Combinatorics Advances

Mathematics and Its Applications

Volume 329

Combinatorics Advances

edited by

Charles J. Colbourn
University of Waterloo,
Canada

and

Ebadollah S. Mahmoodian
Sharif University of Technology,
Tehran, Iran

KLUWER ACADEMIC PUBLISHERS
DORDRECHT / BOSTON / LONDON

Library of Congress Cataloging-in-Publication Data

Combinatorics advances / edited by Charles J. Colbourn, Ebadollah S. Mahmoodian.
p. cm. -- (Mathematics and its applications ; v. 329)
Papers from a conference held Mar. 28-31, 1994, at Sharif University of Technology in Tehran, Iran.
ISBN 0-7923-3574-0 (HB : alk. paper)
1. Combinatorial analysis--Congresses. I. Colbourn, C. J. (Charles J.), 1953- . II. Mahmoodian, Ebadollah S. III. Series: Mathematics and its applications (Kluwer Academic Publishers) ; v. 329.
QA164.C6633 1995
511'.6--dc20 95-20159

ISBN-13: 978-1-4613-3556-6 e-ISBN-13: 978-1-4613-3554-2
DOI: 10.1007/978-1-4613-3554-2

Published by Kluwer Academic Publishers,
P.O. Box 17, 3300 AA Dordrecht, The Netherlands.

Kluwer Academic Publishers incorporates
the publishing programmes of
D. Reidel, Martinus Nijhoff, Dr W. Junk and MTP Press.

Sold and distributed in the U.S.A. and Canada
by Kluwer Academic Publishers,
101 Philip Drive, Norwell, MA 02061, U.S.A.

In all other countries, sold and distributed
by Kluwer Academic Publishers Group,
P.O. Box 322, 3300 AH Dordrecht, The Netherlands.

Printed on acid-free paper

CONTENTS

CONTRIBUTORS

Shahin Ajoodani-Namini
Department of Mathematics 253-37
California Institute of Technology
Pasadena, CA 91125
USA

Mehdi Behzad
Department of Mathematics
Shahid Beheshti University
Tehran, IRAN

Charles J. Colbourn
Combinatorics and Optimization
University of Waterloo
Waterloo, Ontario
CANADA N2L 3G1

Amir Daneshgar
Department of Mathematical Sciences
Sharif University of Technology
P.O. Box 11365-9415
Tehran, IRAN

O. Delgado Friedrichs
Universität Bielefeld
FSPM - Strukturbildungsprozesse
Postfach 10 01 31
33501 Bielefeld
GERMANY

Andreas Dress
Universität Bielefeld
FSPM - Strukturbildungsprozesse
Postfach 10 01 31
33501 Bielefeld
GERMANY

Mohammed Reza Emamy-K.
University of Puerto Rico, and
Institute for Studies in Theoretical
Physics and Mathematics,
Tehran, IRAN

Richard K. Guy
Department of Mathematics and
Statistics
The University of Calgary
2500 University Drive N.W.
Calgary, Alberta, CANADA
T2N IN4

D. Huson
Universität Bielefeld
FSPM - Strukturbildungsprozesse
Postfach 10 01 31
33501 Bielefeld
GERMANY

Zhike Jiang
Combinatorics and Optimization
University of Waterloo
Waterloo, Ontario
CANADA N2L 3G1

Bahman Kalantari
Department of Computer Science,
Rutgers University,
New Brunswick, NJ 08903
USA

Iraj Kalantari
Department of Mathematics
Western Illinois University,
Macomb, IL 61455
USA

Hadi Kharaghani
Mathematics & Computer Science
University of Lethbridge
Lethbridge, Alberta
CANADA T1K 3M4

Gholamreza B. Khosrovshahi
Institute for Studies in Theoretical
Physics and Mathematics, and
University of Tehran
Tehran, IRAN

Ebadollah S. Mahmoodian
Department of Mathematical Sciences
Sharif University of Technology
P.O. Box 11365–9415
Tehran, IRAN

Maryam Mirzakhani
Department of Mathematical Sciences
Sharif University of Technology
P.O. Box 11365–9415
Tehran, IRAN

Dara Moazzami
Shahid Beheshti University
Tehran, IRAN

Javad Mohajeri
Electronic Research Center
Sharif University of Technology
Tehran, IRAN

Yaghout Nourani
Ørsted Laboratory
Niels Bohr Institute
University of Copenhagen
Universitetsparken 5
DK-2100 Copenhagen
DENMARK

Dietmar Schweigert
Universität Kaiserslautern
Fachbereich Mathematik
Postfach 3049
D 6750 Kaiserslautern
GERMANY

Mohammed A. Shokrollahi
Universität Bonn
Informatik
GERMANY

Nasrin Soltankhah
Department of Mathematical Sciences
Sharif University of Technology
P.O. Box 11365–9415
Tehran, IRAN

Heidi J. Strayer
Department of Computer Science
University of Waterloo
Waterloo, Ontario
CANADA N2L 3G1

Anne Penfold Street
Centre for Combinatorics,
Department of Mathematics
The University of Queensland,
Brisbane 4072
AUSTRALIA

L. Zhu
Department of Mathematics,
Suzhou University,
Suzhou 215006
CHINA

PREFACE

On March 28-31, 1994 (Farvardin 8-11, 1373 by Iranian calendar), the Twenty-fifth Annual Iranian Mathematics Conference (AIMC25) was held at Sharif University of Technology in Tehran, Islamic Republic of Iran. Its sponsors included the Iranian Mathematical Society, and the Department of Mathematical Sciences at Sharif University of Technology. Among the keynote speakers were Professor Dr. Andreas Dress and Professor Richard K. Guy. Their plenary lectures on combinatorial themes were complemented by invited and contributed lectures in a Combinatorics Session. This book is a collection of refereed papers, submitted primarily by the participants after the conference.

The topics covered are diverse, spanning a wide range of combinatorics and allied areas in discrete mathematics. Perhaps the strength and variety of the papers here serve as the best indications that combinatorics is advancing quickly, and that the Iranian mathematics community contains very active contributors.

We hope that you find the papers mathematically stimulating, and look forward to a long and productive growth of combinatorial mathematics in Iran.

We would like to dedicate this volume to Mohammad Khaarazmi, a mathematician of the ninth century who is from the same region that the conference was held in. Khaarazmi is known as Muhammad Ibn Musa al-Khowarizmi in the world. The word algorithm comes from his name. He wrote his famous book "Al-djabr wal mukabala" around 830. This book and its translation to English was published by Frederic Rosen in 1831, in London. The word Algebra also is taken from the title of his book.

Waterloo, Canada — Tehran, Iran

February, 1995 — Bahman, 1373

ON A CONJECTURE OF A. HARTMAN

S. Ajoodani-Namini
* G.B. Khosrovshahi

Department of Mathematics 253-37,
California Institute of Technology
Pasadena, CA 91125, USA

* *Institute for Studies in Theoretical Physics and Mathematics, and Department of Mathematics, University of Tehran, Tehran, Iran*

ABSTRACT

We denote the complete design D (or the so-called trivial design) by $S\left(\binom{v-t}{k-t};\ t,k,v\right)$. A conjecture of Hartman states that one can partition D into two $S\left(\binom{v-t}{k-t}/2;\ t,k,v\right)$ designs if and only if $\binom{v-i}{k-i}$ is even for $i=0,\dots t$. In this paper, some progress in support of the conjecture is reported.

1 NOTATIONS AND PRELIMINARIES

Let v,k,t, and λ be positive integers such that $t \le k \le v$. The set of i-subsets of a set X is denoted by $P_i(X)$. A t-design $S(\lambda;t,k,v)$ is a pair $(X,\mathcal{B})$ in which X is a finite set with $|X|=v$ and $\mathcal{B}$ is a family of elements of $P_k(X)$, called blocks, such that every element of $P_t(X)$ appears in exactly λ blocks of $\mathcal{B}$. The parameter k is the *block size*, and t is the *strength*. We denote by $D=(X,P_k(X))$ the complete design, sometimes called the trivial design of block size k. In fact, D is a t-design $S\left(\binom{v-t}{k-t};t,k,v\right)$.

The question addressed in this paper is this: When can one partition the complete design into two designs with $|\mathcal{B}|=\binom{v}{k}/2$?

This partitioning has been referred to as "halving the complete design" by A. Hartman [4], who made the following conjecture: Halving the complete design is possible if and only if $\binom{v-i}{k-i}$ is even for $i=0,\dots,t$.

C. J. Colbourn and E. S. Mahmoodian (eds.), Combinatorics Advances, 1–12.

Some progress in settling this conjecture has been made. For a complete report on the subject, the reader may consult [1,2,4]. In this paper, first we state three theorems and outline their proofs, and then provide some examples which could be helpful in clarifying the underlying constructions. But to do this, first we have to introduce a combinatorial object which is called a "trade".

2 THE STRUCTURE OF TRADES

Let $X = \{x_n | n \geq 1\}$ be a set of distinct indeterminates. Then a subset B of X can be identified with the monomial $\Pi_{x_i \in B} x_i$ (and the empty set is identified with constant polynomial 1), and then every collection of the subsets of X is a polynomial $f \in Z[X]$ with nonnegative coefficients. Let $f \in Z[X]$. We say f is *regular* of degree (or blocksize) k if $f = \sum_{i=1}^{m} n_i X_i$ in which n_i's are nonzero integers, and X_i's are distinct k-subsets of $P_k(X)$. Therefore, 0-regular polynomials are just integers. f is called simple if $|n_i| = 1$ for $i \leq m$. Let f be a regular polynomial of degree k. We define found(f) to be the set of all x_i's such that $\deg_{x_i}(f) = 1$, and the order of f to be the cardinality of found(f). We also define

$$\begin{aligned} f^+ &= \textstyle\sum_{i=1}^{m} \max(n_i, 0) X_i, \\ f^- &= \textstyle\sum_{i=1}^{m} \min(n_i, 0) X_i, \\ \text{supp(f)} &= f^+ + f^-. \end{aligned}$$

Then $f = f^+ - f^-$. For $t \leq k$, we define

$$\phi_t(f) = \sum_{i=1}^{m} n_i P_t(X_i).$$

A regular polynomial f of order v and blocksize k is called a *signed design* with parameters t, k, v, λ (denoted by $SS(\lambda; t, k, v)$) if $\phi_t(f) = \lambda P_t(\text{found(f)})$, and it is called a t-design, (denoted by $S(\lambda; t, k, v)$) if $f^- = 0$. A *(t, k)-trade* is a $SS(0; t, k, v)$ design. It is well known that for any nonzero (t, k)-trade T we have $|\text{found(T)}| \geq k + t + 1$, and $\phi_0(T) \geq 2^{t+1}$. $\frac{1}{2}\phi_0(T)$ is called *volume* of T and is denoted by vol(T). If both equalities hold, then T is called a *minimal (t, k)-trade*. Now it is easy to check that if f is of blocksize k and $t \leq s < k$, then

$$\phi_t \phi_s(f) = \left[\binom{k}{s}\binom{s}{t} \Big/ \binom{k}{t}\right] \phi_t(f).$$

Therefore, a (t,k)-trade is also an (i,k)-trade for $0 \le i \le t$. Now if f and g are regular polynomials such that found $(f) \cap$ found(g) $= \emptyset$, then fg is also regular, and

$$\phi_s(fg) = \sum_{i=0}^{s} \phi_i(f)\phi_{s-i}(g).$$

Therefore, if T_1 is a (t_1, k_1)-trade, T_2 is a regular polynomial of blocksize k_2, and found$(T_1) \cap$ found$(T_2) = \emptyset$, then T_1T_2 is a $(t_1, k_1 + k_2)$-trade, and if T_2 is also a (t_2, k_2)-trade, then T_1T_2 is a $(t_1 + t_2 + 1, k_1 + k_2)$-trade, and in both cases, we have supp(T_1T_2) = supp(T_1)supp(T_2). Now, it is obvious that if $y_1, \ldots, y_{k+t+1}$ are distinct elements of X, then

$$T = (y_1 - y_2)\ldots(y_{2t+1} - y_{2t+2})y_{2t+3}\cdots y_{k+t+1}$$

is a minimal (t,k)-trade. It is well known that any minimal trade is of this form.

Remarks. (i) Let V be a finite subset of X. The set of all (t,k)-trades whose support is a subset of V form a **Z**-module. (ii) If $|V| > k+t$, then this module is of dimension $\binom{v}{k} - \binom{v}{t}$in which $v = |V|$. (iii) In [8], a triangular basis consisting of minimal trades for this module is given. For a comprehensive review on trades and related topics, the reader is referred to [5].

3 BLOCK SIZE 3 AND STRENGTH 2

In [8], an algorithm to produce a $(2,3)$-trade T with vol$(T) = \binom{v}{3}/2$, for $v = 4n+2, n = 1, 2, \cdots$, is given. Clearly this is equivalent to halving the complete design for this case. The idea behind the algorithm is the fact that every simple trade (a trade with no repeated blocks) can be viewed as a halved complete design.

To describe the algorithm, first we call the basis elements produced in [8] as $T_1, T_2, \cdots, T_{\binom{v}{3}-\binom{v}{2}}$. Now the algorithm is as follows:

—**BEGIN**

— *T:=empty;*

— *i:=o;*

— **1:**

— *Find first block "B" in lexicographical order, $B \notin T$;*

```
—       IF NOT found THEN stop
—       ELSE BEGIN
—               i:=i+1;
—               WHILE B ≠ first block of T_i DO i := i + 1;
—               IF T + T_i is simple THEN
—                       T := T + T_i
—               ELSE
—                       T := T − T_i;
—       END;
—       GOTO 1;
—END.
```

Theorem 3.1 *The algorithm produces $S(2n; 2, 3, 4n+2)$ designs.*

4 A POSSIBLE APPROACH TO THE PROBLEM

In this section we describe a possible approach to the problem of halving complete designs. The nature of this method is basically different from that of Theorem 3.1. Here trades are used in a recursive manner. From one hand the construction does not depend on a specific basis, but on the other hand the construction heavily relies on the previous design on the list.

To describe the approach, it is more convenient to reformulate our problem in terms of trades.

Let T be a simple (t, k)-trade of order v (i.e., |found(T)| = v) such that $\text{supp}(T) = P_k(\text{found(T)})$, then $\phi_t(T^+ - T^-) = 0$ and

$$\phi_t(T^+ + T^-) = \binom{v-t}{k-t} P_t(\text{found(T)}).$$

Therefore, both T^+ and T^- are $S(\binom{v-t}{k-t}/2; t, k, v)$ designs. On the other hand, if f is a $S(\binom{v-t}{k-t}/2; t, k, v)$ design, then $2f - P_k(\text{found(f)})$ is a simple (t, k)-trade with $\text{supp}(T) = P_k(\text{found(T)})$ and $T^+ = f$. Therefore, Hartman's conjecture can be reformulated in the following way:

Conjecture 4.1 *A $(t, k, v, 1/2)$-trade, i.e. a simple (t, k)-trade T of order v and volume $\binom{v}{k}$, exists if and only if $\binom{v-i}{k-i}$ is even for $0 \le i \le t$.*

Therefore, to prove this conjecture we must discuss which polynomials can be a support of a simple trade. We already have seen that if f and g are supports of simple trades, then fg is also a support of a simple trade if and only if it is of degree at most one for each x_i. Also, it is clear that $f+g$ is a support of a trade if and only if f and g have the same blocksize and they are disjoint (i.e., $f+g$ is simple). Therefore, if $f_0,\ldots,f_n,g_0,\ldots,g_n$ are regular simple polynomials such that each f_i(and g_i, as well) is the support of a simple trade, then $f_0+g_0+\sum_{i=1}^{n} f_i g_i$ is support of a simple trade if and only if it is regular and simple. In particular, if V is a finite subset of X, and $x_i \in X \setminus V$, then $P_{k+1}(V)+x_iP_k(V) = P_{k+1}(x_iV)$ which is regular and simple, therefore we have the following lemma.

Lemma 4.2 *If a* $(t,k+1,v,1/2)$*-trade and a* $(t,k,v,1/2)$ *exist then a* $(t,k+1,v+1,1/2)$ *also exists.*

Now, let u, v, and k be two positive integers, and let $Y=\{x_i|i\le v+u-t\}$. In [2], it is shown that we can write

$$P_k(Y)=P_k(Y_1)+P_k(Y_2)+\sum_{i=1}^{k-1}P_{k-i}(Z_i)P_i(T_i),$$

in which Y_1, Y_2, Z_i's and T_i's are subsets of X with $|Y_1|=v$, $|Y_2|=u$, $|T_i|=u-k+i$, $Z_i=Y_1$ for $i<k-t-1$, and $|Z_{k-t-1+j}|=v-j$. Therefore we have the following theorem.

Theorem 4.3 *Let* u *and* v *be positive integers such that a* $(t,k,u,1/2)$*-trade exists, and for* $t+1\le i\le k$ *a* $(t,i,v,1/2)$*-trade exist, then a* $(t,k,v+u-t,1/2)$*-trade exists.*

An immediate consequence of this theorem is that if v and u satisfy the conditions of the theorem, then a $(t,k,v,1/2)$-trade exists, where $v_1\equiv u \pmod{v-t}$ and $v_1>u$. Therefore, it is more convenient to interpret the necessary conditions as a system of congruence relations. A triple (v,k,t) is called feasible if $\binom{v-i}{t-i}$ is even for $i=0,\ldots,t$. Let $k<2^{n_k}\le 2k$. In [1], it is shown that a triple (v,k,t) is feasible if and only if either $v\equiv i \pmod{2^{n_k}}$ for some $t\le i<k$ or $v\equiv v_0 \pmod{2^{n_k}}$ in which $k<v_0<2^{n_k}$ and (v_0,v_0-k,t) is feasible.

Theorem 4.3 together with the above remarks suggests the following recursive approach to Hartman's conjecture: Let $B(t,k)$ be the set of all v's for which a $(t,k,v,1/2)$-trade exists. Then $B(t,k) \subset C(t,k) = \{v|(t,k,v) \text{ is feasible }\}$. Let $2 \le t < k$ and $l = n_k$. We would like to prove that $B(t,k) = C(t,k)$. Let $k > t+1$, and assume that $B(t,i) = C(t,i)$ for $t < i < k$. Utilizing Theorem 4.3 to prove $B(t,k) = C(t,k)$, we must prove that (i) $2^l + t \in \cap_{i=t+1}^{k} B(t,i)$, (ii) if either $v_0 < 2k$ and (v_0,t,k) is feasible or $2^l + t \le v_0 < 2^l + k$, then a $(t,k,v_0,1/2)$-trade exists. First we discuss (ii). If $v_0 < 2^l$, then $(v_0, v_0 - k, t)$ is feasible and $v_0 - k < k$, so $v_0 \in B(t, v_0 - k)$ which implies that $v_0 \in B(t,k)$. On the other hand, if (i) is true, then by Lemma 4.2 we have $\{2^l+t,\ldots,2^l+k-1\} \subset B(t,k)$. Therefore (i) implies (ii), and all we have to do is to prove (i). Also by Theorem 4.3, if $2^n + t \in B(t,t+1)$ (in which $t+1 < 2^n < 2t+3$), then $B(t,t+1) = C(t,t+1)$. Therefore to prove $B(t,k) = C(t,k)$ we only need to establish existence of $(k-t)$-designs.

We see in the next section that for $t = 2$, this approach actually works. But now, we just mention an immediate consequence of the above remarks to show how powerful this approach can be.

It is well known that $6 \in B(2,3)$, $10 \in B(2,i)$ for $3 < i < 8$, and $18 \in B(2,i)$ for $7 < i < 15$. Therefore we have the following theorem.

Theorem 4.4 *[2] $B(2,k) = C(2,k)$ for $3 \le k < 15$.*

5 THE CASE OF STRENGTH 2

In this section, we use the method which was briefly sketched in Section 4, to prove that for $t = 2$, the necessary conditions for halving the complete designs are in fact sufficient. As we discussed before, to do this, we must show that for every $k \ge 3$, $2^{n_k} + 2 \in B(2,k)$ in which 2^{n_k} is the smallest power of 2 which is larger than k. For a moment assume that we have proved $2^{l+1} + 2 \in B(2, 2^l)$ for all l, and let k be the smallest integer such that $k > 2$ and $2^{n_k} + 2 \notin C(2,k)$. Then by the remarks before Theorem 4.4, we have $B(2,i) = C(2,i)$ for $2 < i < k$, and $k > 15$. For simplicity let $n_k = n$, and $v = 2^n + 2$. Now if $v < 2k$, then $v - k < k$, therefore $B(2, v-k) = C(2, v-k)$ and in particular $v \in B(2, v-k)$ which implies that $v \in B(2,k)$. Therefore, $v \ge 2k$, but we have $v - 2 = 2^n \le 2k$. Therefore either $v = 2k$ or k is a power of 2. In the first case it is proved by Alltop [3] that $v \in B(2,k)$ and in the second case by our assumption $v \in B(2,k)$.

Now, we describe briefly an outline of the proof of $2^{l+1}+2 \in B(2, 2^l)$ for $l > 3$. For simplicity, let $m = k+1 = 2^l+1$ and $v = 2m$. Let $V = \{x_i | 1 \leq i \leq v\}$ and $V_1 = \{x_{2i-1} | 1 \leq i \leq m\}$. For $1 \leq i \leq m$, we define $x^*_{2i-1} = x_{2i-1}x_{2i}$. Then we can extend this, in a natural way to the subsets of V_1 and then to all regular polynomials whose foundation is a subset of V_1. For example $(x_1x_3 + x_3x_5)^* = x_1x_2x_3x_4 + x_3x_4x_5x_6$. Let $1 \leq t \leq k/2$, and let T be a minimal $(2t-1, k)$-trade with found$(T) \subset V$. Then T is said to be normalized if there exists a permutation $\sigma \in S_m$ such that (i) $\sigma(i) < \sigma(j)$ if $1 \leq i < j \leq 2t$, and (ii)

$$T = (y_1 - y_2) \dots (y_{4t-1} - y_{4t}) y_{4t+1} \cdots y_{k+2t},$$

in which $y_{2i-1} = x_{2\sigma(i)-1}$ and $y_{2i} = x_{2\sigma(i)}$. Let f_t be the sum of the supports of all normalized $(2t-1, k)$-trades, and define $f_0 = \sum_{B \in P_{k/2}(V_1)} B^*$. Then

$$P_k(V) = \sum_{i=0}^{k/2} f_t.$$

Now, it is easy to check that any two normalized trades are disjoint, so each f_t is the support of a simple $(2t-1, k)$-trade, therefore if $t > 1$, then f_t is the support of a simple $(2, k)$-trade. Now, if S is a 1-trade whose support is $P_{k/2}(V_1)$, then S^* is also a 1-trade, and $\text{supp}(S^*) = f_0$. Therefore, if we can find a simple 1-trade T whose support is f_1, and $\phi_2(T) = \phi_2(S^*)$, then $T - S^*$ is a 2-trade whose support is $f_0 + f_1$. To construct T we need a further description of normalized 1-trades and $\phi_2(S)$.

For $1 \leq i < j \leq k+1$, we define

$$T_{ij} = (x_{2i-1} - x_{2i})(x_{2j-1} - x_{2j}).$$

Then if $B \in P_{k/2-1}(V_1 \setminus \{x_{2i-1}, x_{2j-1}\})$, then $T_{ij}B^*$ is a normalized 1-trade, and in fact every normalized 1-trade is of this form. Clearly $\phi_2(T_{ij}B^*) = T_{ij}$. Therefore, if $B, C \in P_{k/2-1}(V_1 \setminus \{x_{2i-1}, x_{2j-1}\})$, then $T_{ij}B^* - T_{ij}C^*$ is a 2-trade.

It is straightforward to check that $\phi_2(S^*)$ can be written as a sum of, say n_S, 1-trades of the form $h = \text{supp}(\text{T}_{\text{ir}} + \text{supp}(\text{T}_{\text{js}}) - \text{supp}(\text{T}_{\text{is}} - \text{supp}(\text{T}_{\text{jr}}))$ (with i, j, r, s distinct). On the other hand, if B is a $(k/2-1)$-subset of V_1 such that $i, j, k, l \notin B$ then hB^* is a 1-trade whose support is the sum of four distinct normalized 1-trades. Putting these informations together, we can show that if $l > 3$, then we can find S in such a way that $n_S < \binom{k-3}{k/2-1}$, and then we can find a 1-trade T with desired property which proves $f_0 + f_1$ (and so $P_k(V)$) is the support of a simple 2-trade. Thus, we have the following theorem.

Theorem 5.1 *For* $k \geq 3$, $B(2, k) = C(2, k)$.

6 SOME EXAMPLES

In this section we present some examples which clarify some of the constructions which are used in Theorems 3.1, 4.3, and 5.1.

Example 1. Let $v = 6, k = 3, t = 2$. Then $\binom{6}{3} - \binom{6}{2} = 5$, and the following 5 minimal trades form a basis:

	T_1	T_2	T_3	T_4	T_5
	123	124	125	134	135
	145	135	136	156	146
T^+	246	256	246	235	236
	356	346	345	246	245
	124	125	126	135	136
	135	134	135	146	145
T^-	236	246	245	234	235
	456	356	346	256	246

Applying the algorithm, we obtain the following result:

$$\begin{aligned} T &= T_1 - T_3 + T_4, T = T^+ \cup T^-, \\ T^+ &= T_1^+ \cup T_3^- \cup T_4^+, T^- = T_1^- \cup T_3^+ \cup T_4^-, \\ T^+ &= \{123, 126, 134, 145, 156, 235, 245, 246, 346, 356\}, \\ T^- &= \{124, 125, 135, 136, 146, 234, 236, 256, 345, 456\}. \end{aligned}$$

Obviously,

$$T^+ = S(2; 2, 3, 6), \; T^- = S(2; 2, 3, 6)$$

Example 2. First, we like to consider a special case of Theorem 4.3 in which $k = t + 1$. In this case Theorem 4.3 says that if just a halved design of order v and block size k exists, then an infinite family of them exists, and in particular for $k = t + 1$, halving of complete designs is equivalent to the existence of first design in the family. For this, we would like to mention this particular case as a theorem.

Theorem 6.1 *If a simple $(v, t+1, t, 1/2)$ trade exists, then for a positive integer n, a simple $(v + n(v-t), t+1, t, 1/2)$ trade also exists.*

Now, suppose we know that the design $S(4;2,3,6)$ is halvable and by utilizing Theorem 4.3, we wish to construct $S(4;2,3,10)$. Here $k = 2+1$ and $v = 6+6-2 = 10$ and

$$X = \{1,\cdots,10\}.$$

We define the following sets

$$\begin{array}{ll} X_0 = \{1,\cdots,6\}, & \\ X_1 = \{1,\cdots,5\}, & Y_1 = \{7,\cdots,10\}, \\ X_2 = \{1,\cdots,4\}, & Y_2 = \{6,\cdots,10\}, \\ & Y_3 = \{5,\cdots,10\}. \end{array}$$

Now we partition the set $P_3(X)$ into $2+2=4$ classes:

$$\begin{array}{rcl} \text{First Class} & : & D_1 = \{\{x_1,x_2,x_3\}|x_1 < x_2 < x_3 < 7\}, \\ \text{Second Class} & : & D_2 = \{\{x_1,x_2,x_3\}|x_1 < x_2 < 6 < x_3\}, \\ \text{Third Class} & : & D_3 = \{\{x_1,x_2,x_3\}|x_1 < 5 < x_2 < x_3\}, \\ \text{Fourth Class} & : & D_4 = \{\{x_1,x_2,x_3\}|5 < x_1 < x_2 < x_3\}. \end{array}$$

Now it suffices to construct simple and disjoint $(2,3)$-trades $T_0,\cdots,T_3$ based on blocks belonging to corresponding classes. Then our big trade is

$$T = \sum_{i=0}^{3} T_i.$$

$$\text{vol(T)} = \binom{10}{3}/2 = 60.$$

Now the 4 trades are as follows:

First Class, T_0. We use the existence of $S(2;2,3,6)$ based on X_0 to construct the trade

$$T_0 = S(2;2,3,6) - S^c(2;2,3,6),$$

where

$$S\ (2;2,3,6) = \{123,124,135,146,156,234,245,256,345,346\},$$

$$S^c(2;2,3,6) = \{125,126,134,136,156,235,236,246,356,456\}.$$

Second Class, T_1. First we construct a 2-factorization of K_5 based on X_1, $F = \{F_1, F_2\}$, where

$$F_1 = \{12, 13, 24, 35, 45\}, \quad F_2 = \{14, 15, 23, 25, 34\}.$$

Now we add the elements of Y_1 to the factors of F in an appropriate way to construct $T_1 = T_1^+ \cup T_1^-$ as follows, where 'a' denotes 10:

$$\begin{aligned} T_1^+ = \{&127, 137, 247, 357, 457,\\ &128, 138, 248, 358, 458,\\ &149, 159, 239, 259, 349, \quad 14a, 15a, 23a, 25a, 34a\}. \end{aligned}$$

$$\begin{aligned} T_1^- = \{&129, 139, 249, 359, 459,\\ &12a, 13a, 24a, 35a, 45a,\\ &147, 157, 237, 257, 347,\\ &148, 158, 238, 258, 348\}. \end{aligned}$$

Third Class, T_2. This trade is constructed exactly like T_1, namely we construct a 2-factorization of K_5 based on Y_2, $G = (G_1, G_2)$ and add the elements of X_2 in an appropriate way to G_1 and G_2.

$$G_1 = \{67, 68, 79, 89, 9a\}, \quad G_2 = \{69, 6a, 78, 7a, 89\}.$$

Fourth Class, T_3. This trade is also constructed exactly like T_0, but based on Y_3.

If $k > t + 1$, then the smallest design which can be constructed by Theorem 4.3 has at least 1530 blocks, so we are unable to discuss it in details here, but as Example 2 shows the main step in this construction is partitioning $P_k(X)$ into some subsets such that each of them is a support of a t-trade. Therefore we bring an example to show how this partitioning looks like if $k > t + 1$ (for $k = t + 1$, the partitioning is very much like Example 1).

Example 3. Let $k = 4$, $t = 2$. It is well known that $(28; 2, 4, 10)$ is halvable and by utilizing Theorem 4.3, we wish to construct an $S(120; 2, 4, 18)$. Let

$$X = \{1, \cdots, 10\}.$$

We define the following sets

$$\begin{array}{ll} X_0 = \{1, \cdots, 10\} & \\ X_1 = \{1, \cdots, 10\} & Y_1 = \{11, \cdots, 18\} \\ X_2 = \{1, \cdots, 9\} & Y_2 = \{11, \cdots, 18\} \\ X_3 = \{1, \cdots, 8\} & Y_3 = \{10, \cdots, 18\} \\ & Y_4 = \{9, \cdots, 18\} \end{array}$$

Now we partition $P_4(X)$ into the following classes:

$$\begin{aligned}
D_0 &= P_4(X_0),\\
D_1 &= \{B \in P_4(X) \| B \cap X_1| = 3 \ \& |B \cap Y_1| = 1\},\\
D_2 &= \{B \in P_4(X) \| B \cap X_1| = 2 \ \& |B \cap Y_2| = 2\},\\
D_3 &= \{B \in P_4(X) \| B \cap X_1| = 1 \ \& |B \cap Y_3| = 3\},\\
D_4 &= P_4(Y_4)
\end{aligned}$$

Now, for $0 \leq i \leq 5$, we must show that D_i is the support of a 2-trade. For $i = 0, 5$, it is part of our assumption. For $i = 2$, we form a halving of $S(4; 2, 3, 10)$, say E_1, E_2, on X_1, then if D_{1i} $(i = 1, 2)$ consists of all 4-subsets of the form $B \cup \{j\}$ in which $B \in E_i$ and $j \in Y_1$, then $D_{11} - D_{12}$ is a 2-trade whose support is D_1. Notice that here cardinality of D_1 is odd, so we can't use the method we used in Example 2, and this justifies why in Theorem 4.3 we need a halving of $S(\binom{v-t}{i-t}; t, i, v)$ for $t < i \leq k$. For $i = 2$, we form two 4-factorizations $\{F_1, F_2\}$ and $\{G_1, G_2\}$ of K_9 on X_2 and Y_2, respectively. Then

$$D_{2,1} = \{A \cup B | A \in F_j \text{ and } \mathrm{B} \in \mathrm{G_j},\ \mathrm{j} = 1, 2\},$$

and $D_{22} = D_2 \setminus D_{21}$ form a 2-trade with support D_2, finally case $i = 3$ is essentially the same as second class in Example 2.

7 CONCLUDING REMARKS

In spite of what is offered in Theorem 5.1, the establishment of the conjecture is far from complete and although it seems intuitively correct but otherwise it is a hard one to be proved in its totality.

In closing, G.B. Khosrovshahi would like to state 3 conjectures which are related to the Hartman's conjecture. Conjectures 7.1 and 7.2 are old ones, but Conjecture 7.3 is new.

Conjecture 7.1 *Let λ_* be the smallest feasible λ. If a $D = S(\lambda_*; t, k, v)$ design exists, then a $D' = S(\lambda_*; t, k, v)$ design also exists such that $D \cap D' = \emptyset$.*

Conjecture 7.2 *There exists a partition of complete design $S\left(\binom{v-t}{k-t}; t, k, v\right)$ into four $S\left(\binom{v-t}{k-t}/4; t, k, v\right)$ designs if and only if $4|\binom{v-i}{k-i}$ for $i = 0, \cdots, t$.*

Conjecture 7.3 *For any given positive integers v, k, and t such that $v \geq k \geq t$ and λ_*, the smallest feasible λ, there exists a λ, $\lambda_* \leq \lambda < \binom{v-t}{k-t}$ for which a simple $S(\lambda; t, k, v)$ exists.*

REFERENCES

[1] S. Ajoodani-Namini, *All block designs with $b = \binom{v}{k}/2$ exist*, 1994, submitted.

[2] —— and G.B. Khosrovshahi, *More on halving the complete designs*, 1992, Discrete Math., in press.

[3] W.O. Alltop, *Extending t-designs*, J. Combin. Theory, Ser. A **18** (1975), 177-186.

[4] A.Hartman, *Halving the complete design*, Ann. Discrete Math. **34** (1987), 207-224.

[5] A.S. Hedayat, *The theory of trade-off for t-design*, in: Coding Theory and Design Theory, Part II, Design Theory (D.Ray-Chaudhuri, ed.), IMA Vol Math. Appl. **21**, Springer-Verlag, 1990, 101-126.

[6] A.S. Hedayat, G.B. Khosrovshahi, and D. Majumdar, *A prospect for a general method for constructing t-designs*, Discrete Appl. Math. **42** (1993), 31-50.

[7] G.B. Khosrovshahi and S. Ajoodani-Namini, *An infinite family of 6-designs exists*, Sankhyā **54** (1992), 259-264.

[8] —— and S. Ajoodani-Namini, *A new basis for trades*, SIAM J. Discrete Math. **3** (1990), 364-372.

SOME PROBLEMS IN TOTAL GRAPH THEORY

M. Behzad

Beheshti University
Tehran, Iran
arashbeh@irearn.bitnet

ABSTRACT

Thirty years ago the total chromatic number and the total graph of a graph was introduced and a conjecture was stated in the author's Ph.D. dissertation. This conjecture is known as the Total Chromatic Conjecture (TCC). At this time numerous results concerning these and other total concepts such as total groups, total crossing numbers, and total Ramsey numbers exist in the literature. More specifically, during these decades results concerning different properties of total graphs, such as planarity, reconstructibility, and traversability have been obtained, and some parameters, including (vertex) connectivity, edge connectivity, arboricity, and eigenvalues, of such graphs have been studied. Some total concepts have been generalized, and some have been applied to other areas. In recent years a lot of attention has been paid to TCC and tremendous effort is being used toward its settlement. In this brief expository article we confine ourselves to some concepts and questions which are tangible even by some bright undergraduate students. As far as references are concerned, as usual, there are a lot of duplications. For the sake of briefness we have to be selective, and, with apology, we eliminate a lot of articles and results.

1 INTRODUCTION AND PRELIMINARIES

We consider (simple) graphs which are finite, undirected, with no loops or multiple edges. The minimum number of colors required to color the elements of a graph G so that no two adjacent or incident elements have the same color

C. J. Colbourn and E. S. Mahmoodian (eds.), Combinatorics Advances, 13–26.

is called the *total chromatic number* of G and is denoted by $\chi''(G)$. For every graph G a graph $T(G)$, called the *total graph* of G, is defined in such a way that $\chi''(G)$ equals the chromatic number of $T(G)$. The author introduced these two notions in 1965 is his Ph.D. dissertation [3] in which he also stated the so-called,

Total Chromatic Conjecture–TCC. *For every graph G we have $\chi''(G) \leq \Delta(G) + 2$, where $\Delta(G)$ is the maximum degree of G.*

This conjecture was published for the first time in the Proceedings of the International Symposium on Graph Theory, Rome, 1966 [8]. See also [7].

The inequality $\Delta(G) + 1 \leq \chi''(G)$ is obviously true and best possible. For clarification we present the following figures.

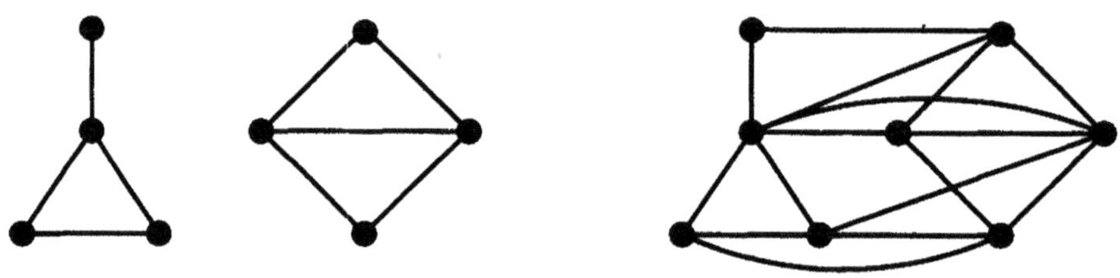

G $L(G)$ $T(G)$

$$\chi(G) = \chi'(G) = \chi(L(G)) = 3; \; \chi''(G) = \chi(T(G)) = 4$$

Clearly, $T(G)$ is the square of the subdivision graph $S(G)$ which is obtained by "adding" a new vertex on every edge of G. Since G is an induced subgraph of $T(G)$, no "forbidden subgraph characterization" of total graphs can exist. Structure of regular total graphs is presented in [14], and a characterization of nonregular total graphs appears in [5] which is algorithmic in nature. In fact, based on this characterization computer programs have been devised (see for example, [25]). These characterizations reveal that total graphs, in general, are not regular, yet they are very regularly structured.

The majority of notions and notations are the same as those of [10].

2 TOTAL RAMSEY NUMBERS

For positive integers r and s, the Ramsey number $R(r,s)$ is the smallest positive integer ν such that every graph of order ν contains K_r or $\overline{K_s}$ as a subgraph

[26]. Here the class of graphs can be restricted to that of connected ones, except for the trivial case in which $r = 2$. Similarly, we define $R''(r, s)$ to be the smallest positive integer ν such that every connected graph of order ν has r mutually associated elements or s mutually nonassociated elements. Although this Ramsey type problem seems much harder than the original problem which is still wide open, the exact values of $R''(r, s)$ for a relatively big portion of the first quadrant of the (r, s)-plane is known [15].

Theorem 2.1 *For all pairs of positive integers (r, s) outside the region*

$$\{(r,s) \mid r > 4,\ s > r^2 - 5r + 8\}$$

the values of $R''(r, s)$ are as follows:

$$\begin{aligned}
&R''(1,s) = R''(r,1) = 1, \quad \textit{for all } r \textit{ and } s,\\
&R''(2,s) = R''(3,s) = 2, \quad s \geq 2,\\
&R''(r,2) = 3, \quad r \geq 4,\\
&R''(4,s) = \lfloor (3s+1)/2 \rfloor, \quad s \geq 2,\\
&R''(r,s) = 2s - 1, \quad 2 < s < r, \textit{ and}\\
&R''(r,s) = 2s - 2, \quad 4 \leq s \leq r^2 - 5r + 8.
\end{aligned}$$

It is desirable to know the exact values of $R''(r, s)$ for the remaining part of the first quadrant. A partial solution, or bounds would be appreciated, too.

3 VERTEX RECONSTRUCTIBILITY OF TOTAL GRAPHS

The (vertex) Reconstruction Conjecture of Kelly and Ulam is over fifty years old, and has a vast literature; see, for example, [18] and [41]. One of its several formulations is as follows: Every graph with at least three vertices can be reconstructed from the deck of its vertex-deleted subgraphs.

Let C be a class of graphs. A *graph* $G \in C$ is *class reconstructible* from any n of the vertex-deleted subgraphs of G if no other member of C contains n such vertex-deleted subgraphs. The *class* C is *class-reconstructible* if every member of C is class-reconstructible. The Reconstruction Conjecture may now be stated as follows: The class of all graphs of order $\nu \geq 3$ is class-reconstructible from the ν vertex-deleted subgraphs.

Theorem 3.1 *The class of total graphs is class-reconstructible from any one of its vertex-deleted subgraphs.*

This interesting theorem [2] provides another indication of the fact that total graphs are highly structured. In fact, $T(G)$ effectively contains two copies of the graph G, and so one vertex-deleted subgraph suffices to determine G, and $T(G)$ uniquely within the class. In this regard the following problem is still open: Are total graphs recognizable from the given deck of vertex- deleted subgraphs?

4 EDGE RECONSTRUCTIBILITY OF TOTAL GRAPHS

In 1964 F. Harary stated the Edge Reconstruction Conjecture: Every graph with at least four edges is reconstructible. Again the literature is quite vast. In these studies algebraic tools such as groups are heavily used. We refer the interested reader to [18].

In 1990 the following theorem was proved [35].

Theorem 4.1 *The square of a graph is edge reconstructible if its average degree is at least 9.5.*

One can immediately conclude that the total graph of a graph is edge reconstructible provided its average degree is at least 9.5. Are total graphs which are not included in the above statement edge reconstructible? The answer ought to be yes. One might want to tackle the whole problem directly.

5 THE SPECTRUM OF TOTAL GRAPHS

The eigenvalues of the adjacency matrix of a graph G of order ν constitute the spectrum of G: $\lambda_1 \geq \lambda_2 \geq \cdots \lambda_\nu$. In 1973 D. M. Cvetković considered the relationship between the eigenvalues of a connected r-regular graph G and the eigenvalues of $T(G)$ [24]. If $r = 1$ then $G = K_2$ has eigenvalues 1 and -1, while $T(G) = K_3$ has eigenvalues $2, -1$, and -1. For $r > 1$ we have

Theorem 5.1 *If a connected r-regular graph G, $r > 1$, has order ν and size ϵ, then $T(G)$ has $\epsilon - \nu$ eigenvalues equal to -2, and 2ν eigenvalues of the form $(2\lambda_i + r - 2 \pm \sqrt{4\lambda_i + r^2 + 4})/2$ where $\lambda_i, 1 \leq i \leq \nu$, is an eigenvalue of G.*

For $r = 2$, $G = C_\nu$, $\nu \geq 3$; the eigenvalues of G are $2\cos\frac{2\pi i}{\nu}$, $1 \leq i \leq \nu$. In this case the least eigenvalue for $T(G)$ in greater than $-9/4$.

For $r > 2$, the eigenvalues of $T(G)$ lie in $[-r, 2r]$. The maximum eigenvalue of G is r and the maximum eigenvalue of $T(G)$ is $2r$. The number -2 is an eigenvalue of $T(G)$ with multiplicity $\epsilon - \nu + p_{-r} + p_{-1}$ where p_λ is the multiplicity of the eigenvalue λ of G. For example, $K_\nu, \nu > 1$, has eigenvalues $\nu - 1$ of multiplicity 1 and -1 of multiplicity $\nu - 1$, while $T(K_\nu)$ has eigenvalues $2\nu - 2, \nu - 3$, and -2 of multiplicity $1, \nu$, and $(\nu^2 - \nu - 2)/2$, respectively.

Ultimately one would like to see such a theorem for an arbitrary connected graph G; this seems to be a very difficult task. Are the following problems easier? How about some special cases? Find the relationship between the eigenvalues of a nonempty connected graph G and the eigenvalues of:

i) $S^e(G)$, where $S^e(G)$ is an elementary subdivision of G.

ii) $S(G)$, where $S(G)$ is the subdivision of G obtained by "inserting" a vertex of degree two on every edge of G.

iii) $T(G) = (S(G))^2$, and, in general, G^2 which is the square of G.

6 GROUPS, AND POLYNOMIALS OF GRAPHS

Five groups are associated with every nonempty graph G: the degree preserving group $\Gamma_d(G)$, the (vertex) group $\Gamma(G)$, the induced edge group $\Gamma_i(G)$, the edge group $\Gamma'(G)$, and the total group $\Gamma''(G)$. The total group of G is the group of all association preserving permutations of the elements of G. For the isomorphism of any two of these five groups necessary and sufficient conditions for G exist. For example, we have the following theorem [13].

Theorem 6.1 *For any graph $G \neq K_1, \Gamma(G) \cong \Gamma''(G)$ if and only if no component of G is either a cycle or a complete graph.*

In spite of the fact that this algebraic tool is vital for solving different graph theoretical problems, no systematic study, easy to use knowledge, about such groups exist. Indeed, a book should have been written on this topic. Another interesting question is the following. What is the relationship between the chromatic polynomial of G and that of the $T(G)$? The book [16] contains a whole chapter on this topic.

7 RELATIONSHIPS BETWEEN SOME PARAMETERS OF G AND THOSE OF $T(G)$

The (edge) arboricity $a_1(G)$ of a nonempty graph G is the minimum number of edge disjoint spanning forests into which G can be partitioned. By using a well known theorem of Nash-williams, Akiyama and Hamada [1] proved the following

Theorem 7.1 *For any r-regular graph $G, r \geq 1, a_1(T(G)) = r + 1$.*

This is the only result about the arboricity of total graphs.

The crossing number $cr(G)$ of a graph G is the minimum number of crossings of the edges of G among all drawings of G in the plane. There are three main results concerning crossing numbers of total graphs.

(i) $cr(T(G)) = 0$ if and only if $\Delta(G) \leq 3$, and if $\deg_G v = 3$, then v is a cut vertex of G [4].

(ii) $cr(T(G)) = 1$ of and only if $\Delta(G) = 4$, G has exactly one vertex of degree 4, and every vertex of degree 4 or 3 of G is a cut vertex of G [36]

(iii) Finally, Patil [42] established a characterization of total graphs with crossing number one in terms of forbidden subgraphs. Are characterizations of graphs G with the property that $cr(T(G)) = 2$, or 3 very hard? How about graphs G which are arbitrary members of special classes of graphs? For arboricity, too, there are quite a few tangible questions. How about thickness and genus which are untouched, and binding numbers on which there is only one article [45]?

8 SOME GENERALIZATIONS AND APPLICATIONS

The following generalization of total graphs is defined in [37]. Let $k \geq 2$ be a positive integer. The *k-total graph* $T_k(G)$ of G has complete subgraphs K_{k-1} and K_k of G as vertices; 2 distinct vertices S and S' are adjacent in $T_k(G)$ if at least one is a k-simplex and $|S \cap S'| = k - 1$, or both are $(k-1)$ simplicies such that $S \cup S'$ is a k-simplex. Note that $T_2(G)$ is the same as $T(G)$. A graph G is *perfect* if for each induced subgraph G' of G, $\chi(G') = \omega(G')$, the clique number of G'. The graph C_{2n+1}, $n \geq 2$, is an *odd hole* and is not perfect. A graph is *Berge* if it does not contain an odd hole or the complement of an odd hole. Berge Perfect Graph Conjecture (BPGC) states that every Berge graph is perfect. In 1977 it was proved that BPGC holds for total graphs [43]. In 1993 Van Bang Lê [37] extended this result to

Theorem 8.1 *BPGC holds for 3-total graphs.*

As far as we know there is only one article on total directed graphs [19], as this is the case for k-total graphs, and the total chromatic numbers of hypergraphs [40]. Certainly, in these areas there are plenty of interesting, worthwhile problems.

As for applications, total graphs of balanced incomplete block designs have been introduced [11]. To test the isomorphism of designs the two invariants: the eccentric sequences and the triangle sequences of such graphs have shown to be quite effective [12]. We ask the interested design theorists to see these two articles.

The remaining part is devoted to total colorings. There are numerous papers written on this subject. Several are expository and their titles include expressions such as "recent results", "recent developments", "a summary of", "progress report", and all have been published in the last few years. In Table 1 we simply tabulate some, and definitely not all, of the classes of graphs for which TCC holds. Section 9 deals with several upper bounds for total chromatic numbers. Finally, we end the article with a few comments, and the references.

Name	Condition	χ''	Comments and References
Complete Graphs K_ν	$\begin{cases} \nu \text{ odd} \\ \nu \text{ even} \end{cases}$	$=\Delta+1$ $=\Delta+2$	[9]
Cycles C_ν	$\begin{cases} \nu \equiv 0 \pmod 3 \\ \text{otherwise} \end{cases}$	$=3$ $=4$	[6]
Complete Bigraphs $K_{m,n}$	$\begin{cases} m \neq n \\ m = n \end{cases}$	$=\Delta+1$ $=\Delta+2$	[9]
Bigraphs		$\leq \Delta+2$	
Complete 3-Partite Graphs		$\leq \Delta+2$	[44]
Husimi Trees H	$\begin{cases} H = K_\nu, \nu \text{ even} \\ \text{otherwise} \end{cases}$	$=\Delta+2$ $=\Delta+1$	[3] (includes all trees)
Any graph with	$\Delta = 3$	≤ 5	[44] and [46]
Any graph with	$\Delta = 4$	≤ 6	[33]
Any graph with	$\Delta \geq \nu - 4$	$\leq \Delta+2$	[48]
Any graph with	$\Delta \geq \frac{3}{4}\nu$	$\leq \Delta+2$	[28]
Any graph with odd ν	$\delta \geq \frac{5}{6}(\nu+1)$	$\leq \Delta+2$	[23]
Any graph with even ν	$\delta \geq \frac{3}{4}(\nu+1)$	$\leq \Delta+2$	
r-Regular Graphs with odd ν	$r \geq \frac{6}{7}\nu$	$\leq \Delta+2$	(See [20].) To prove TCC it is sufficient to prove it for 2-connected regular graphs [6].
r-Regular Graphs with even ν	$r \geq \frac{3}{4}\nu$	$\leq \Delta+2$	
Planar Graphs	$\Delta \geq 8$	$\leq \Delta+2$	[47].

Table 1 Several Classes of Graphs for Which TCC Holds (The order, the maximum degree, and the minimum degree of a graph is denoted by ν, Δ and δ, respectively).

9 UPPER BOUNDS FOR $\chi''(G)$

Here we present just a few of the known upper bounds for graphs which are not very specific.

i) One obvious upper bound is

$$\chi''(G) \leq 2\Delta(G) + 1.$$

ii) Another obvious upper bound is

$$\chi''(G) \leq \chi'(G) + \chi(G).$$

For the graph $K_{n,n}$ the equality holds. In fact, one can easily prove that if $\chi''(G) = \chi'(G) + \chi(G)$, then G is bipartite [9].

iii) Kostochka [33, 34] established that

$$\text{if } \Delta(G) \geq 6, \text{ and } \Delta(G) \neq 9, 15, \text{ then } \chi''(G) \leq \tfrac{3}{2}\Delta(G).$$

iv) In 1985, Bollobas and Harris [17], as a corollary to a theorem on list colorings, proved that

$$\text{if } \Delta(G) \geq 3919, \text{ then } \chi''(G) \leq \tfrac{11}{6}\Delta(G).$$

Obviously, the bound given in (iii) is better than this one.

v) In 1990 H. R. Hind [29] showed that

$$\text{for every } 1 \leq k \leq \Delta(G), \chi''(G) \leq \chi'(G) + 2\lceil \tfrac{\Delta(G)}{k} \rceil + k.$$

vi) Then by taking $k = \lceil \sqrt{\Delta(G)} \rceil$ Hind observed that

$$\chi''(G) \leq \chi'(G) + 2\lceil \sqrt{\Delta(G)} \rceil.$$

Clearly, if $\Delta(G) \geq 23$, then this upper bound in better than the one given in (iii).

vii) Eventually, as the last result of this paper the following bound was obtained

$$\chi''(G) \leq \chi'(G) + 2\lceil \sqrt{\chi(G)} \rceil.$$

Note that, in general, this bound, referred to as the Hind bound, is by far better than the ones given thus far.

viii) From the articles [21], and [38] one can conclude that if t is an integer such that $t! > \nu(G)$, then $\chi''(G) \leq \chi'(G)+t$. Proof of this result is probabilistic and we urge the interested reader to see these articles.

ix) In 1992 the following upper bound was established by Hind [31].

$$\chi''(G) \leq \Delta(G) + 2\lceil\frac{\nu(G)}{\Delta(G)}\rceil + 1.$$

This result shows that for those graphs with large Δ relative to ν, TCC is not too far off.

x) Finally, in 1993 McDiarmid and Sanchez-Arroyo [39] proved that

$$\chi''(G) \leq \frac{7}{5}\Delta(G) + \frac{13}{5}.$$

If $\Delta(G) = 26$, then the two bounds given in (x) and (iii) are the same. As Δ increases (x) provides a much better bound compared to the one given by (iii).

10 REMARKS

i) As can be observed this article is by no means a survey on total concepts on graphs. In fact, a whole book can be written on this subject. There are several excellent articles with plenty of open problems and conjectures which deal with the recent developments and results concerning the total chromatic numbers of graphs and related coloring numbers [20], [27], [30] and [32].

ii) A graph G is of *type* 1 if $\chi''(G) = \Delta(G) + 1$, and is of *type* 2 if $\chi''(G) = \Delta(G) + 2$. If TCC is correct, then every graph is either of type 1 or of type 2. In fact, in 1993 McDiarmid and Reed [38] have proved that almost every graph is of type 1. Thus graphs for which TCC does not hold are scarce. For this categorization of graphs, and some elegant results and open problems see [22] and [27].

iii) A graph G in called *uniquely total colorable* if $V(G) \cup E(G)$ can be partitioned into $\chi''(G)$ color classes in exactly one way. For example, cycles of order $\nu, \nu \equiv 0(mod\ 3)$, empty graphs, and paths are uniquely total colorable. In this line, a few other classes of graphs have been examined, and several results have been obtained by S. Akbari, M. Behzad, H.

Haji-Abolhasan and E.S. Mahmoodian all of which confirm the following conjecture (elsewhere in this volume).

Conjecture *Aside from the cycles of order* $\nu, \nu \equiv 0(mod\ 3)$, *empty graphs, and paths there are no graphs which are uniquely total colorable.*

iv) In 1969 while the author was busy presenting the article [6], Paul Erdős said something like this, "Boy, this is a very hard problem. It (TCC) is not going to be solved in this century."

The twenty–first century is not too far away!

REFERENCES

[1] J. Akiyama and T. Hamada, *The decomposition of the line graphs, middle graphs and total graphs*, Discrete Math., 26 (1979), pp. 203–208.

[2] D. W. Bange, A. E. Barkauskas, and L. H. Host, *Class-reconstruction of total graphs*, J. Graph Theory, 11 (1987), pp. 221–230.

[3] M. Behzad, *Graphs and their chromatic numbers*, PhD dissertation, Michigan State University, Department of Mathematics, 1965.

[4] ——, *A criterion for the planarity of a total graph of a graph*, Camb. Phil. Soc., 63 (1967), pp. 679–681.

[5] ——, *A characterization of total graphs*, Proc. Amer. Math. Soc., 26 (1970), pp. 383–389.

[6] ——, *The total chromatic number of a graph, a survey*, in Comb. Math. and its Applications, D. J. A. Welsh, ed., Acad. Press, 1971, pp. 1–9.

[7] M. Behzad and G. Chartrand, *Total graphs and traversability*, Proc. Edinburgh Math. Soc., 15 Ser. II (1966), pp. 117–120.

[8] ——, *An introduction to total graphs*, in Proc. of the Inter. Symposium on Theory of Graphs, Rome, 1967, pp. 31–33.

[9] M. Behzad, G. Chartrand, and J. K. Cooper Jr., *The colour numbers of complete graphs*, J. London Math. Soc., 42 (1967), pp. 226–228.

[10] M. Behzad, G. Chartrand, and L. Lesniak-Foster, *Graphs and Digraphs*, Wadsworth International Group, Belmont,California, 1979.

[11] M. Behzad and E. S. Mahmoodian, *Graphs versus designs: A quasisurvey*, in Graph Theory, Combinatorics, and Applications, Y. Alavi and et al, eds., John Wiley and Sons, Inc., 1991, pp. 125–141.

[12] ——, *Eccentric sequences and triangle sequences of block designs*, Discrete Math., 127 (1994), pp. 47–56.

[13] M. Behzad and H. Radjavi, *The total group of a graph*, Proc. American Math. Soc., 19 (1968), pp. 158–163.

[14] ——, *Structure of regular total graphs*, J. London Math. Soc., 44 (1969), pp. 433–436.

[15] ——, *Another analog of Ramsey numbers*, Math. Ann., 186 (1970), pp. 228–232.

[16] N. Biggs, *Algebraic Graph Theory*, Camb. Univ. Press, 1974.

[17] B. Bollobás and A. J. Harris, *List-colourings of graphs*, Graphs and Comb., 1 (1985), pp. 115–127.

[18] J. A. Bondy, *A graph reconstructor's manual*, London Math. Soc. Lecture Note Series, 166 (1991), pp. 221–252.

[19] G. Chartrand and M. J. Stewart, *Total digraphs*, Canadian Math. Bull., 9 (1966), pp. 171–176.

[20] A. G. Chetwynd, *Total colorings of graphs-a progress report*, in Graph Colorings, R. Nelson and R. J. Wilson, eds., Research Notes in Mathematics, Pitman, 1990, pp. 65–77.

[21] A. G. Chetwynd and R. Häggkvist, *Some upper bounds on the total and list chromatic numbers of multigraphs*, J. Graph Theory, 16 (1992), pp. 503–516.

[22] A. G. Chetwynd and A. J. W. Hilton, *Some refinements of the total chromatic number conjecture*, Congressus Numerantium, 66 (1988), pp. 195–216.

[23] A. G. Chetwynd, A. J. W. Hilton, and Zhao Cheng, *The total chromatic number of graphs of high minimum degree*, J. London Math. Soc. (2), 44 (1991), pp. 193–202.

[24] D. M. Cvetković, *Spectrum of the total graph of a graph*, Publications de L'Institut Mathematique, 16 (1973), pp. 49–52.

[25] F. Gavril, *A recognition algorithm for the total graphs*, Networks, 8 (1978), pp. 121–133.

[26] R. L. Graham, B. L. Rothschild, and J. H. Spencer, *Ramsey Theory*, John Wiley & Sons, 1990.

[27] A. J. W. Hilton, *Recent results on the total chromatic number*, Discrete Math., 111 (1993), pp. 323–331.

[28] A. J. W. Hilton and H. R. Hind, *The total chromatic number of graphs having large maximum degree*, Discrete Math., 117 (1993), pp. 127–140.

[29] H. R. Hind, *An upper bound for the total chromatic number*, Graphs and Comb., 6 (1990), pp. 153–159.

[30] ——, *A summary of total coloring results*. Preprint, May 29 1991.

[31] ——, *An upper bound for the total chromatic number of dense graphs*, J. Graph Theory, 16 (1992), pp. 197–203.

[32] ——, *Recent developments in total coloring*, Discrete Math., 125 (1994), pp. 211–218.

[33] A. V. Kostochka, *The total colouring of a multigraph with maximal degree 4*, Discrete Math., 17 (1977), pp. 161–163.

[34] A. V. Kostochka, *An analogue of Shannon's estimate for complete colorings* (Russian), Diskret. Analiz. 30 (1977), pp. 13–22.

[35] I. Krasikŏv and Y. Roditty, *Recent applications of Nash-Williams lemma to the edge-reconstruction conjecture*, Ars Comb., Ser. A, 29 (1990), pp. 215–224.

[36] V. R. Kulli and N. S. Annigeri, *Total graphs with crossing number 1*, J. Math. Phys. Sci., 12 (1978), pp. 615–617.

[37] Van Bang Lê, *Perfect k-line graphs and k-total graphs*, J. Graph Theory, 17 (1993), pp. 65–73.

[38] C. J. H. McDiarmid and B. Reed, *On total colourings of graphs*, J. Comb. Theory, Ser. B, 57 (1993), pp. 122–130.

[39] C. J. H. McDiarmid and A. Sanchéz-Arroyo, *An upper bound for total colouring of graphs*, Discrete Math., 111 (1993), pp. 389–392.

[40] J. C. Meyer, *Nombre chromatique total d'un hypergraphe*, J. Comb. Theory, Ser. B, 24 (1978), pp. 44–50.

[41] C. St. J. A. Nash-Williams, *The reconstruction problem*, in Selected Topics in the Theory of Graphs, L. W. Beineke and R. J. Wilson, eds., Academic Press, 1978.

[42] N. P. Patil, *Forbidden subgraphs and total graphs with crossing number 1*, J. Math. Phys. Sci., 17 (1983), pp. 293–295.

[43] S. B. Rao and G. Ravindra, *A characterization of perfect total graphs*, J. Math phys. Sci., 11 (1977), pp. 25–26.

[44] M. Rosenfeld, *On the total chromatic number of certain graphs*, Israel J. Math., 9 (1971), pp. 396–402.

[45] A. Saito and Tian Songlin, *The binding number of line graphs and total graphs*, Graphs and Comb., 1 (1985), pp. 351–356.

[46] N. Vijayaditya, *On total chromatic number of a graph*, J. London Math. Soc., 3 (1971), pp. 405–408.

[47] H. P. Yap, *On the total chromatic number of a graph*, Research Report No. 343, 1988.

[48] H. P. Yap, Wang Jian-Fang, and Zhang Zhongfu, *Total chromatic number of graphs of high degree*, J. Aust. Math. Soc., Ser. A, 47 (1989), pp. 445–452.

CONSTRUCTION TECHNIQUES FOR MUTUALLY ORTHOGONAL LATIN SQUARES

Charles J. Colbourn

Combinatorics and Optimization
University of Waterloo
Waterloo, Ontario, CANADA N2L 3G1

ABSTRACT

Recent developments concerning the construction of mutually orthogonal latin squares (MOLS) and incomplete MOLS are discussed. Some improvements in the number of MOLS are presented using Greig's line–flip technique, and some new sets of incomplete MOLS found by computer search are presented. Finally, a variant of Wilson's theorem is developed.

1 BACKGROUND

A *transversal design* $(V, \mathcal{G}, \mathcal{B})$ of *order* kn and *blocksize* k ($TD(k,n)$) is a kn-set V which is partitioned into k equivalence classes $\mathcal{G} = \{G_1, \ldots, G_k\}$ called *groups*, and each having n elements of V, together with a set $\mathcal{B}$ of k-subsets of V, called *blocks*. Every unordered pair from V occurs together either in a group, or in precisely one block, but not both. A transversal design is *idempotent* if and only if it contains a set of n blocks that are pairwise disjoint.

A *latin square* of *side* n is an $n \times n$ array in which each entry contains a single element from an n-set S, such that each element occurs exactly once in each row and exactly once in each column. Two latin squares L and L' are *orthogonal* if their rows and columns are indexed by the same sets, and if $L(a,b) = L(c,d)$ and $L'(a,b) = L'(c,d)$, then $a = c$ and $b = d$. A set of latin squares $L_1, \ldots, L_m$ is *mutually orthogonal*, or a set of *MOLS*, if for every $1 \le i < j \le m$, L_i and L_j are orthogonal.

C. J. Colbourn and E. S. Mahmoodian (eds.), Combinatorics Advances, 27–48.

An *orthogonal array* of *size* N, with k *constraints* (or of *degree* k), s *levels* (or of *order* s), and *strength* t, denoted $OA(N,k,s,t)$, is a $k \times N$ array with entries from a set of $s \geq 2$ symbols, having the property that in every $t \times N$ submatrix, every $t \times 1$ column vector appears the same number $\lambda = \frac{N}{s^t}$ times. The parameter λ is the *index* of the orthogonal array. An $OA(N,k,s,t)$ is also denoted by $t - OA_\lambda(k,s)$; in this notation, if t is omitted it is understood to be 2, and if λ is omitted it is understood to be 1. In this paper, we are concerned only with $OA(k,s)$.

Let $\{L_i : 1 \leq i \leq k\}$ be a set of k MOLS on symbols $\{1,\ldots,n\}$. Form a $(k+2) \times n^2$ array $A = (a_{ij})$ whose columns are

$$(i, j, L_1(i,j), L_2(i,j), \ldots, L_k(i,j))^T$$

for $1 \leq i,j \leq k$. Then A is an $OA(k+2,n)$. This process can be reversed to recover k MOLS of side n from an $OA(k+2,n)$, by choosing any two rows of the OA to index the rows and columns of the k squares.

Let A be an $OA(k,n)$ on the n symbols in X. On $V = X \times \{1,\ldots,k\}$ (a set of size kn), form a set $\mathcal{B}$ of k-sets as follows. For $1 \leq j \leq n^2$, include $\{(a_{i,j}, i) : 1 \leq i \leq k\}$ in $\mathcal{B}$. Then let $\mathcal{G}$ be the partition of V whose classes are $\{X \times \{i\} : 1 \leq i \leq k\}$. Then $(V, \mathcal{G}, \mathcal{B})$ is a $TD(k,n)$. This process can be reversed to recover an $OA(k,n)$ from a $TD(k,n)$.

Thus, k MOLS of side n, a $TD(k+2,n)$, and an $OA(k+2,n)$ are all equivalent. In each of these disguises, mutually orthogonal latin squares have been extensively studied and are central in combinatorial design theory and in experimental design theory.

Recent advances have come, for the most part, from the study of such structures with various patterns of "holes". An *incomplete transversal design* of *order* or *groupsize* n, *blocksize* k, *index* λ, and *holesizes* $b_1,\ldots,b_s$, denoted $ITD_\lambda(k,n;b_1,\ldots,b_s)$, is a quadruple $(V,\mathcal{G},\mathcal{H},\mathcal{B})$, where

1. V is a set of kn elements;
2. $\mathcal{G}$ is a partition of V into k classes (called *groups*), each of size n;
3. $\mathcal{H}$ is a set of disjoint subsets $H_1,\ldots,H_s$ of V, with the property that, for each $1 \leq i \leq s$ and each $G \in \mathcal{G}$, $|G \cap H_i| = b_i$;
4. $\mathcal{B}$ is a collection of k–subsets of V (called *blocks*);

5. every unordered pair of elements from V is
   - contained in a hole, and contained in no blocks; or
   - contained in a group, and contained in no blocks; or
   - contained in neither a hole nor a group, and contained in λ blocks.

When $\lambda = 1$, it can be omitted from the notation. An $ITD(k, n; b_1, \ldots, b_s)$ is also denoted as a $TD(k, n) - \sum_{i=1}^{s} TD(k, b_i)$.

Analogous definitions can be made for incomplete MOLS and for incomplete orthogonal arrays. When a single hole of size h is present, we write $N(n, h)$ for

$$\max(k : \exists ITD(k+2, n; h)).$$

When the set of holes is spanning, suppose that there are t_i holes each of size g_i, for $1 \leq i \leq s$. We then employ exponential notation and write $N(g_1^{t_1} \ldots g_s^{t_s})$ for

$$\max(k : \exists TD(k+2, n) - \sum_{i=1}^{s} t_i TD(k+2, g_i)).$$

For example, $N(1^n)$ is the number of idempotent MOLS of side n.

The wide variety of constructions available for transversal designs and incomplete transversal designs precludes the possibility of a complete survey here. Thus we content ourselves to discuss three directions in which substantial progress has been made recently, after giving a brief history.

2 HISTORY AND SMALL ORDERS

Euler [13] was apparently the first to study mutually orthogonal latin squares. He advanced the conjecture that $N(n) = 1$ for $n \equiv 2 \pmod 4$ on the evidence that $N(2) = 1$ and his unsuccessful attempt to construct 2 MOLS of side 6. Remarkably, Euler did succeed in constructing an $ITD(4, 6; 2)$.

Kirkman [15] established in 1850 that a projective plane of order n exists whenever n is a prime power. A finite projective plane of order n is a set of n^2+n+1 lines on n^2+n+1 points; each line has length $n+1$ and each point lies on $n+1$ lines. Finally, each pair of points occurs on a unique line, and every pair of lines meet in a unique point.

Evidently, removing one point from a projective plane, and treating the truncated lines as groups and the other lines as blocks, one obtains a $TD(n+1,n)$. An easy counting argument shows that in a $TD(k,n)$ one has $k \leq n+1$. Hence Kirkman implicitly determined that $N(q) = q-1$ for q a prime power.

Tarry [13] in 1900 returned attention to the Euler conjecture, and by systematic hand search established that $N(6) = 1$. In 1922, MacNeish [17] established the important recursion that $N(nm) \geq \min(N(n), N(m))$. He mistakenly claimed equality and thus purported to prove the Euler conjecture. Primarily as orthogonal arrays, transversal designs were widely examined for applications in experimental design and coding theory. However, the next major advance for existence came in the period 1957-1960, culminating with Bose, Shrikhande and Parker's disproof of the Euler conjecture [7]. They developed techniques to construct transversal designs from pairwise balanced designs, setting the stage in the process for many of the techniques of modern design theory.

However, it was Wilson's 1974 paper [30] that advanced techniques using incomplete transversal designs, and gave the most powerful technique for asymptotic existence available today. Variants of Wilson's technique have a predominant place in the existence of MOLS for large sides. Both the techniques of Wilson and of Bose, Shrikhande and Parker are recursive in nature, and so rely on the construction of "small" examples for their success.

Thus computational methods to construct transversal designs and incomplete transversal designs are fundamental construction techniques for the existence of MOLS. In Table 1, we present the best current bounds for $N(n)$ for $6 \leq n \leq 50$ when n is not a prime power.

A few features of Table 1 warrant comment. First, we observe that MacNeish's construction has been improved upon in all but two cases (35 and 45) in this range. Secondly, many of the constructions are quite recent and all of these have employed computational methods. Thirdly, and perhaps most disappointingly, the lower bounds given here give us little clue about what the true values are. For instance, in the case of order 12, although the lower bound of 5 is striking for being relatively large, we do not as yet know whether or not $N(12) = 11$. Nevertheless, improvements in these small sides remain of substantial interest.

Order	Number	Year	Reference(s)
6	1	1900	[13]
10	2	1960	[7]
12	5	1961	[16]
14	3	1985	[25]
15	4	1978	[23]
18	3	1978	[28]
20	4	1989	[26]
21	5	1991	[20]
22	3	1978	[28]
24	5	1992	[3, 21, 32]
26	4	1994	[9]
28	4	1991	[1, 22]
30	4	1993	[4]
33	5	1994	[2]
34	4	1994	[2]
35	4	1922	[17]
36	5	1994	[22]
38	4	1993	[4]
39	4	1977	[18]
40	7	1994	[3]
42	3	1978	[28]
44	4	1993	[4]
45	4	1922	[17]
46	4	1974	[29]
48	5	1994	[3, 22]
50	6	1974	[29]

Table 1 Best Current Lower Bounds on $N(n)$

3 PAIRWISE BALANCED DESIGNS AND GREIG'S LINE–FLIP

Bose, Shrikhande and Parker [7] pioneered the use of pairwise balanced designs in the construction of MOLS. A *pairwise balanced design* (*PBD*) of *order* v and *blocksizes* K, is a pair $(V, \mathcal{B})$. V is a v-set of elements, and $\mathcal{B}$ is a collection of subsets (*blocks*) of V for which $\{|B| : B \in \mathcal{B}\} \subseteq K$. Moreover, every pair of distinct elements occurs together in exactly one block. A *clear set* in a PBD is a set of pairwise disjoint blocks. Suppose that a PBD of order v exists, having a clear set with block sizes in C and all other blocks of sizes in K. Then the number of MOLS of order v is at least

$$\min(\min(N(1^k) : k \in K), \min(N(k) : k \in C)).$$

See [6] for details.

When a clear set in a PBD contains all elements, it is a *parallel class*. Thus a parallel class partitions the elements of the PBD. This leads naturally to the notion of PBD with a distinguished parallel class. A *group divisible design* (*GDD*) of *order* v, *blocksizes* K, and *group-type* $g_1^{t_1} \dots g_s^{t_s}$, is a triple $(V, \mathcal{G}, \mathcal{B})$. V is a set of v elements, and $\mathcal{G}$ is a partition of V into $\sum_{i=1}^{s} t_i$ classes, where t_i classes have size g_i for each $1 \leq i \leq s$. These classes are called *groups*. $\mathcal{B}$ is a set of subsets of V with sizes from K, so that every unordered pair of elements of V either occurs in a group together, or in a block together, but not both. Using GDDs, one can obtain MOLS in a manner analogous to using PBDs; observe, for example, that from a GDD of order v, blocksizes K and group-type $g_1^{t_1} \dots g_s^{t_s}$, one obtains

$$N(v+h, h) \geq \min(\min(N(1^k) : k \in K), \min(N(g_i + h, h) : 1 \leq i \leq s)).$$

Moreover, if $t_1 = 1$, then

$$N(v+h, g_1 + h) \geq \min(\min(N(1^k) : k \in K), \min(N(g_i + h, h) : 2 \leq i \leq s)).$$

The construction of pairwise balanced designs and group-divisible designs with large block sizes poses many challenging problems, however. Recently, Greig [14] developed an important technique that has substantial consequences for the existence of MOLS.

Suppose that a $\{k\}$-GDD of type k^k exists; this is an affine plane of order k, and hence is known whenever k is a prime power. Now choose an integer x

with $1 \leq x < k$, and choose one block B. Delete all points from x of the groups *except* those on block B. Next delete all points on block B in the remaining $k - x$ groups. The resulting PBD has blocks of five types:

1. a single block on x points, which is the truncation of B;
2. $k - 1$ disjoint blocks each of size $k - x$, which are the truncations of the blocks disjoint from B in the affine plane;
3. $k - x$ disjoint blocks each of size $k - 1$, which are the truncations of the remaining groups;
4. blocks of size $k - x - 1$ that do not intersect the truncation of B (in the affine plane, they did intersect B);
5. blocks of size $k - x + 1$ that do intersect the truncation of B.

Types (1) and (2) form a parallel class; so also do types (1) and (3). Thus adding a point at infinity to the blocks of type (1) and (2), called a *type A extension*, gives a $\{k - x - 1, k - x + 1\}$-GDD of type $(k - 1)^{k-x}(x + 1)^1$. On the other hand, adding a point at infinity to blocks of types (1) and (3) gives a $\{k-x-1, k, k-x+1\}$-GDD of type $(k-x)^{k-1}(x+1)^1$ (a *type B extension*). Greig [14] observes that either GDD can be extended with a further point at infinity to form a PBD on $(k-x)(k-1)+x+2$ with block sizes $\{k, k-x-1, k-x+1, x+2^\star\}$. The superscript $\star$ indicates that a block of size $x + 2$ is present, and that all other blocks have sizes from $\{k, k - x - 1, k - x + 1\}$.

Take, for example, $q = 8$ and $x = 2$. Initially, we obtain a PBD on 44 points with block sizes $\{7, 6, 5, 2^\star\}$. A type A extension gives a GDD of order 45 with blocksizes $\{5, 7\}$ and group-type $7^6 3^1$. Since $N(1^5) = 3$ and $N(1^7) = 5$, we obtain $N(45, 3) \geq 3$. Adding a point at infinity and noting that $N(4, 1) = 3$ and $N(8, 1) = 7$, we also obtain $N(46, u) \geq 3$ for $u \in \{4, 5, 7, 8\}$. These settle cases marked unknown in Du's list [12].

However, the more significant applications arise in the construction of MOLS. Here we take $k = q$, q a prime power, so that the required affine plane exists. We illustrate the application of Greig's technique to the construction of MOLS of orders up to 10000 in Table 2.

For order 9229, note that the PBD constructed is a GDD with block sizes $\{25, 27\}$ of type $342^{26}318^1$. Since $ITD(20, 342 + 19; 19)$ and $TD(20, 318 + 19)$ exist, we can add 19 points at infinity to obtain the desired $TD(20, 9229)$.

q	x	num	order	blocksizes	Remark
16	6	7	158	$\{16, 9, 11, 8^*\}$	
29	11	12	517	$\{29, 17, 19, 13^*\}$	
32	14	15	574	$\{32, 17, 19, 16^*\}$	
41	17	18	979	$\{41, 23, 25, 19^*\}$	
41	15	16	1057	$\{41, 25, 27, 17^*\}$	
43	17	18	1111	$\{43, 25, 27, 19^*\}$	
53	23	24	1585	$\{53, 29, 31, 25^*\}$	
128	110	13	2398	$\{128, 17, 19, 112^*\}$	$N(112) \geq 13$ [3]
81	39	39	3401	$\{81, 41, 43, 41^*\}$	
128	102	23	3405	$\{25, 27, 127, 103^*\}$	type A extension
83	41	39	3487	$\{83, 41, 43, 43^*\}$	
128	100	25	3657	$\{27, 29, 127, 101^*\}$	type A extension
89	47	39	3745	$\{89, 41, 43, 49^*\}$	
83	35	36	3973	$\{83, 47, 49, 37^*\}$	
89	41	42	4267	$\{89, 47, 49, 43^*\}$	
113	71	39	4777	$\{113, 41, 43, 73^*\}$	
121	79	39	5121	$\{121, 41, 43, 81^*\}$	
107	59	45	5149	$\{107, 47, 49, 61^*\}$	
137	95	39	5809	$\{137, 41, 43, 97^*\}$	
125	77	45	6031	$\{125, 47, 49, 79^*\}$	
127	79	45	6129	$\{127, 47, 49, 81^*\}$	
128	80	45	6177	$\{47, 49, 127, 81^*\}$	type A extension
149	107	39	6325	$\{149, 41, 43, 109^*\}$	
107	47	48	6409	$\{107, 59, 61, 49^*\}$	
167	125	39	7099	$\{167, 41, 43, 127^*\}$	
101	29	30	7231	$\{101, 71, 73, 31^*\}$	
179	137	39	7615	$\{179, 41, 43, 139^*\}$	
137	77	57	8239	$\{137, 59, 61, 79^*\}$	
139	79	57	8361	$\{139, 59, 61, 81^*\}$	
109	29	30	8671	$\{109, 79, 81, 31^*\}$	
343	317	18	9229	$\{25, 27, 342, 318^*\}$	see note
233	191	39	9937	$\{233, 41, 43, 193^*\}$	

Table 2 Applications of Greig's Line-flip

4 DIFFERENCE MATRICES: SOME DIRECT CONSTRUCTIONS

Let $(\Gamma, +)$ be a group of order n. A *quasi-difference matrix* with parameters n, u, k, λ and μ is an array $Q = (q_{ij})$ having k rows, in which each entry is either empty or contains an element of Γ; moreover,

1. each row contains λu empty entries;
2. each column contains at most one empty entry; and
3. for two distinct rows h and i, the list of differences $q_{hj} - q_{ij}$ contains each nonzero element of Γ exactly λ times, and contains 0 exactly μ times (differences with empty entries are undefined).

Wilson [29] established the following:

Lemma 4.1 *If a quasi-difference matrix with parameters n, u, $\lambda = 1$, $0 \leq \mu \leq 1$ and k exists, then there exist $k-2$ MOLS of side $n+u$ with a hole of side u.*

To see this, append $\lambda - \mu$ columns consisting of zeroes. Then develop each column into n columns under the action of the group Γ.

Constructions of quasi-difference matrices form the main tool in the direct construction of incomplete transversal designs. Recently, a large number of quasi-difference matrices over cyclic groups have been found [10]. We present one new quasi-difference matrix as an illustration of the method. Consider the following matrix:

$$\begin{pmatrix} 0 & 0 & 0 & - & - & - \\ 0 & 2 & 7 & 0 & 0 & 0 \\ 1 & 8 & 3 & 3 & 5 & 10 \\ - & - & - & 6 & 9 & 4 \\ 0 & 9 & 11 & 6 & 5 & 8 \\ 2 & 7 & 5 & 1 & 8 & 7 \end{pmatrix}.$$

We employ this matrix over the group Z_{12}. Replacing every column by three columns obtained by applying the row permutation (1 2 3)(4 5 6) gives the 18 columns of a quasi–difference matrix with $n = 12$, $k = 6$, $u = 3$, $\lambda = 1$ and $\mu = 1$. Hence we obtain an $ITD(6,15;3)$.

Wilson [29] went on to describe a class of quasi–difference matrices that can be succinctly presented. Let $q = mt + 1$ and let ω be a primitive element of $GF(q)$. Now suppose that a vector $(a_1, \ldots, a_{m+1})$ exists for which, for each $1 \le k < m$, the differences

$$\{a_{i+k} - a_i : 1 \le i \le m+1, i+k \ne m+2\}$$

represent the m cyclotomic classes of $GF(mt+1)$ (compute subscripts modulo $m+2$ as needed). In other words, for a fixed k, if $a_{i+k} - a_i = \omega^{mx+\alpha}$ and $a_{j+k} - a_j = \omega^{my+\beta}$, we find that $\alpha \not\equiv \beta \pmod{m}$. Then form a single column of length $m+2$ whose first entry is empty, and whose remaining entries are $(a_1, \ldots, a_{m+1})$. Form t columns by multiplying this column by the powers of ω^m. From each of these t columns, form $m+2$ columns by taking the $m+2$ cyclic shifts of the column. To the resulting $mt+2t$ columns, append a column of all zeroes. The result is a quasi–difference matrix with parameters $n = q$, $u = t$, $\lambda = 1$, $\mu = 1$ and $k = m+2$. It follows that $N(q+t,t) \ge m$. Such a vector m is termed a *V(m,t) vector* in [19] and [8].

Recent progress on the existence of $V(m,t)$ vectors [8, 10, 27] has established the following:

Theorem 4.2 *A $V(m,t)$ vector exists if m and t are not both even,*

1. *whenever $m = 2$ and $mt+1$ is a prime or prime power;*
2. *whenever $m = 3$ and $mt+1$ is a prime;*
3. *whenever $mt+1 \le 5000$, $m-1 \le t$, $m \le 10$ and $mt+1$ is prime, except when $m = 9$ and $t = 8$, as no $V(9,8)$ exists.*
4. *whenever $mt+1 \le 5000$, $m-1 \le t$, $m \le 6$ and $mt+1$ is a prime power, except when $m = 3$ and $t = 5$, as no $V(3,5)$ exists.*

The restriction that m and t are not both even is necessary. To see this, consider the differences between elements at distance $\frac{m+2}{2}$ when m is even. As a result of the cyclic shifts, each difference and its additive inverse appear; however, when t is even these also appear in the same cyclotomic class.

When m and t are both even, one can instead consider the partition of the nonzero elements into $2m$ cyclotomic classes, each having $\frac{t}{2}$ elements. In this case, one requires a pair of vectors that together represent all $2m$ cyclotomic classes. We call such a configuration a $V^{(2)}(m,t)$ *matrix*. Observe that such a matrix can exist only when $t \equiv 2 \pmod 4$. We have systematically searched for $V^{(2)}(m,t)$ matrices whenever $p = mt + 1$ is a prime less than 5000, $t \equiv 2 \pmod 4$, and $m \in \{2,4,6\}$. Under the condition that $t \geq m$, we found $V^{(2)}(m,t)$ matrices in each case; here we present solutions for $m \in \{4,6\}$, since the solutions with $m = 2$ are trivial to find. We also present solutions for $m = 8$ for suitable primes p in the range $1777 \leq p < 5000$. For each of the solutions presented, we omit the first entry of each vector, which is always 0 in these solutions.

m	t	p	$V^{(2)}(m,t)$ matrix	
4	10	41	1 3 2 8	5 34 4 19
4	18	73	1 4 11 3	11 45 24 51
4	22	89	1 4 13 23	11 30 61 47
4	34	137	1 3 2 5	5 15 52 109
4	58	233	1 4 11 16	9 7 74 108
4	70	281	1 3 6 21	5 2 101 65
4	78	313	1 3 7 2	5 15 68 175
4	102	409	1 3 2 5	7 21 14 35
4	114	457	1 3 2 10	5 15 85 419
4	130	521	1 3 2 8	3 10 18 95
4	142	569	1 3 2 5	6 28 156 59
4	150	601	1 6 20 5	11 5 114 483
4	154	617	1 3 8 27	6 27 57 165
4	190	761	1 3 2 8	3 11 46 150
4	202	809	1 3 2 5	6 4 21 501
4	214	857	1 3 2 7	3 9 18 770
4	234	937	1 4 11 28	5 28 71 267
4	238	953	1 3 2 5	5 14 90 341
4	258	1033	1 3 6 12	5 15 38 178
4	262	1049	1 3 6 11	6 1 29 312
4	274	1097	1 3 6 11	6 1 30 122
4	282	1129	1 3 2 17	11 33 52 661
4	298	1193	1 3 6 11	6 13 49 36
4	322	1289	1 4 11 21	9 22 16 3
4	330	1321	1 3 7 20	5 12 88 1282
4	358	1433	1 4 13 40	5 57 10 287
4	370	1481	1 3 6 11	6 1 16 1443
4	402	1609	1 4 13 32	7 49 76 473
4	414	1657	1 3 7 2	5 13 128 901

continued on next page

m	t	p	$V^{(2)}(m,t)$ matrix	
4	430	1721	1 4 11 7	9 31 66 464
4	438	1753	1 3 6 11	6 1 16 237
4	450	1801	1 4 14 44	11 24 117 1446
4	478	1913	1 4 11 16	9 7 34 604
4	498	1993	1 3 2 8	5 15 26 401
4	522	2089	1 4 11 7	5 36 47 414
4	534	2137	1 3 2 10	5 15 10 351
4	538	2153	1 3 2 8	3 8 84 379
4	570	2281	1 3 6 19	5 26 61 215
4	574	2297	1 3 2 7	3 9 26 1557
4	594	2377	1 3 2 29	5 15 10 312
4	598	2393	1 3 6 17	5 18 7 346
4	610	2441	1 4 13 32	7 34 105 219
4	618	2473	1 3 8 18	7 20 85 460
4	630	2521	1 3 2 17	11 33 22 237
4	654	2617	1 3 2 5	5 15 10 132
4	658	2633	1 3 2 5	5 13 10 119
4	678	2713	1 3 2 7	8 23 46 958
4	682	2729	1 3 2 5	6 4 77 461
4	694	2777	1 3 2 5	5 19 40 1785
4	714	2857	1 3 6 11	6 1 16 2002
4	738	2953	1 3 2 8	5 15 30 824
4	742	2969	1 4 13 3	11 28 112 1682
4	750	3001	1 3 2 17	7 26 19 221
4	762	3049	1 3 6 33	11 33 99 2586
4	802	3209	1 3 2 8	3 11 52 480
4	814	3257	1 4 13 3	5 27 25 326
4	858	3433	1 3 2 5	5 15 10 387
4	862	3449	1 4 13 3	7 26 63 875
4	882	3529	1 3 8 23	13 39 104 679
4	898	3593	1 3 2 5	6 4 46 312
4	918	3673	1 3 8 21	11 33 70 505
4	942	3769	1 3 2 10	7 21 63 1843
4	958	3833	1 3 6 19	5 14 1 322
4	970	3881	1 3 2 5	6 14 28 329
4	982	3929	1 3 2 8	3 15 30 763
4	1014	4057	1 4 11 21	15 46 89 106
4	1018	4073	1 3 2 5	5 12 3 178
4	1038	4153	1 3 8 22	7 40 20 1935
4	1050	4201	1 4 13 26	19 42 7 443
4	1054	4217	1 3 6 11	6 1 32 553

continued on next page

m	t	p	$V^{(2)}(m,t)$ matrix	
4	1074	4297	1 6 31 64	13 11 142 586
4	1102	4409	1 4 13 3	7 18 44 2131
4	1110	4441	1 3 2 10	7 21 63 1013
4	1114	4457	1 3 6 11	6 1 39 523
4	1162	4649	1 3 2 7	3 15 30 453
4	1182	4729	1 3 2 14	11 33 77 351
4	1198	4793	1 3 2 5	6 4 28 244
4	1222	4889	1 3 2 8	3 11 42 285
4	1234	4937	1 3 6 30	7 16 61 1263
4	1242	4969	1 3 7 14	8 19 121 234
6	6	37	1 3 8 15 19 36	9 8 3 26 18 2
6	10	61	1 3 2 13 9 28	13 42 24 29 12 35
6	18	109	1 3 2 5 17 38	8 59 44 103 31 89
6	26	157	1 3 2 5 10 62	11 34 10 117 142 48
6	30	181	1 3 2 5 9 69	6 41 85 109 20 52
6	38	229	1 3 2 5 11 67	12 202 148 92 211 51
6	46	277	1 3 2 5 11 50	11 53 133 42 130 32
6	58	349	1 3 2 5 12 30	6 65 212 206 183 41
6	62	373	1 3 2 5 11 49	4 58 67 206 109 214
6	66	397	1 3 2 5 10 16	8 28 146 267 155 132
6	70	421	1 3 2 5 9 25	6 4 28 66 252 294
6	90	541	1 3 2 5 9 51	5 55 492 175 383 329
6	102	613	1 3 2 7 15 38	4 25 469 130 414 34
6	110	661	1 3 2 7 15 26	6 23 383 299 497 263
6	118	709	1 3 2 5 10 20	4 10 590 125 627 588
6	122	733	1 3 2 5 13 62	6 15 650 24 370 445
6	126	757	1 3 7 14 6 30	5 15 29 430 142 409
6	138	829	1 3 2 5 11 15	5 13 124 333 317 699
6	142	853	1 3 2 7 13 5	4 14 46 167 760 55
6	146	877	1 3 2 5 14 31	6 19 292 590 580 635
6	166	997	1 3 2 8 22 69	7 28 166 855 778 159
6	170	1021	1 3 2 8 23 28	10 35 710 185 927 755
6	178	1069	1 3 2 5 10 19	6 4 67 11 606 207
6	182	1093	1 3 7 2 15 87	5 19 594 1032 711 831
6	186	1117	1 3 7 2 17 54	7 5 187 653 912 841
6	202	1213	1 3 2 5 10 40	4 12 248 26 1135 415
6	206	1237	1 3 2 5 11 31	4 12 335 395 420 574
6	230	1381	1 3 2 5 10 48	4 12 69 672 662 706
6	238	1429	1 3 2 5 11 45	4 18 36 344 735 578
6	242	1453	1 3 2 5 11 9	4 15 61 1373 445 22
6	258	1549	1 3 2 9 15 23	4 18 42 394 734 177

continued on next page

m	t	p	$V^{(2)}(m,t)$ matrix	
6	266	1597	1 3 7 14 19 33	11 28 77 1366 1237 96
6	270	1621	1 3 7 2 17 47	7 4 60 722 1570 762
6	278	1669	1 3 7 2 22 13	7 24 179 667 1145 1404
6	282	1693	1 3 2 5 12 33	6 15 122 323 504 1305
6	290	1741	1 3 2 5 11 22	10 6 23 198 270 1655
6	298	1789	1 3 2 5 13 22	5 19 77 87 1080 1682
6	310	1861	1 3 2 9 25 49	4 12 42 628 872 932
6	322	1933	1 3 2 8 15 25	5 20 55 1838 1202 1890
6	338	2029	1 3 7 2 15 12	5 12 91 1949 1639 1006
6	342	2053	1 3 2 5 10 19	4 10 21 2 81 183
6	370	2221	1 3 7 15 12 47	10 38 51 1013 1583 1849
6	378	2269	1 3 2 7 15 50	4 2 50 2212 822 1290
6	382	2293	1 3 2 5 11 15	8 6 22 350 101 445
6	390	2341	1 3 7 2 15 22	5 13 35 2023 387 635
6	398	2389	1 3 2 9 15 53	8 19 40 313 1201 1823
6	406	2437	1 3 2 5 11 15	8 6 28 2233 2251 1656
6	426	2557	1 3 2 5 12 17	6 14 32 847 2126 418
6	446	2677	1 3 2 5 11 19	4 13 35 1221 1650 1877
6	458	2749	1 3 2 5 14 32	6 17 29 1094 950 336
6	466	2797	1 3 2 5 11 18	4 12 10 1827 1277 230
6	486	2917	1 3 2 5 10 18	6 17 3 216 383 2743
6	506	3037	1 3 2 5 10 14	7 4 25 478 1894 198
6	510	3061	1 3 2 5 11 39	7 4 18 419 2277 1623
6	518	3109	1 3 2 7 13 32	4 12 60 1091 2073 1012
6	530	3181	1 3 7 2 15 22	8 19 33 368 923 1436
6	538	3229	1 3 2 5 11 18	8 19 12 3193 2550 2624
6	542	3253	1 3 7 2 26 17	5 24 49 132 972 2113
6	550	3301	1 3 2 5 10 25	6 14 12 915 1101 679
6	562	3373	1 3 7 2 26 100	5 15 40 161 1558 260
6	578	3469	1 3 2 5 10 62	4 14 1 379 2159 1195
6	586	3517	1 3 2 7 17 30	4 20 12 744 1003 3267
6	590	3541	1 3 2 5 12 75	8 5 38 1063 2975 3282
6	602	3613	1 3 2 5 10 22	6 20 11 1473 3235 1188
6	606	3637	1 3 2 5 10 49	4 10 8 608 713 1154
6	618	3709	1 3 2 5 11 32	4 11 22 714 3648 2864
6	622	3733	1 3 7 15 2 31	10 5 30 1269 2465 567
6	642	3853	1 3 7 2 15 12	5 38 25 105 1701 2883
6	646	3877	1 3 2 5 10 20	15 9 40 1325 3838 1589
6	670	4021	1 3 2 5 9 18	6 4 24 610 2879 54
6	682	4093	1 3 7 2 15 25	5 25 60 29 453 149
6	710	4261	1 3 2 5 11 22	7 5 14 655 1127 3137

continued on next page

m	t	p	$V^{(2)}(m,t)$ matrix	
6	726	4357	1 3 2 5 10 26	6 20 15 1716 363 1755
6	758	4549	1 3 2 5 11 18	8 5 14 766 144 2986
6	766	4597	1 3 2 5 10 24	6 4 48 315 120 2192
6	770	4621	1 3 2 5 10 23	4 27 9 805 599 4122
6	798	4789	1 3 2 5 15 43	4 10 8 1079 876 301
6	802	4813	1 3 2 5 10 48	4 10 50 1686 276 3601
6	810	4861	1 3 2 5 9 50	6 4 36 2199 4387 2266
6	818	4909	1 3 2 5 11 29	7 21 19 342 2732 3884
6	822	4933	1 3 7 2 21 108	8 4 37 1175 542 2493
6	826	4957	1 3 7 2 23 46	5 13 42 373 748 4146
8	222	1777	1 3 6 11 4 21 31 60	9 126 1429 278 576 1232 1683 441
8	234	1873	1 3 2 5 15 10 73 102	4 190 1585 499 326 1428 766 142
8	266	2129	1 3 2 5 11 9 115 54	7 58 1160 1142 1389 658 1832 2052
8	270	2161	1 3 2 5 10 19 34 95	7 77 581 39 602 949 221 1943
8	302	2417	1 3 2 5 9 16 6 29	5 132 141 1383 2313 1601 1612 1236
8	326	2609	1 3 2 5 9 17 8 53	5 28 684 41 134 497 491 2220
8	350	2801	1 3 2 5 11 9 22 49	4 60 729 463 1066 1870 1703 1329
8	354	2833	1 3 6 11 4 21 31 226	6 68 1826 1475 2032 682 2319 1124
8	362	2897	1 3 2 5 9 15 25 49	7 18 539 1008 1802 840 709 2795
8	386	3089	1 3 2 5 11 9 19 38	5 27 601 1155 2625 2889 1976 2878
8	390	3121	1 3 2 5 13 19 34 71	7 38 2881 1364 173 220 2758 395
8	402	3217	1 3 2 5 13 10 30 170	5 26 603 160 2110 580 465 1292
8	414	3313	1 3 2 5 10 16 27 40	8 5 105 2868 1614 850 1016 141
8	462	3697	1 3 7 2 17 13 43 138	8 5 1861 854 2534 3338 1517 708
8	470	3761	1 3 2 5 10 16 40 87	7 22 3314 527 2042 2321 2883 385
8	474	3793	1 3 2 5 9 15 8 46	5 15 40 295 824 2444 3087 2744
8	486	3889	1 3 7 15 4 26 14 117	10 8 3137 2379 762 3383 1823 1592
8	506	4049	1 3 6 11 4 26 13 75	6 19 3006 833 3114 309 3157 3486
8	522	4177	1 4 11 2 23 17 77 220	5 14 1534 2572 3915 2724 1335 848
8	530	4241	1 3 2 5 9 15 6 45	8 20 568 2524 1273 3565 351 597
8	534	4273	1 3 6 11 4 21 96 120	9 24 1140 2911 205 2857 2044 4108
8	542	4337	1 3 2 5 9 19 6 46	8 4 235 2715 2052 657 2489 810
8	570	4561	1 3 7 15 4 23 66 92	10 39 1741 2475 215 1809 3819 2012
8	582	4657	1 3 2 5 10 23 20 60	8 23 382 3922 2475 1535 4349 2629
8	590	4721	1 4 13 2 22 7 76 250	19 15 150 3235 337 2144 4291 1314
8	602	4817	1 3 2 5 10 23 52 20	6 24 593 4247 3200 476 1013 4784

One can push this idea further, by considering $t \equiv 4 \pmod 8$, and considering the $4m$ cyclotomic classes of size $\frac{t}{4}$ in $GF(mt+1)$. The configuration of four vectors representing all $4m$ cyclotomic classes is a $V^{(4)}(m,t)$ *matrix*. The computation gets substantially more difficult. However, we were able to complete the search for $V^{(4)}(m,t)$ matrices, when $p = mt+1$ is a prime less than 5000, $t \equiv 4 \pmod 8$ and $m \in \{2,4\}$. Under the condition that $t \geq m$, all of these exist; we give solutions for $m = 4$. Again, we omit the first entry of each of the vectors, which is always 0.

m	t	p	$V^{(4)}(m,t)$ matrix			
4	4	17	1 3 2 7	3 14 5 12	4 13 10 5	6 16 14 10
4	28	113	1 3 2 5	5 11 9 38	9 22 61 48	11 57 17 103
4	60	241	1 3 2 5	4 9 15 26	7 21 42 85	19 143 192 148
4	84	337	1 4 11 3	5 14 29 23	17 55 12 152	23 286 146 135
4	100	401	1 3 2 5	4 11 1 7	12 4 31 65	18 200 124 53
4	108	433	1 3 7 2	5 4 12 23	7 5 15 180	31 225 117 290
4	148	593	1 3 2 5	5 12 3 25	8 5 25 43	9 50 43 348
4	220	881	1 4 11 3	5 14 27 20	11 34 61 190	25 343 733 310
4	244	977	1 3 2 5	4 10 3 56	7 4 24 148	13 7 969 673
4	252	1009	1 3 2 9	4 12 1 45	11 24 22 55	14 172 947 916
4	300	1201	1 3 2 5	5 15 10 36	11 26 63 156	29 165 505 275
4	324	1297	1 3 2 10	4 2 12 26	5 20 79 42	24 178 1277 1215
4	340	1361	1 3 2 5	4 10 1 15	7 16 14 78	12 47 480 1155
4	372	1489	1 3 2 5	4 10 17 27	8 25 54 199	13 39 816 622
4	388	1553	1 3 6 19	5 23 1 38	9 31 54 131	26 55 405 1309
4	444	1777	1 3 6 11	6 1 15 38	7 25 46 82	38 15 254 1050
4	468	1873	1 3 2 5	4 9 1 24	9 29 15 46	15 66 345 1233
4	532	2129	1 3 2 8	3 10 18 45	11 29 22 124	12 43 279 1480
4	540	2161	1 3 2 7	3 7 5 12	9 23 69 110	21 58 583 331
4	604	2417	1 3 2 5	4 9 1 7	7 5 14 34	10 58 418 2062
4	652	2609	1 3 2 5	4 10 1 15	5 15 3 241	8 30 517 737
4	700	2801	1 3 2 8	3 11 15 9	9 6 40 28	12 4 366 1497
4	708	2833	1 3 6 11	6 1 16 50	7 21 63 163	13 34 1287 1576
4	724	2897	1 3 2 5	4 10 1 9	7 4 14 37	11 5 503 1011
4	772	3089	1 3 2 5	5 11 9 24	7 19 12 34	13 37 83 451
4	780	3121	1 3 2 8	3 10 18 47	11 24 46 135	12 63 516 1220
4	804	3217	1 3 2 5	5 11 26 50	8 5 46 186	10 30 463 1683
4	828	3313	1 3 2 5	5 11 19 16	10 25 3 40	11 30 804 362
4	924	3697	1 3 7 2	5 4 15 29	10 8 30 70	13 123 18 3429
4	940	3761	1 3 2 5	5 11 9 32	9 27 21 126	12 67 457 620
4	948	3793	1 3 2 5	4 9 15 23	10 5 25 11	12 42 757 2197
4	972	3889	1 3 7 18	5 4 28 38	13 11 68 145	26 130 1465 3309
4	1012	4049	1 3 6 11	6 1 31 59	13 27 52 150	23 98 523 765
4	1044	4177	1 4 11 3	5 14 29 50	13 30 65 198	19 136 3183 1570
4	1060	4241	1 3 2 5	4 10 1 45	8 20 14 99	13 61 316 1649
4	1068	4273	1 3 6 11	6 15 29 58	15 6 57 180	17 206 2085 4039
4	1084	4337	1 3 2 5	4 10 8 17	8 20 33 60	21 85 243 3775
4	1140	4561	1 3 7 17	5 13 24 47	13 33 85 306	22 109 1090 3460
4	1164	4657	1 3 2 5	5 11 23 47	8 23 46 241	10 29 1800 651

continued on next page

m	t	p	$V^{(4)}(m,t)$ matrix			
4	1180	4721	1 4 13 3	5 20 39 15	11 9 77 305	25 100 1200 700
4	1204	4817	1 3 2 5	5 11 9 26	9 22 42 289	10 67 1847 3742

Similar techniques can be used to find incomplete transversal designs with a spanning set of holes. A *holey quasi-difference matrix* over the group Γ with subgroup Γ' with parameters n, u, λ and k is a $k \times s$ array whose entries are either empty or consist of a single entry of Γ. Moreover,

1. each row contains u empty entries;
2. each column contains at most one empty entry; and
3. for two distinct rows h and i, the list of differences $q_{hj} - q_{ij}$ contains each element of $\Gamma \setminus \Gamma'$ exactly λ times, and does not contain any element of Γ' (differences with empty entries are undefined).

If Γ' has order h and has index m in Γ, the quasi-difference matrix gives an incomplete TD establishing that $N(h^m u^1) \geq k-2$.

Again, to illustrate this we take the groups Γ and Γ' to be cyclic, and give the columns of some holey quasi-difference matrices. Each column of length k develops cyclically into k columns. We first give solutions in which each column is to be first developed by adjoining the column obtained by multiplying by -1 in Z_{hm}. The column *num* specifies the number of incomplete MOLS obtained:

num	h	m	u	Columns of holey qdm		
3	2	7	4	− 0 1 4 6	− 0 4 12 3	
3	2	10	6	− 0 1 3 6	− 0 4 9 16	− 0 6 18 7
3	2	12	4	0 1 3 7 16	− 0 3 8 22	− 0 6 23 10
3	2	13	8	− 0 1 3 6	− 0 4 9 19	− 0 6 14 2
				− 0 9 16 1		
3	3	5	4	− 0 1 3 7	− 0 3 11 2	
3	3	7	6	− 0 1 3 16	− 0 4 1 12	− 0 6 11 2
3	3	9	8	− 0 1 3 6	− 0 4 11 17	− 0 8 23
				− 0 10 15 2		
3	3	11	10	− 0 1 3 6	− 0 4 9 16	− 0 6 19 10
				− 0 8 25 15	− 0 12 31 13	
3	3	11	0	0 1 3 26 7	0 3 12 16 21	0 6 14 1 16
5	3	11	6	− 0 1 3 8 2 20	− 0 7 31 21 4 25	− 0 8 5 24 28 15

We also give some solutions for the columns of holey quasi–difference matrices with no additional automorphism assumed, but with columns developed cyclically.

num	h	m	u	Columns of holey qdm		
3	2	8	3	0 1 3 2 13	− 0 4 2 9	− 0 5 11 4
				− 0 10 6 3		
4	2	9	4	− 0 1 3 15 2	− 0 3 13 11 7	− 0 4 1 8 16
				− 0 6 5 16 11		
3	2	11	5	0 1 3 2 5	− 0 4 16 8	− 0 6 19 16
				− 0 7 5 14	− 0 8 13 1	− 0 15 9 5
3	3	6	5	− 0 1 3 8	− 0 3 13 17	− 0 7 16 9
				− 0 13 11 10	− 0 14 4 1	

One further example is given here, where the the array

$$\begin{array}{cccccc} 0 & 0 & 0 & - & - & - \\ 1 & 9 & 11 & 0 & 0 & 0 \\ 4 & 8 & 2 & 2 & 4 & 8 \\ - & - & - & 8 & 2 & 4 \\ 3 & 13 & 5 & 12 & 10 & 6 \\ 9 & 11 & 1 & 11 & 1 & 9 \end{array}$$

is inflated by having each column replaced by three columns under the row permutation (1 2 3)(4 5 6). This establishes that $N(2^7 3^1) \geq 4$.

Of course, the difficulty here is in producing solutions for a large number of MOLS, and the computational methods employed seem to be suited only for quite small problems, unless the cyclotomic structure of the finite field is used. A most challenging area is to find effective computational techniques for larger problems, possibly by using the algebraic structure of finite fields in another way.

5 A VARIANT OF WILSON'S THEOREM

In the preceding two sections, we have seen two direct methods for producing incomplete transversal designs. Here we examine a recursive technique, which is a variant of the central construction in the field due to Wilson [30].

Theorem 5.1 *If there exists a* $ITD(k, n+h; h)$ *for which* $(k-2)h = n$*, and there exists a* $TD(k, m)$*, then there exists an* $ITD(k, mn+(m-1)h; n^m(h(m-1))^1)$.

Proof: Delete the kh points of the $ITD(k, n+h; h)$ which are in the hole. Since $(k-2)h = n$, all blocks have one point deleted in this way, and so we obtain a GDD of block size $k-1$ and type n^k. This GDD has a partition into holey parallel classes, so that for each group there are h of the holey parallel classes missing that group. Now give every point of this GDD weight m, and place a copy of a resolvable $ITD(k-1, m; 1^m)$ on the $(k-1)m$ points arising from each block of the GDD, arranging the placement so that, for $1 \leq i \leq m$, the ith hole of size 1 in the ITD coincides with the ith copy of each point in the block of the GDD. After carrying out this placement for each block of the GDD, we obtain an incomplete GDD with k groups each of size mn, and m holes each of size kn, with every hole intersecting every group in exactly n points. All blocks are of size $k-1$, and they are partitioned into holey parallel classes, with $h(m-1)$ holey parallel classes missing each group. Adding $h(m-1)$ points to each group, each attached to the blocks of a holey parallel class missing that group, gives the required ITD. □

One useful application of this theorem employs a $TD(p+1, p)$ for p a prime power. Using $ITD(p+1, (p-1)+1; 1)$ and $TD(p+1, p)$, we obtain a $ITD(p+1, p^2-1; (p-1)^{p+1})$.

Numerous constructions of this type are known, and are the main technique by which the incomplete designs constructed directly are used to produce larger transversal designs. Examples of other variants of Wilson's theorem appear in [6, 8, 11, 12].

6 CONCLUDING REMARKS

The construction of MOLS remains a challenging problem. In this paper, we have seen two techniques that can be used with some success in constructing small transversal designs and incomplete transversal designs for use in the recursive constructions. Most existence questions in combinatorial design theory are solved, if at all, by an effective marriage of direct and recursive methods. We have not explored the recursive techniques here, and refer the interested reader to [6, 8].

Let us close with a few open problems. Constructions from pairwise balanced designs afford a number of dramatic improvements on other recursive constructions. Thus it would be particularly interesting to extend the battery of constructions for pairwise balanced designs with large (and hopefully prime power) block sizes. Greig [14] has made progress in this direction, and it appears to be a very important direction for future research.

Turning to quasi–difference matrices, the main obstacle appears to be the limitation to small numbers of MOLS. Even for $V(m,t)$ vectors, no information at all is available for $m > 10$ (except for some asymptotic existence results). Direct constructions of $V(m,t)$ vectors for large m are very much to be desired; even for $m = 3$, however, the existence of $V(3,t)$ vectors is not completely settled.

REFERENCES

[1] R.J.R. Abel, "Four mutually orthogonal latin squares of orders 28 and 52", *J. Combinat. Theory (A)* 58 (1991), 306–309.

[2] R.J.R. Abel, e–mail communications, 1994.

[3] R.J.R. Abel and Y.W. Cheng, "Some new MOLS of order $2^n p$ for p a prime power", *Austral. J. Combin.* 10 (1994), 175–186.

[4] R.J.R. Abel and D.T. Todorov, "Four MOLS of order 20, 30, 38 and 44", *J. Combinat. Theory (A)* 64 (1993), 144–148.

[5] R.D. Baker, "Whist tournaments", *Proc. Sixth Southeastern Conf. Combin., Graph Theory, Computing* Utilitas Math. (1975) 89–100.

[6] T. Beth, D. Jungnickel and H. Lenz, *Design Theory*, Cambridge University Press, Cambridge, 1986.

[7] R.C. Bose, S.S. Shrikhande and E.T. Parker, "Further results on the construction of mutually orthogonal latin squares and the falsity of Euler's conjecture", *Canad. J. Math.* 12 (1960), 189–203.

[8] A.E. Brouwer and G.H.J. van Rees, "More mutually orthogonal latin squares", *Discrete Math.* 39 (1982), 263–281.

[9] C.J. Colbourn, "Four MOLS of order 26", *J. Comb. Math. Comb. Comput.*, to appear.

[10] C.J. Colbourn, "Some direct constructions for incomplete transversal designs", *J. Stat. Plan. Infer.*, to appear.

[11] C.J. Colbourn, J.H. Dinitz and M. Wojtas, "Thwarts in transversal designs", *Des. Codes Crypt.*, to appear.

[12] B. Du, "On the existence of incomplete transversal designs with block size 5", *Discrete Math.*, to appear.

[13] L. Euler, "Recherches sur une nouvelle espèce de quarrés magiques", *Verh. Zeeuw. Gen. Weten. Vlissengen* 9 (1782), 85–239.

[14] M. Greig, "Designs from configurations in projective planes", preprint, 1992.

[15] T.P. Kirkman, "On the perfect r–partitions of $r^2 - r + 1$", *Transactions of the Historic Society of Lancashire and Cheshire* (1850), 127–142.

[16] D.M. Johnson, A.L. Dulmage and N.S. Mendelsohn, "Orthomorphisms of groups of orthogonal latin squares", *Canad. J. Math.* 13 (1961), 356–372.

[17] H.F. MacNeish, "Euler squares", *Ann. Math. (NY)* 23 (1922), 221–227.

[18] W.H. Mills, "Some mutually orthogonal latin squares", *Proc. Eighth S.E. Conf. Combin. Graph Theory Computing*, 1977, pp. 473–487.

[19] R.C. Mullin, P.J. Schellenberg, D.R. Stinson and S.A. Vanstone, "Some results on the existence of squares", *Ann. Discrete Math.* 6 (1980), 257–274.

[20] A.V. Nazarok, "Five pairwise orthogonal latin squares of order 21", *Issled. oper. i ASU*, 1991, pp. 54–56.

[21] C.E. Roberts Jr., "Sets of mutually orthogonal latin squares with 'like subsquares' ", *J. Combinat. Theory (A)* 61 (1992) 50–63.

[22] C.E. Roberts Jr., "Sets of mutually orthogonal latin squares with 'like subsquares' II", preprint.

[23] P.J. Schellenberg, G.H.J. van Rees and S.A. Vanstone, "Four pairwise orthogonal latin squares of order 15", *Ars Combinat.* 6 (1978), 141–150.

[24] G. Tarry, "Le problème de 36 officiers", *Ass. Franc. Av. Sci.* 29 (1900), 170–203.

[25] D.T. Todorov, "Three mutually orthogonal latin squares of order 14", *Ars Combinat.* 20 (1985), 45–48.

[26] D.T. Todorov, "Four mutually orthogonal latin squares of order 20", *Ars Combinat.* 27 (1989), 63–65.

[27] G.H.J. van Rees, private communication, 1994.

[28] S.P. Wang, *On self-orthogonal latin squares and partial transversals of latin squares*, Ph.D. thesis, Ohio State University, 1978.

[29] R.M. Wilson, "A few more squares", *Proc. Fifth Southeastern Conf. Combin. Graph Theory Computing* (1974), pp. 675–680.

[30] R.M. Wilson, "Concerning the number of mutually orthogonal latin squares", *Discrete Math.* 9 (1974), 181–198.

[31] M. Wojtas, "Some new matrices–minus–diagonal and MOLS", *Discrete Math.* 76 (1989) 291–292.

[32] M. Wojtas, "Five mutually orthogonal latin squares of orders 24 and 40", preprint.

THE SPECTRUM OF R-ORTHOGONAL LATIN SQUARES

Charles J. Colbourn
*** L. Zhu**

Combinatorics and Optimization
University of Waterloo
Waterloo, Ontario, CANADA N2L 3G1

** Department of Mathematics,*
Suzhou University, Suzhou 215006, China

ABSTRACT

Two latin squares of side n are r-orthogonal if, when superimposed, there are exactly r distinct ordered pairs. In this paper, it is established that for all $n \geq 27$, r-orthogonal latin squares of side n exist if and only if $n+2 \leq r \leq n^2-2$ or $r \in \{n, n^2\}$. An almost complete solution is given for smaller sides.

1 LATIN SQUARES AND R-ORTHOGONALITY

A *latin square* of *side* n is an $n \times n$ array L, with rows indexed by an n-set $\rho(L)$ and columns indexed by an n-set $\gamma(L)$, in which each entry contains a single element of an n-set $\sigma(L)$. Every element of $\sigma(L)$ occurs precisely once in each row and in each column. Two latin squares of side n, $L = (\ell_{ij})$ and $M = (m_{ij})$, are *r-orthogonal* if they have the same row, column and symbol sets and

$$|\{(\ell_{ij}, m_{ij}) : i \in \rho(L),\ j \in \gamma(L)\}| \ = \ r.$$

For the most part, we treat latin squares of side n in which the row, column and symbol sets each equal $\{0, 1, \ldots, n-1\}$.

The natural question, apparently first treated systematically by Belyavskaya [3], is: For which integers n and r does a pair of r-orthogonal latin squares of side n exist? Evidently, $n \leq r \leq n^2$, and an easy argument [3] establishes that

C. J. Colbourn and E. S. Mahmoodian (eds.), Combinatorics Advances, 49–75.

Side n	Genuine Exceptions	Possible Exceptions
2	2	
3	2, 3, 4	
4	3, 6, 7, 9, 10	
5	3, 4, 15, 17, 18	
6	30	27
7, 8, 10, 11, 13, 14, 16, 17, 18, 19, 20, 22, 23, 25, 26		$n^2 - n - 3$

Table 1 Genuine and Possible Exceptions Remaining

$r \notin \{n+1, n^2-1\}$. We define two sets of interest:

$$\mathcal{P}_n = \{0, \ldots, n^2 - n\} \setminus \{1, n^2 - n - 1\}, \text{ and}$$

$$\mathcal{E}_n = \{r - n : \exists\ r - \text{orthogonal latin squares of side } n\}.$$

Using this notation, $\mathcal{E}_n \subseteq \mathcal{P}_n$. In this paper, we establish that in fact equality holds, except for certain small values of n. We also provide a fairly complete solution for small n; in Table 1, we list the numbers in $\mathcal{P}_n \setminus \mathcal{E}_n$ as *genuine exceptions*, and the numbers in $\mathcal{P}_n$ whose membership status in $\mathcal{E}_n$ is unknown as *possible exceptions*. We prove the following theorem:

Main Theorem 1.1 *1. For $n \in \{9, 12, 15, 21, 24\}$ and $n \geq 27$, $\mathcal{E}_n = \mathcal{P}_n$;*

2. For $n \in \{7, 8, 10, 11, 13, 14, 16, 17, 18, 19, 20, 22, 23, 25, 26\}$, $\mathcal{P}_n \setminus \{n^2 - n - 3\} \subseteq \mathcal{E}_n \subseteq \mathcal{P}_n$.

2 SOME BASIC CONSTRUCTIONS

Lemma 2.1 $0 \in \mathcal{E}_n$ *for all* $n \geq 1$.

Proof: Simply take two copies of the same $n \times n$ latin square. □

Lemma 2.2 *$n^2 - n \in \mathcal{E}_n$ for all $n \geq 1$ except when $n \in \{2, 6\}$.*

Proof: Simply take two mutually orthogonal latin squares of order n. □

Belyavskaya [3] established the following result:

Lemma 2.3 *If $2 \leq k \leq \frac{n}{2}$, then $k \in \mathcal{E}_n$.*

We use some well known embedding results to handle numerous cases near the minimum and maximum. We require a technical lemma first.

Lemma 2.4 *If there are two r-orthogonal latin squares of side n, then there are two r-orthogonal latin squares of side n in which the superposition contains every ordered pair in $\{(i,i) : 0 \leq i < n\}$.*

Proof: Let L and M be r-orthogonal latin squares of side n on symbol set $\{0, \ldots, n-1\}$. Form a bipartite multigraph G on vertices $V = \{x_0, \ldots, x_{n-1}\} \cup \{y_0, \ldots, y_{n-1}\}$, where $\{x_i, y_j\}$ is an edge λ times if (i, j) occurs exactly λ times in the superposition of L and M. G is n-regular and bipartite, so G has a perfect matching F (see Corollary 5.2 in [5], for example). Rename the symbols of M so that, if $\{x_i, y_j\} \in F$, symbol j in M is renamed as i. The resulting squares have the required property. □

We suppose henceforth that whenever r-orthogonal latin squares are used, they have all constant pairs (of the form (i, i)) in the superposition.

Lemma 2.5 *$\mathcal{E}_n \subseteq \mathcal{E}_m$ for all $m \geq 2n$, $n \geq 1$.*

Proof: If $r \in \mathcal{E}_n$, consider two $(r+n)$-orthogonal latin squares L_1 and L_2 of side n. Form an $m \times m$ square M that contains a subsquare L of side n on symbols $\{0, \ldots, n-1\}$ in the upper left corner. Form two squares of side m, M_1 and M_2, by replacing the subsquare L in M by L_1 and L_2, respectively. Then M_1 and M_2 are $(r+m)$-orthogonal, so that $r \in \mathcal{E}_m$. □

Lemma 2.6 *If $x, y \in \mathcal{E}_n$, then $x + y \in \mathcal{E}_{2n}$.*

Proof: Given an arbitrary latin square L of side n, let L' be the latin square obtained by adding n to each entry of L. Form the $2n \times 2n$ latin square M as $\left(\begin{smallmatrix} LL' \\ L'L \end{smallmatrix}\right)$. Replacing both occurrences of L by $(x+n)$-orthogonal latin squares L_1 and L_2, and both occurrences of L' by $(y+n)$-orthogonal latin squares L'_1 and L'_2, gives two $(2n+x+y)$-orthogonal latin squares. □

Lemma 2.7 *If $n^2 - n - x \in \mathcal{E}_n$, then $m^2 - m - x \in \mathcal{E}_m$ whenever $m \geq 3n$, $n \geq 2$.*

Proof: $IMOLS(m, n)$ exist whenever $m \geq 3n$, $n \geq 2$ [11]. Filling the hole with two $(n^2 - x)$-orthogonal latin squares of side n gives the desired result. □

3 SMALL SIDES

Belyavskaya [3] performed an exhaustive search for the cases $2 \leq n \leq 5$, and obtained the complete characterization for these sides given in Table 1. We extended her investigation of small orders. The primary computational device that we use is the following. Take two $n \times n$ (not necessarily distinct) latin squares L_1 and L_2. Choose two permutations π_r and π_c, and use them to permute the rows and columns of L_2 to obtain a square L'_2. Then determine the number of distinct ordered pairs when L_1 and L'_2 are superimposed. For side 6, we employ the following squares:

0	1	2	3
0 1 2 3 4 5	0 1 2 3 4 5	0 1 2 3 4 5	0 1 2 3 4 5
1 0 3 2 5 4	1 0 3 2 5 4	1 0 3 4 5 2	1 0 3 4 5 2
2 3 4 5 0 1	2 3 4 5 0 1	2 3 0 5 1 4	2 3 0 5 1 4
3 2 5 4 1 0	3 2 5 4 1 0	3 4 5 0 2 1	3 4 5 0 2 1
4 5 0 1 2 3	4 5 0 1 3 2	4 5 1 2 0 3	4 5 1 2 3 0
5 4 1 0 3 2	5 4 1 0 2 3	5 2 4 1 3 0	5 2 4 1 0 3

Next we specify which squares are used, along with the row and column permutations to be applied to the second square. The permutations are applied as follows. If, say, 4 1 5 0 2 3 is the row permutation, applying it means to take the rows of the second square in the order specified, namely rows 4, 1, 5, 0, 2, and 3 in that order.

$r-n$	First Square	Second Square	Row Permutation	Column Permutation
4	1	1	2 3 5 4 0 1	3 2 5 4 0 1
5	3	3	4 1 5 0 2 3	3 5 2 4 1 0
7	2	0	1 3 0 5 2 4	4 0 5 2 1 3
8	2	1	2 0 1 4 3 5	2 0 1 4 3 5
9	0	1	1 5 2 3 4 0	3 2 5 1 0 4
10	0	2	3 1 0 4 2 5	5 2 4 0 1 3
11	1	1	4 0 5 2 3 1	5 2 3 1 4 0
12	0	0	4 5 1 2 3 0	0 3 2 1 5 4
13	2	2	1 3 2 4 5 0	3 5 4 0 1 2
14	3	3	2 1 5 3 4 0	5 1 4 3 0 2
15	1	3	3 5 2 0 4 1	4 3 0 5 1 2
16	3	3	1 5 4 2 3 0	2 4 3 0 1 5
17	0	2	2 5 1 0 4 3	2 1 3 0 5 4
18	3	2	5 3 0 2 1 4	5 4 3 2 0 1
19	1	1	2 4 1 5 0 3	3 0 5 4 1 2
20	0	1	1 3 5 4 0 2	1 2 4 5 3 0
21	0	3	2 5 1 4 0 3	1 0 4 5 3 2
22	0	2	5 4 3 1 0 2	1 2 5 3 0 4
23	3	1	0 4 1 5 2 3	4 1 2 5 3 0
24	0	0	3 2 1 0 5 4	4 3 5 0 1 2
25	1	2	2 5 0 3 4 1	2 3 5 4 0 1
26	3	3	1 3 2 0 5 4	1 3 0 5 4 2

These, together with Lemma 2.3, 2.5 and 2.7 determine $\mathcal{E}_6$ with the exception of 27 and 30. It has been known since 1900 [13] that two 36-orthogonal latin squares of order 6 do not exist, so 30 is a genuine exception. The case of 27 remains open.

For side 7, we employ three squares:

0	1	2
0 1 2 3 4 5 6	0 1 2 3 4 5 6	0 1 2 3 4 5 6
1 0 3 2 5 6 4	1 0 3 2 5 6 4	1 0 3 2 5 6 4
2 3 4 5 6 0 1	2 3 4 5 6 0 1	2 3 1 4 6 0 5
3 4 5 6 1 2 0	3 4 1 6 0 2 5	3 4 5 6 0 2 1
4 6 1 0 2 3 5	4 6 5 1 2 3 0	4 6 0 5 2 1 3
5 2 6 1 0 4 3	5 2 6 0 1 4 3	5 2 6 1 3 4 0
6 5 0 4 3 1 2	6 5 0 4 3 1 2	6 5 4 0 1 3 2

The squares, and row and column permutations used, are:

$r-n$	First Square	Second Square	Row Permutation	Column Permutation
4	2	2	6 5 2 4 3 0 1	1 0 6 4 2 5 3
5	2	2	4 0 5 3 1 6 2	0 4 3 5 6 2 1
7	0	0	2 5 6 4 1 0 3	6 1 0 3 2 4 5
8	2	2	5 4 1 6 3 0 2	3 0 6 5 1 4 2
9	2	2	2 1 0 3 6 4 5	0 3 1 2 4 6 5
10	2	2	1 0 2 4 3 6 5	0 5 3 2 6 1 4
11	0	0	0 1 4 3 2 6 5	6 4 2 3 1 5 0
12	2	2	4 3 2 1 0 6 5	4 6 3 2 1 0 5
13	2	2	3 4 0 5 1 6 2	2 3 5 6 0 1 4
14	1	1	1 2 6 5 0 4 3	5 6 1 3 4 0 2
15	2	2	6 2 3 1 5 0 4	4 2 6 3 5 0 1
16	2	1	4 1 0 5 3 2 6	0 5 6 1 2 4 3
17	2	2	1 5 2 3 6 0 4	3 1 0 2 4 6 5
18	2	0	1 0 2 5 6 4 3	1 4 2 0 6 5 3
19	1	0	6 2 3 1 4 0 5	3 5 6 0 4 2 1
20	1	2	3 0 1 2 4 6 5	5 4 1 2 0 6 3
21	1	2	0 5 4 6 2 3 1	3 6 5 1 4 0 2
22	1	2	4 0 2 5 3 6 1	3 5 1 2 6 0 4
23	1	0	3 4 2 5 1 6 0	0 5 3 2 4 1 6
24	2	2	2 3 4 0 1 6 5	5 6 0 3 1 4 2
25	0	1	2 0 3 1 5 4 6	2 1 0 5 6 3 4
26	1	0	5 6 2 3 1 0 4	3 4 2 1 6 0 5
27	1	0	6 5 2 4 0 3 1	1 6 2 3 0 5 4
28	1	0	4 0 2 6 1 5 3	1 2 0 6 5 3 4
29	2	0	6 5 4 1 2 0 3	2 0 4 5 6 3 1
30	1	0	2 1 3 0 6 4 5	5 3 6 1 0 4 2
31	2	1	6 5 0 1 2 3 4	0 5 1 4 2 3 6
32	0	1	2 1 0 5 3 4 6	5 6 2 3 4 1 0
33	2	0	5 6 3 2 0 1 4	5 0 1 3 6 4 2
34	2	0	0 1 6 4 5 2 3	4 0 2 5 1 3 6
35	2	2	5 3 0 4 2 6 1	6 0 1 4 2 3 5
36	1	1	2 6 0 5 4 3 1	0 1 2 3 4 5 6
37	1	0	5 4 3 6 2 1 0	4 1 3 0 5 2 6
38	0	0	0 3 4 1 2 5 6	3 1 2 0 4 6 5

Together with Lemmas 2.3, 2.5 and 2.7, this leaves only the possible exception of 39.

For side 8, we employ four symmetric squares:

0	1	2	3
0 1 2 3 4 5 6 7	0 1 2 3 4 5 6 7	0 1 2 3 4 5 6 7	0 1 2 3 4 5 6 7
1 0 3 2 5 4 7 6	1 0 3 2 5 4 7 6	1 0 3 2 5 4 7 6	1 0 3 2 5 4 7 6
2 3 0 1 6 7 4 5	2 3 0 1 6 7 4 5	2 3 0 1 6 7 4 5	2 3 0 1 6 7 4 5
3 2 1 0 7 6 5 4	3 2 1 0 7 6 5 4	3 2 1 0 7 6 5 4	3 2 1 0 7 6 5 4
4 5 6 7 0 1 2 3	4 5 6 7 0 1 2 3	4 5 6 7 0 1 3 2	4 5 6 7 0 1 3 2
5 4 7 6 1 0 3 2	5 4 7 6 1 0 3 2	5 4 7 6 1 0 2 3	5 4 7 6 1 0 2 3
6 7 4 5 2 3 0 1	6 7 4 5 2 3 1 0	6 7 4 5 3 2 0 1	6 7 4 5 3 2 1 0
7 6 5 4 3 2 1 0	7 6 5 4 3 2 0 1	7 6 5 4 2 3 1 0	7 6 5 4 2 3 0 1

Then the required squares, and row and column permutations, are:

$r-n$	First Square	Second Square	Row Permutation	Column Permutation
11	3	3	6 5 0 3 7 4 1 2	4 7 1 2 5 6 0 3
15	0	1	5 4 7 6 1 2 0 3	7 6 4 5 1 3 0 2
18	2	0	1 0 7 6 5 4 2 3	3 6 5 4 7 2 1 0
19	2	1	5 1 3 7 4 0 6 2	0 4 2 3 1 5 6 7
21	0	1	7 0 5 6 3 4 1 2	0 1 6 3 2 5 4 7
22	2	0	6 3 2 7 1 0 5 4	6 0 1 7 5 4 2 3
23	2	2	0 2 3 4 5 7 1 6	2 0 5 3 6 4 7 1
25	1	3	0 6 7 1 4 2 5 3	0 4 6 1 5 2 3 7
26	0	3	5 7 0 6 2 4 1 3	7 1 0 6 4 2 3 5
27	3	3	5 7 0 3 1 6 4 2	5 7 2 4 1 6 0 3
28	1	1	0 5 3 1 6 4 7 2	2 4 0 1 7 6 3 5
29	2	2	6 2 1 3 5 7 0 4	0 3 1 5 2 7 4 6
30	0	0	5 7 6 1 3 0 4 2	0 4 2 1 7 5 3 6
31	0	1	7 1 3 4 6 5 0 2	6 3 4 1 0 2 7 5
32	2	3	4 0 1 2 6 5 7 3	6 7 2 5 1 4 3 0
33	1	1	6 4 2 0 5 1 3 7	1 4 5 7 6 3 2 0
34	1	2	3 6 1 4 2 0 5 7	3 6 5 2 7 1 4 0
35	2	1	2 5 4 0 6 7 3 1	0 7 2 6 3 5 1 4
36	3	3	2 1 6 3 4 5 0 7	4 1 7 0 3 6 2 5
37	2	2	0 2 5 6 3 1 4 7	1 7 2 4 3 5 0 6
38	1	0	4 3 7 2 5 1 6 0	0 6 2 4 7 1 5 3
39	1	1	7 5 0 3 6 4 2 1	7 5 4 6 2 3 1 0

$r-n$	First Square	Second Square	Row Permutation	Column Permutation
40	0	2	5 0 7 3 2 6 1 4	2 1 4 0 3 6 5 7
41	3	3	7 2 1 5 4 0 6 3	3 1 2 0 4 6 7 5
42	2	0	3 1 2 6 7 5 0 4	6 2 0 1 7 3 4 5
43	2	2	4 0 2 1 7 3 6 5	0 3 4 6 2 1 7 5
44	0	3	4 1 5 0 2 7 3 6	6 4 5 7 3 1 0 2
45	1	2	2 1 3 0 4 7 5 6	1 5 4 0 2 6 7 3
46	2	1	0 3 1 2 6 5 4 7	6 1 2 5 3 4 0 7
47	3	3	0 2 7 5 3 1 4 6	2 6 7 3 0 4 5 1
48	1	2	5 1 2 6 0 4 7 3	7 6 2 3 4 5 1 0
49	1	1	6 2 3 7 0 4 5 1	6 0 2 4 5 3 1 7
50	3	2	2 7 3 6 0 5 1 4	5 3 1 7 6 0 2 4
52	0	3	5 3 7 1 6 0 4 2	5 4 6 7 3 2 0 1

The remaining values come from Lemmas 2.6 and 2.7.

For side 9, Belyavskaya [3] establishes that if $0 \leq x \leq 72$, and x is a multiple of 6 or of 9, then $x \in \mathcal{E}_9$. For some of the remaining cases, we employ six squares:

0	1	2
0 1 2 3 4 5 6 7 8	0 2 1 3 5 4 6 8 7	0 1 2 3 4 5 6 7 8
2 0 1 5 3 4 8 6 7	2 1 0 5 4 3 8 7 6	1 2 3 4 5 6 7 8 0
1 2 0 4 5 3 7 8 6	1 0 2 4 3 5 7 6 8	2 3 4 5 6 7 8 0 1
6 7 8 0 1 2 3 4 5	6 8 7 0 2 1 3 5 4	3 4 5 6 7 8 0 1 2
8 6 7 2 0 1 5 3 4	8 7 6 2 1 0 5 4 3	4 5 6 7 8 0 1 2 3
7 8 6 1 2 0 4 5 3	7 6 8 1 0 2 4 3 5	5 6 7 8 0 1 2 3 4
3 4 5 6 7 8 0 1 2	3 5 4 6 8 7 0 2 1	6 7 8 0 1 2 3 4 5
5 3 4 8 6 7 2 0 1	5 4 3 8 7 6 2 1 0	7 8 0 1 2 3 4 5 6
4 5 3 7 8 6 1 2 0	4 3 5 7 6 8 1 0 2	8 0 1 2 3 4 5 6 7

3
0 1 2 3 4 5 6 7 8
1 0 3 2 5 6 7 8 4
2 3 1 0 6 7 8 4 5
3 2 0 1 7 8 4 5 6
4 5 6 7 8 0 1 2 3
5 6 7 8 1 4 2 3 0
6 7 8 4 2 3 5 0 1
7 8 4 5 3 1 0 6 2
8 4 5 6 0 2 3 1 7

4
0 1 2 3 4 5 6 7 8
1 0 3 2 5 6 7 8 4
2 3 1 0 6 7 8 4 5
3 2 0 1 7 8 4 5 6
4 5 6 7 8 0 1 2 3
5 6 7 8 2 4 3 0 1
6 7 8 4 3 2 5 1 0
7 8 4 5 1 3 0 6 2
8 4 5 6 0 1 2 3 7

5
0 1 2 3 4 5 6 7 8
1 0 3 2 5 6 7 8 4
2 3 0 1 6 7 8 4 5
3 2 1 0 7 8 4 5 6
4 5 6 7 8 0 1 2 3
5 6 7 8 1 4 2 3 0
6 7 8 4 2 3 5 0 1
7 8 4 5 3 1 0 6 2
8 4 5 6 0 2 3 1 7

Then the squares, and row and column permutations used, are:

$r-n$	First Square	Second Square	Row Permutation	Column Permutation
10	5	5	2 3 0 1 4 5 6 7 8	2 3 0 1 4 5 6 7 8
13	5	5	2 0 1 3 4 5 6 7 8	2 0 1 3 4 5 6 7 8
14	1	0	2 7 3 8 6 0 5 1 4	3 5 6 7 2 1 4 8 0
15	3	5	1 3 2 0 4 5 6 7 8	1 3 2 0 4 5 6 7 8
16	1	1	4 7 2 3 8 1 5 6 0	1 8 3 0 6 4 2 7 5
17	4	4	1 3 2 0 4 5 6 7 8	2 1 3 0 4 5 6 7 8
19	2	2	2 0 7 1 5 3 8 6 4	5 1 8 6 4 2 0 3 7
20	2	2	4 0 5 7 3 8 2 6 1	1 6 0 4 8 2 7 3 5
21	0	2	0 5 1 7 6 2 3 4 8	0 4 8 2 3 7 5 1 6
22	2	2	0 4 1 7 2 5 8 3 6	5 8 3 0 6 1 4 7 2
23	2	1	2 7 3 6 4 0 8 1 5	5 7 4 8 1 3 2 6 0
25	1	0	3 2 5 1 4 0 6 7 8	5 4 3 8 0 2 1 7 6
26	0	2	2 8 4 0 6 1 5 3 7	3 7 0 5 1 2 8 4 6
28	2	1	2 5 8 6 3 0 4 7 1	0 1 7 4 2 8 5 3 6
29	0	1	6 3 0 5 8 2 7 4 1	5 3 6 7 0 1 4 2 8
31	1	2	1 4 7 0 6 3 8 5 2	7 0 4 5 1 2 6 8 3
32	1	1	4 5 3 0 1 6 8 2 7	1 7 2 3 8 6 5 4 0
33	0	0	6 5 8 4 7 3 1 0 2	2 7 1 6 4 8 3 5 0
34	1	0	1 7 3 0 4 5 2 6 8	8 0 4 6 1 3 7 2 5
35	2	1	1 7 2 8 0 6 4 3 5	4 0 2 7 6 8 1 3 5
37	1	2	3 5 1 2 8 0 6 4 7	8 5 7 0 1 6 2 4 3
38	2	2	3 0 4 8 1 6 5 2 7	2 0 3 4 5 6 1 8 7
39	1	1	6 5 0 1 7 2 8 4 3	7 4 0 3 2 8 6 1 5

$r-n$	First Square	Second Square	Row Permutation	Column Permutation
40	0	1	5 2 7 8 0 1 4 3 6	7 3 6 4 5 8 2 0 1
41	2	0	0 2 3 7 5 6 1 4 8	2 6 5 8 0 7 3 1 4
43	1	0	2 3 4 7 8 0 5 6 1	1 3 8 4 2 5 6 7 0
44	1	2	8 4 1 7 5 0 6 2 3	6 8 0 5 1 3 2 7 4
46	1	2	6 3 5 4 7 1 2 8 0	8 3 4 7 6 5 2 1 0
47	1	1	5 2 3 8 7 0 1 6 4	2 8 0 1 7 6 3 5 4
49	2	0	5 6 3 0 1 7 4 2 8	1 5 4 2 8 7 0 6 3
50	0	1	6 8 4 5 2 3 1 0 7	4 7 3 0 8 2 1 6 5
51	0	2	1 6 2 0 4 7 8 3 5	7 6 8 3 5 4 2 1 0
52	2	2	2 1 5 8 6 4 0 3 7	3 4 5 1 0 6 2 8 7
53	2	0	1 3 2 7 0 5 8 6 4	5 1 0 2 7 8 6 3 4
55	2	2	4 0 3 1 7 6 2 8 5	2 6 1 5 8 3 7 0 4
56	2	2	3 6 5 8 7 1 0 2 4	2 7 8 4 0 5 6 3 1
57	1	0	4 0 8 5 3 6 7 2 1	6 2 0 3 1 8 5 4 7
58	2	2	1 7 4 0 6 3 8 5 2	4 8 2 5 3 6 0 7 1
64	2	2	1 6 3 0 8 5 2 7 4	7 8 2 5 0 3 6 1 4

With Belyavskaya's results and Lemmas 2.3, 2.5 and 2.7, the cases that remain are 7, 11, 59, 61, 62, 65, 67, 68, and 69. First we treat the two remaining small values. Consider the square

$$\begin{array}{|ccccccccc|}
\hline
\underline{0} & \underline{1} & \underline{2} & \underline{3} & 6 & \underline{4} & \underline{5} & 8 & 7 \\
\underline{2} & \underline{3} & \underline{1} & \underline{0} & 8 & \underline{5} & \underline{4} & 7 & 6 \\
\underline{1} & \underline{0} & \underline{3} & \underline{2} & 5 & 6 & 7 & 4 & 8 \\
\underline{3} & \underline{2} & \underline{0} & \underline{1} & 4 & 7 & 8 & 6 & 5 \\
6 & 7 & 8 & 4 & 1 & 2 & 0 & 5 & 3 \\
4 & 5 & 6 & 8 & 7 & 1 & 2 & 3 & 0 \\
5 & 4 & 7 & 6 & 2 & 8 & 3 & 0 & 1 \\
7 & 8 & 4 & 5 & 3 & 0 & 6 & 1 & 2 \\
8 & 6 & 5 & 7 & 0 & 3 & 1 & 2 & 4 \\
\hline
\end{array}$$

A 4×4 subsquare is shown on symbols $\{0, 1, 2, 3\}$, and a 2×2 subsquare is shown on symbols $\{4, 5\}$. For $7 \in \mathcal{E}_9$, replace the 4×4 subsquare by the two 9-orthogonal squares

$$\begin{pmatrix} 0 & 1 & 2 & 3 \\ 1 & 0 & 3 & 2 \\ 2 & 3 & 1 & 0 \\ 3 & 2 & 0 & 1 \end{pmatrix}, \quad \begin{pmatrix} 1 & 3 & 0 & 2 \\ 3 & 0 & 2 & 1 \\ 2 & 1 & 3 & 0 \\ 0 & 2 & 1 & 3 \end{pmatrix}.$$

Although these squares are only 9-orthogonal, their superposition does not yield the pairs (1,1) and (3,3). Since these arise elsewhere in the superposition of the 9×9 squares, we obtain $(9-4)+2 = 7 \in \mathcal{E}_9$. Instead replace the 4×4 subsquare by the two 12-orthogonal squares

$$\begin{pmatrix} 0 & 1 & 2 & 3 \\ 1 & 0 & 3 & 2 \\ 2 & 3 & 0 & 1 \\ 3 & 2 & 1 & 0 \end{pmatrix}, \quad \begin{pmatrix} 1 & 3 & 2 & 0 \\ 2 & 1 & 0 & 3 \\ 0 & 2 & 3 & 1 \\ 3 & 0 & 1 & 2 \end{pmatrix}.$$

These miss only one constant pair, (0,0). Further replace the 2×2 subsquare by the pair $\left(\begin{smallmatrix} 4 & 5 \\ 5 & 4 \end{smallmatrix}\right)$ and $\left(\begin{smallmatrix} 5 & 4 \\ 4 & 5 \end{smallmatrix}\right)$ to obtain $11 \in \mathcal{E}_9$.

Until this point, we have found no value of n for which (n^2-3)-orthogonal latin squares exist. For the purposes of later recursions, we require such a pair for some n. We present a technique here that yields 78–orthogonal latin squares of order 9, and hence establishes that $69 \in \mathcal{E}_9$.

Suppose that there are two mutually orthogonal latin squares L and M of order n such that L contains a 3×3 subsquare $\begin{pmatrix} 0 & 2 & 1 \\ 2 & 1 & 0 \\ 1 & 0 & 2 \end{pmatrix}$ at the top left, and M contains a 3×3 subarray $\begin{pmatrix} a & b & x \\ y & a & b \\ b & z & a \end{pmatrix}$ at the top left. Suppose further that x, y and z are all distinct. Then replacing the subsquare in L by $\begin{pmatrix} 1 & 0 & 2 \\ 0 & 2 & 1 \\ 2 & 1 & 0 \end{pmatrix}$ yields another latin square L'.

Now in the superposition of L' and M, all pairs except $\{(1,x),(2,y),(0,z)\}$ appear, so L' and M are (n^2-3)–orthogonal. It remains to exhibit such squares; we do so here for $n = 9$:

0	1	7	2	3	6	4	5	8
8	0	1	7	2	3	6	4	5
1	6	0	3	8	2	5	7	4
2	3	6	4	5	8	0	1	7
7	2	3	6	4	5	8	0	1
3	8	2	5	7	4	1	6	0
4	5	8	0	1	7	2	3	6
6	4	5	8	0	1	7	2	3
5	7	4	1	6	0	3	8	2

1	0	2	3	5	4	6	8	7
0	2	1	5	4	3	8	7	6
2	1	0	4	3	5	7	6	8
6	8	7	0	2	1	3	5	4
8	7	6	2	1	0	5	4	3
7	6	8	1	0	2	4	3	5
3	5	4	6	8	7	0	2	1
5	4	3	8	7	6	2	1	0
4	3	5	7	6	8	1	0	2

In fact, we prove a more general result:

Lemma 3.1 *If there are two mutually orthogonal latin squares of side k, and α, β and γ nonnegative integers for which $0 \le \alpha+\beta+\gamma \le k$, then $(3k)^2 - 3k - 3\alpha - 2\beta - 6\gamma \in \mathcal{E}_{3k}$.*

Proof: Let $A = \begin{pmatrix} 0 & 2 & 1 \\ 2 & 1 & 0 \\ 1 & 0 & 2 \end{pmatrix}$ and let $B = \begin{pmatrix} 0 & 1 & 2 \\ 2 & 0 & 1 \\ 1 & 2 & 0 \end{pmatrix}$. A and B form MOLS(3). Let $L = (\ell_{ij})$ and $M = (m_{ij})$ be MOLS(k). Let R be the latin square of side $3k$ obtained by substituting, for each entry ℓ_{ij} of L, the array $A + 3\ell_{ij}$ (that is, add $3\ell_{ij}$ to each entry of A).

Let C be the partial array $C = \begin{pmatrix} 0 & 1 & \\ & 0 & 1 \\ 1 & & 0 \end{pmatrix}$. Form a partial latin square S of side $3k$ by replacing each entry m_{ij} of M by the partial latin square $C + 2m_{ij}$. Now let $T_i = \{(a,b) : m_{ab} = i\}$ for $0 \le i < k$. The entries T_i form a transversal of L for each $0 \le i < k$ since M is orthogonal to L. Now place in S the element $2k + j$ in cells $\{(3a, 3b+2) : (a,b) \in T_j\}$, $\{(3a+1, 3b) : (a,b) \in T_{j+1}\}$ and $\{(3a+2, 3b+1) : (a,b) \in T_{j+2}\}$, for $0 \le j < k$, with subscripts reduced modulo k.

Then the latin squares R and S are orthogonal. Select $\alpha+\beta$ 3×3 subsquares of R (indexed by rows $\{0,1,2\}$ and columns $\{3b, 3b+1, 3b+2\}$ for $0 \le b < \alpha+\beta+\gamma$. Replace each of the first α subarrays $\begin{pmatrix} a & c & b \\ c & b & a \\ b & a & c \end{pmatrix}$ by $\begin{pmatrix} b & a & c \\ a & c & b \\ c & b & a \end{pmatrix}$; call this a *type I switch*. For the next β subarrays, replace each by $\begin{pmatrix} b & c & a \\ c & a & b \\ a & b & c \end{pmatrix}$; call this a *type II switch*. Finally, for the γ remaining subarrays, replace each by $\begin{pmatrix} a & b & c \\ c & a & b \\ b & c & a \end{pmatrix}$; call this a *type III switch*. Call the resulting square R'.

It is routine to verify that R' and S are $((3k)^2 - 3\alpha - 2\beta - 6\gamma)$-orthogonal. □

Lemma 3.1 shows $\{61, 62, 65, 67, 68\} \subset \mathcal{E}_9$ (and more). It remains to treat 59. We explicitly exhibit 68-orthogonal latin squares to handle this last case.

1	3	2	5	4	6	8	7	0
3	2	1	4	6	5	7	0	8
2	1	3	6	5	4	0	8	7
7	8	0	2	3	1	4	6	5
0	7	8	3	1	2	6	5	4
8	0	7	1	2	3	5	4	6
4	6	5	7	0	8	1	3	2
6	5	4	0	8	7	3	2	1
5	4	6	8	7	0	2	1	3

1	2	8	3	4	7	5	6	0
0	1	2	8	3	4	7	5	6
2	7	1	4	0	3	6	8	5
3	4	7	5	6	0	1	2	8
8	3	4	7	5	6	0	1	2
4	0	3	6	8	5	2	7	1
5	6	0	1	2	8	3	4	7
7	5	6	0	1	2	8	3	4
6	8	5	2	7	1	4	0	3

Thus no exceptions remain for side 9.

4 A GDD CONSTRUCTION

We employ a construction using group divisible designs to handle the majority of the cases, and use Lemmas 2.5 and 2.7 with the small cases done in the previous section to handle values near 0 and near $n^2 - n$, respectively.

We require an auxiliary notion in providing ingredients for the *GDD* construction. A latin square is *idempotent* if, for each $0 \le i < n$, the entry (i, i) of the square is i (in fact, since we are free to rename symbols, it suffices to require that on the main diagonal all entries are distinct). Define

$$\mathcal{I}_n = \{r - n : \exists\ r - \text{orthogonal } \textit{idempotent} \text{ latin squares of side } n\}.$$

It is an easy matter to verify that $\mathcal{I}_2 = \emptyset$, $\mathcal{I}_3 = \{0\}$ and $\mathcal{I}_4 = \{0, 12\}$. In addition to this, we require $\mathcal{I}_5$. Consider the three *idempotent* latin squares:

0	1	2
0 3 1 4 2	0 2 4 1 3	0 2 4 1 3
4 2 0 3 1	4 1 3 0 2	3 1 0 4 2
3 1 4 2 0	3 0 2 4 1	4 3 2 0 1
2 0 3 1 4	2 4 1 3 0	2 4 1 3 0
1 4 2 0 3	1 3 0 2 4	1 0 3 2 4

Here we apply the *same* permutation to the rows and columns of the second square, to obtain the following members of $\mathcal{I}_5$:

$\mathcal{I}_5$	First Square	Second Square	Permutation
0	0	0	4 0 1 2 3
6	1	2	2 0 4 3 1
8	1	0	3 1 0 4 2
9	2	2	0 3 1 4 2
12	2	1	1 2 4 3 0
14	2	2	1 4 3 2 0
16	2	2	1 2 0 3 4
20	1	0	0 4 3 2 1

Next we give the basic *GDD* construction:

Theorem 4.1 *Let $\mathcal{G} = \{G_1, \ldots, G_g\}$ and $\mathcal{B} = \{B_1, \ldots, B_b\}$ be the groups and blocks of a GDD, $(V, \mathcal{G}, \mathcal{B})$, having*

1. *$\alpha_i \in \mathcal{E}_{|G_i|}$ for $1 \le i \le g$; and*
2. *$\beta_i \in \mathcal{I}_{|B_i|}$ for $1 \le i \le b$,*

then $\sum_{i=1}^{g} \alpha_i + \sum_{i=1}^{b} \beta_i \in \mathcal{E}_{|V|}$.

Moreover, if

1. $\gamma_i \in \mathcal{E}_{|G_i|+1}$ *for* $1 \leq i \leq g$; *and*

2. $\beta_i \in \mathcal{I}_{|B_i|}$ *for* $1 \leq i \leq b$,

then $\sum_{i=1}^{g} \gamma_i + \sum_{i=1}^{b} \beta_i \in \mathcal{E}_{|V|+1}$.

Proof: Place β_i-orthogonal idempotent latin squares of side $|B_i|$ on the elements of B_i for $1 \leq i \leq b$, omitting the entries on the diagonals. Place α_i-orthogonal latin squares of side $|G_i|$ on the elements of G_i for $1 \leq i \leq g$. For the latter, use Lemma 2.4 to ensure that each square contains all constant pairs (of the form (i,i)). The resulting squares are easily checked to have the required number of pairs in their superposition.

For the construction of side $|V|+1$, employ the blocks as before. Add one extra row, column and symbol, indexed by ∞; then use each group G_i, together with ∞, to form γ_i-orthogonal latin squares of side $|G_i|+1$. Arranging each such selection so that entry (∞,∞) of each square is ∞ can be done without loss of generality to obtain the desired result. □

Of course, the use of Theorem 4.1 requires that we find suitable *GDD*. Hence we employ the following corollary:

Corollary 4.2 *If a* $\{4,5\}-GDD$ *of type* $4^n h^1$ *exists having* $h < 13$ *and having at least two blocks of size 5, then for any* r *satisfying* $17 \leq r \leq (4n+h)^2 - (4n+h) - 13$, $r \in \mathcal{E}_{4n+h}$.

Proof: Suppose there are α blocks of size 4, and $\beta + 2$ blocks of size 5 in the $\{4,5\} - GDD$. From two blocks of size 5, we obtain the a contribution from

$$\{0,6,8,9,12,14,15,16,17,18,20,21,22,23,24,25,26,28,29,30,32,34,36,40\}$$

to the sum. This contains all even integers from 12 to 36. Thus the contribution from the blocks can be any even integer of the form $12a + 20b + c$ where $0 \leq a \leq \alpha$, $0 \leq b \leq \beta$, and $12 \leq c \leq 36$, c even. In particular, the contribution of the blocks to the sum can be any even number from 12 to $(4n+h)^2 - (4n+h) - 12n - (h^2-h) - 4$. For each group of size 4, we may choose any number in $\mathcal{E}_4$, so that after considering groups of size four, we can cover all even numbers from 12 to $(4n+h)^2 - (4n+h) - (h^2-h) - 4$. By choosing $5 \in \mathcal{E}_4$ for one group, we also obtain all odd numbers from 17 to $(4n+h)^2 - (4n+h) - (h^2-h) - 11$.

Unless $h \in \{2, 6\}$, we may choose 0 or $h^2 - h$ for the contribution of the group of size h; when $h \in \{2, 6\}$, our selection is limited to 0 or $h^2 - h - 2$. This covers the interval of odd numbers from 17 to $(4n + h)^2 - (4n + h) - 13$. □

This corollary can be used to settle further cases outside the specified range, depending upon the value of h.

Our next task is to apply this corollary, by producing suitable $\{4, 5\} - GDD$:

Lemma 4.3 *For $u \geq 17$, $r \in \mathcal{E}_u$ if $17 \leq r \leq u^2 - u - 13$.*

Proof: Write $u = 4m + h$, $1 \leq h \leq 4$. Let $v = 4(m + 1) + 1$. If $m \notin \{6, 7, 11\}$, we form a $\{5, k^\star\}$-PBD on v points using one of:

1. $k = 5$ and $v \equiv 1, 5 \pmod{20}$;
2. $k = 9$ and $v \equiv 9, 17 \pmod{20}$, $v \neq 17, 29, 49$; or
3. $k = 13$ and $v \equiv 13 \pmod{20}$, $v \neq 33$.

Such PBD all exist [9, 8]. Deleting $v - u$ points from the long block gives a $\{4, 5\} - GDD$ of order u and group-type $4^n h^1$ having at least two blocks of size 5; apply Corollary 4.2.

If $m = 11$, start with a $\{5\} - GDD$ of type $4^{10} 12^1$, obtained by deleting one point from a $\{5, 13^\star\}$-PBD on 53 points [8]; delete $52 - u$ points from the long group (note that $4 \leq 52 - u \leq 7$). Apply Corollary 4.2 to the resulting $\{4, 5\} - GDD$.

If $m = 7$, start with a $TD(5, 7)$ and delete one point to form a $\{5, 7\} - GDD$ of type $4^7 6^1$. Delete $34 - u$ points from the long group ($2 \leq 34 - u \leq 5$). The result is a $\{4, 5, 7\} - GDD$ of type $4^7 h^1$; apply Corollary 4.2, observing that the four blocks of size 7 can be filled yielding a contribution of 0 or 42 (each).

For $m = 6$, we start with a resolvable $\{4\} - GDD$ of type 3^8 [4]. Add points at infinity to three parallel classes, and a point at infinity to the groups. Using a parallel class as groups, form a further group containing the four points at infinity, to obtain a $\{4, 5\} - GDD$ of type 4^7. Deleting 1, 2 or 3 points and applying Corollary 4.2 handles $u \in \{26, 27, 28\}$. For $u = 25$, we employ

Theorem 4.1 directly on a $TD(5,5)$, which is a $\{5\} - GDD$ of type 5^5 (the details are similar to Corollary 4.2). □

Now we prove the Main Theorem for $n \geq 27$. For $n \geq 12$, we have $\mathcal{E}_6 \subseteq \mathcal{E}_n$ by Lemma 2.5; see [10] for the required characterization of latin squares with subsquares. For $n \geq 18$, we have $n^2 - n - x \in \mathcal{E}_n$ whenever $30 - x \in \mathcal{E}_6$ by Lemma 2.7. Together with Lemma 2.2, this ensures that for $x = 0$, $x = 2$ and $4 \leq x \leq 28$, $n^2-n-x \in \mathcal{E}_n$ provided $n \geq 18$. With Lemma 4.3, this ensures that the only possible exception for $n \geq 18$ is $n^2 - n - 3$. Since $69 = 9^2 - 9 - 3 \in \mathcal{E}_9$, we obtain $n^2 - n - 3 \in \mathcal{E}_n$ for all $n \geq 27$.

5 INTERMEDIATE SIDES

It remains to treat the cases in the range $10 \leq n < 27$, although they are not used in the general solution in any way. Nonetheless, it is of interest to determine the spectrum for these intermediate values. We treat them in turn, after giving a useful lemma.

Lemma 5.1 *If HMOLS of type* $(h_1, \ldots, h_n)$ *exist, with* $v = \sum_{i=1}^{n} h_i$, *and* $\alpha_i \in \mathcal{E}_{h_i+1}$, *then*

$$v^2 - \sum_{i=1}^{n} h_i^2 + \sum_{i=1}^{n} \alpha_i \in \mathcal{E}_{v+1}.$$

Proof: similar to the second part of the proof of Theorem 4.1. □

For side 10, Lemma 2.6 establishes that $x \in \mathcal{E}_{10}$ for all $2 \leq x \leq 40$ except possibly 3, 35, 37, 38, and 39. Lemma 2.3 handles $x = 3$. Lemma 2.7 handles $x = 84$ and $x = 88$. We employed the same permutation technique as before, using the single square defined by $\ell_{i+1,j} = \ell_{i,j-1}$, subscripts modulo 10. We present the row and column permutations applied to the second copy of the square here:

$r - n$	Row Permutation	Column Permutation
35	9 2 5 8 1 7 0 4 3 6	6 9 5 2 1 8 7 0 4 3
37	9 6 1 0 7 4 3 8 5 2	0 7 3 1 6 5 8 9 4 2
38	0 3 9 6 1 5 4 8 7 2	7 1 8 5 3 6 2 9 4 0

$r-n$	Row Permutation	Column Permutation
39	1 2 6 0 4 5 7 3 8 9	9 3 7 6 0 1 8 2 4 5
41	6 4 0 7 3 2 8 9 5 1	0 1 7 6 8 3 2 9 5 4
42	8 0 1 5 9 2 7 6 3 4	6 0 2 8 9 1 5 7 3 4
43	0 8 7 6 2 3 5 9 4 1	8 0 5 6 4 3 7 2 9 1
44	9 0 6 8 1 2 3 4 5 7	2 6 7 8 4 5 9 3 1 0
45	2 3 4 9 7 0 1 6 5 8	4 7 9 0 1 6 5 2 8 3
46	3 5 4 8 6 0 7 1 9 2	5 6 2 7 4 1 8 0 9 3
47	2 5 4 6 8 7 1 3 9 0	0 1 9 3 5 4 6 8 2 7
48	7 5 8 3 2 0 1 9 4 6	0 4 6 7 5 9 1 8 3 2
49	7 6 3 0 8 4 2 5 9 1	2 9 5 4 0 1 6 8 7 3
50	4 2 8 6 7 1 0 5 3 9	4 7 0 6 3 9 2 5 1 8
51	3 0 7 5 9 6 8 1 2 4	2 9 4 8 6 0 7 3 1 5
52	8 5 0 4 3 9 7 2 6 1	2 3 0 7 6 8 5 9 4 1
53	8 4 3 5 7 2 0 6 1 9	3 7 2 5 1 0 9 6 4 8
54	6 2 4 0 8 3 7 9 5 1	4 6 9 8 2 7 0 3 1 5
55	3 8 7 9 5 6 1 4 0 2	8 0 7 9 2 3 4 6 1 5
56	8 0 6 4 2 3 1 9 5 7	1 6 9 2 0 4 3 7 8 5
57	9 4 5 8 7 6 2 1 0 3	3 8 4 2 9 7 5 0 6 1
58	3 6 9 2 7 8 1 0 4 5	6 5 4 7 2 9 8 0 1 3
59	1 3 4 9 7 2 8 6 5 0	5 3 0 2 9 1 7 4 6 8
60	3 6 7 9 4 5 8 2 1 0	2 8 0 7 9 6 4 1 5 3
61	8 2 9 7 5 4 6 1 3 0	9 0 3 6 2 8 4 7 5 1
62	4 0 7 5 6 8 1 9 3 2	7 0 5 4 9 6 1 2 8 3
63	1 0 9 8 7 6 4 5 3 2	0 1 6 8 2 3 5 7 9 4
64	0 9 1 8 5 7 6 4 3 2	7 2 0 3 4 9 8 5 1 6
65	0 3 5 2 6 8 1 9 7 4	9 6 7 4 3 8 2 1 5 0
66	7 5 9 2 6 4 8 1 0 3	9 5 2 3 1 7 6 8 4 0
67	4 1 0 3 9 5 8 7 6 2	8 7 2 4 9 3 5 6 0 1
68	1 7 2 5 6 0 4 8 9 3	9 7 6 4 3 1 0 5 2 8
69	2 8 3 9 1 0 6 4 7 5	4 1 3 6 5 8 7 0 2 9
70	3 2 1 0 9 8 7 6 5 4	1 5 8 4 2 9 3 6 7 0
72	3 1 4 2 5 8 6 9 7 0	3 0 6 9 5 1 2 7 8 4
75	3 0 9 5 2 7 4 6 1 8	7 0 1 4 5 8 9 2 3 6
80	0 1 2 3 4 5 6 7 8 9	4 3 7 1 0 2 6 5 9 8

Using HMOLS of type $2^4 1^1$ [14], apply Lemma 5.1 to establish that $\{76, 82\} \subset \mathcal{E}_{10}$.

Now consider the two mutually orthogonal squares:

0	x	y	6	$\underline{z}$	3	5	$\underline{1}$	2	4
6	1	x	y	0	z	4	2	3	5
5	0	2	x	y	1	z	3	4	6
z	6	1	3	x	y	2	4	5	0
3	z	$\underline{0}$	2	4	x	y	$\underline{5}$	6	1
y	4	z	1	$\underline{3}$	5	x	6	0	$\underline{2}$
x	y	$\underline{5}$	z	$\underline{2}$	4	6	$\underline{0}$	1	$\underline{3}$
2	3	4	5	6	0	1	x	z	y
4	5	6	0	$\underline{1}$	2	3	$\underline{z}$	y	x
1	2	3	4	5	6	0	y	x	z

0	6	$\underline{5}$	z	$\underline{3}$	x	y	2	4	1
y	1	0	6	z	4	x	3	5	2
x	y	2	1	0	z	5	4	6	3
6	x	$\underline{y}$	3	2	1	z	5	0	$\underline{4}$
z	0	x	y	4	3	2	6	1	5
3	z	1	x	y	5	4	0	2	6
5	4	z	2	x	y	6	1	3	0
1	2	$\underline{3}$	4	$\underline{5}$	6	0	$\underline{x}$	$\underline{y}$	$\underline{z}$
2	3	$\underline{4}$	5	6	0	1	$\underline{z}$	$\underline{x}$	$\underline{y}$
4	5	6	0	1	2	3	$\underline{y}$	$\underline{z}$	$\underline{x}$

Let A_1, A_2 and A_3 be the subsquares of the first square indicated underlined on symbol sets {1,z}, {2,3} and {0,5}, respectively. Let B_1, B_2 and B_3 be the subsquares indicated underlined in the second square on symbol sets {3,5}, {x,y,z} and {y,4}, respectively. We alter one or both squares by interchanging the last two columns in some of these subsquares. Here we indicate which subsquares are altered to obtain a number of different members of $\mathcal{E}_{10}$:

$\in \mathcal{E}_{10}$	Subsquares Altered
86	A_1
83	A_1, B_1
79	A_1, B_1, A_2
74	A_1, B_1, A_2, B_2
81	A_1, B_2
78	A_1, B_1, B_2
77	A_1, A_2, B_2
73	A_1, A_2, B_2, A_3
71	A_1, A_2, B_1, A_3, B_3

To obtain $85 \in \mathcal{E}_{10}$, consider the two squares

6 9 8 1 7 5 2 0 3 4
2 7 9 8 1 4 6 3 0 5
8 3 4 9 0 2 5 7 1 6
9 8 0 5 4 1 3 6 2 7
4 6 1 3 8 7 0 9 5 2
0 5 7 2 9 8 4 $\underline{1}$ $\underline{6}$ 3
3 1 6 4 2 9 8 5 7 0
5 0 2 7 6 3 9 8 4 1
1 2 3 0 5 6 7 $\underline{4}$ $\underline{8}$ 9
7 4 5 6 3 0 1 2 9 8

8 1 7 9 0 3 5 4 2 6
9 8 2 4 5 1 0 6 3 7
5 9 8 3 7 6 2 1 0 4
0 6 9 8 2 4 7 3 1 5
3 2 6 5 1 9 8 7 4 0
6 0 3 7 4 2 9 $\underline{8}$ $\underline{5}$ 1
4 7 1 0 8 5 3 9 6 2
1 5 4 2 9 8 6 0 7 3
2 3 0 1 6 7 4 $\underline{5}$ $\underline{8}$ 9
7 4 5 6 3 0 1 2 9 8

These squares are $(10^2 - 2)$-orthogonal. Replacing the subsquare on symbols $\{5,8\}$ in the second by the other 2×2 square yields $(10^2 - 5)$-orthogonal squares. The only remaining possible exception is 87 for side 10.

For side 11, apply Theorem 4.1 with one point at infinity to a $\{3\} - GDD$ of type $2^3 4^1$, obtained by deleting one point from a $\{3, 5^\star\}$-PBD on 11 points [9]. This establishes that $x \in \mathcal{E}_{11}$ for $5 \le x \le 32$, and for $x \in \{34, 38\}$. Lemma 2.7 establishes $\{104, 108\} \subset \mathcal{E}_{11}$. There exist incomplete MOLS of side 11 with two holes, one each of size 2 and 3 [15]; filling the holes establishes that $102 \in \mathcal{E}_{11}$. There exist HMOLS of type 2^5; applying Lemma 5.1, we obtain $98 \in \mathcal{E}_{11}$.

For some of the remaining values, we employ a single square defined by $\ell_{i+1,j} = \ell_{i,j-1}$, subscripts modulo 11. We give the permutations applied to the second copy of the square:

$r - n$	Row Permutation	Column Permutation
33	7 3 10 6 1 9 5 2 8 4 0	4 10 6 0 9 5 2 8 3 1 7
35	0 8 1 4 9 3 7 2 6 10 5	9 2 10 5 0 4 8 3 7 1 6
36	7 9 3 0 2 10 4 1 6 5 8	0 5 2 7 1 6 3 8 10 4 9
37	0 2 8 5 4 10 7 1 9 6 3	10 7 4 1 9 6 5 0 8 3 2
39	5 2 1 0 8 9 10 7 6 3 4	8 0 10 5 6 9 4 3 2 1 7
40	7 1 5 3 10 8 6 4 2 9 0	2 6 0 9 5 7 3 1 8 10 4
41	9 2 5 3 10 1 0 7 6 8 4	7 6 2 1 3 5 4 0 10 9 8
42	6 8 5 7 4 3 0 2 10 1 9	2 7 6 0 8 10 4 9 1 3 5
43	10 9 5 1 0 4 8 2 6 7 3	0 7 9 4 3 2 6 5 1 8 10
44	1 9 4 2 10 7 5 3 0 8 6	6 4 1 9 5 7 2 10 8 3 0
45	0 3 10 4 7 8 1 5 9 2 6	7 0 4 3 8 2 9 1 6 10 5
46	0 8 2 3 9 1 6 4 7 10 5	1 4 7 10 2 5 8 0 3 9 6

$r-n$	Row Permutation	Column Permutation
47	5 0 8 4 9 3 2 10 6 1 7	6 0 8 2 9 5 3 10 4 1 7
48	2 1 6 5 9 3 0 7 4 10 8	2 6 10 9 3 7 1 0 5 4 8
49	3 5 6 9 4 2 1 10 0 7 8	7 6 8 2 4 10 1 3 5 0 9
50	8 2 6 3 0 1 9 5 10 7 4	2 9 0 5 4 6 3 10 7 1 8
51	7 8 3 2 10 4 0 9 6 5 1	0 7 3 1 8 6 2 10 9 5 4
52	9 2 1 6 4 7 0 10 5 8 3	3 7 4 9 1 10 5 0 6 2 8
53	7 3 8 5 9 10 4 0 1 6 2	8 1 10 3 7 4 5 2 9 6 0
54	4 5 10 1 2 6 8 7 0 9 3	0 1 6 4 5 7 2 9 3 8 10
55	10 0 1 6 5 2 7 9 3 8 4	6 0 8 2 10 9 1 5 3 4 7
56	4 2 9 8 6 5 10 7 1 3 0	4 7 10 9 5 8 2 1 6 3 0
57	0 3 6 10 7 9 5 8 4 1 2	5 1 4 0 2 7 6 8 9 10 3
58	4 1 8 2 7 9 6 10 3 0 5	9 10 7 6 5 1 3 8 2 4 0
59	2 4 8 5 0 9 7 3 1 6 10	5 3 9 6 10 1 2 7 8 0 4
60	10 0 2 1 9 3 5 6 8 4 7	10 2 4 1 6 5 7 9 0 3 8
61	3 8 2 0 10 4 5 6 1 7 9	6 1 9 3 0 8 2 4 5 7 10
62	2 9 1 5 6 4 7 10 8 3 0	8 1 9 7 0 10 2 5 6 4 3
63	5 1 2 9 8 4 3 0 6 10 7	9 6 3 4 10 5 2 7 0 1 8
64	0 5 2 10 3 4 7 9 1 8 6	9 10 2 6 7 4 8 5 3 0 1
65	8 4 6 1 0 7 5 3 9 10 2	3 7 2 10 0 1 4 9 5 8 6
66	7 10 3 0 2 8 1 4 9 5 6	8 2 7 5 3 10 4 9 0 6 1
67	1 7 5 3 8 2 0 9 10 4 6	5 1 3 0 6 10 4 9 2 8 7
68	3 5 10 2 7 6 0 9 1 4 8	8 1 0 6 2 10 4 5 7 9 3
69	3 6 5 4 8 0 10 7 9 2 1	8 4 9 10 2 3 6 5 7 0 1
70	9 4 5 0 1 2 7 6 3 8 10	3 1 7 9 10 5 6 2 4 0 8
71	2 8 6 1 3 10 7 9 5 4 0	9 1 0 8 6 3 5 7 4 2 10
72	7 8 4 6 2 9 1 5 10 3 0	1 7 3 2 8 5 4 9 6 0 10
73	9 2 7 8 5 6 1 3 4 0 10	9 5 0 6 3 1 8 7 2 10 4
74	2 4 10 3 1 6 9 0 7 8 5	3 8 6 9 10 2 1 7 5 0 4
75	4 7 1 10 8 2 0 9 6 3 5	7 9 5 2 3 6 8 4 1 10 0
76	7 1 9 10 4 5 2 8 3 0 6	8 6 10 2 1 4 0 3 5 7 9
77	5 6 7 10 2 0 3 4 8 9 1	1 9 3 10 0 2 8 5 7 4 6
78	4 1 8 6 10 0 9 5 3 7 2	1 2 8 0 3 4 9 7 6 5 10
79	2 4 8 10 6 1 9 5 7 3 0	0 5 10 2 3 4 9 1 6 7 8
80	7 1 4 2 5 3 6 9 10 8 0	10 3 5 1 8 7 6 0 2 4 9
81	5 4 2 7 6 0 9 8 3 1 10	6 9 4 5 2 3 10 1 7 0 8

$r-n$	Row Permutation	Column Permutation
82	7 3 5 2 0 9 6 8 4 1 10	8 9 0 6 5 7 2 3 1 4 10
83	0 2 3 4 6 1 7 9 8 10 5	3 6 10 1 8 5 2 9 7 4 0
84	9 10 7 8 2 3 5 4 0 1 6	2 9 1 7 5 3 0 10 8 6 4
85	1 7 4 9 6 3 5 8 0 10 2	5 6 2 0 7 8 4 3 1 10 9
86	5 4 1 7 0 6 2 9 8 3 10	8 2 6 10 0 5 9 3 4 7 1
87	7 9 3 5 8 1 2 6 10 0 4	8 5 3 9 6 4 1 0 7 2 10
88	9 8 7 6 5 4 3 2 1 0 10	6 4 5 8 0 7 9 3 1 10 2
89	5 4 3 0 10 1 7 2 6 9 8	10 4 2 7 5 0 9 3 8 6 1
90	2 5 10 0 9 3 8 1 4 7 6	5 0 6 8 7 2 10 3 4 1 9
91	4 1 10 7 5 3 2 0 9 8 6	9 1 7 3 6 8 5 0 4 10 2
92	7 6 10 3 2 1 5 9 8 0 4	9 0 2 10 1 3 7 4 6 8 5
94	4 2 9 5 3 7 6 10 8 1 0	3 5 1 2 9 0 7 8 4 6 10

Now consider the two mutually orthogonal latin squares:

0	z	y	x	7	3	2	1	4	6	5
2	1	z	y	x	0	4	3	5	7	6
4	3	2	z	y	x	1	5	6	0	7
6	5	4	3	z	y	x	2	7	1	0
3	7	6	5	4	z	y	x	0	2	1
x	4	0	7	6	5	z	y	1	3	2
y	x	5	1	0	7	6	z	2	4	3
z	y	x	6	2	1	0	7	3	5	4
$\underline{1}$	$\underline{2}$	3	$\underline{4}$	$\underline{5}$	$\underline{6}$	7	$\underline{0}$	$\underline{x}$	$\underline{y}$	$\underline{z}$
$\underline{5}$	$\underline{6}$	7	$\underline{0}$	$\underline{1}$	$\underline{2}$	3	$\underline{4}$	$\underline{z}$	$\underline{x}$	$\underline{y}$
7	0	1	2	3	4	5	6	$\underline{y}$	$\underline{z}$	$\underline{x}$

0	2	4	6	3	x	y	$\underline{z}$	1	$\underline{5}$	7
z	1	3	5	7	4	x	y	2	6	0
y	z	2	4	6	0	5	x	3	7	1
$\underline{x}$	y	z	3	5	7	1	6	$\underline{4}$	0	2
7	x	y	z	4	6	0	2	5	1	3
3	0	x	y	z	5	7	1	6	2	4
2	4	1	x	y	z	6	0	7	3	5
1	3	5	2	x	y	z	7	0	4	6
$\underline{4}$	5	6	7	0	1	2	3	$\underline{x}$	$\underline{y}$	$\underline{z}$
6	7	0	1	2	3	4	$\underline{5}$	$\underline{y}$	$\underline{z}$	$\underline{x}$
5	6	7	0	1	2	3	4	$\underline{z}$	$\underline{x}$	$\underline{y}$

Let A_1, A_2, A_3 and A_4 be the subsquares of the first square indicated underlined on symbol sets {1,5}, {2,6}, {0,4}, and {x,y,z}, respectively. Let B_1, B_2 and B_3 be the subsquares of the second square indicated underlined on symbol sets {4,x}, {5,z}, and {x,y,z}, respectively. Again we alter some of the subsquares by interchanging the last two columns. Here we tabulate the new values in $\mathcal{E}_{11}$ so obtained:

$\in \mathcal{E}_{11}$	Subsquares Altered
106	A_1
103	A_1, B_1
99	A_1, A_2, B_1
96	A_1, A_2, B_1, B_2
95	A_1, A_2, A_3, B_1
100	A_1, B_3
97	A_2, B_1, B_3
101	A_4, B_1
93	A_2, A_3, A_4, B_1

To treat $11^2 - 11 - 5$, consider the square

0	x	6	1	7	4	2	$\underline{\text{y}}$	5	3	$\underline{8}$
6	1	x	7	2	8	5	3	y	4	0
y	7	2	x	8	3	0	6	4	5	1
5	y	8	3	x	0	4	1	7	6	2
8	6	y	0	4	x	1	5	2	7	3
3	0	7	y	1	5	x	2	6	8	4
7	4	1	8	y	2	6	x	3	0	5
4	8	5	2	0	y	3	7	x	1	6
x	5	0	6	3	1	y	4	8	2	7
1	2	3	4	5	6	7	$\underline{8}$	0	x	$\underline{\text{y}}$
2	3	4	5	6	7	8	0	1	y	x

This square is $(11^2 - 2)$-orthogonal to its transpose. Replacing the subsquare on {8,y} yields $105 \in \mathcal{E}_{11}$. The remaining exception is 107 for side 11.

For side 12, Belyavskaya [3] establishes that

$$\mathcal{P}_{12} \setminus \{23, 27, 119, 121, 123, 126, 127, 129\} \subseteq \mathcal{E}_{12}.$$

Values 23 and 27 are handled by Lemma 2.6. Value 126 is handled by Lemma 2.7. Using HMOLS of type $2^4 3^1$ from [12], apply Lemma 5.1 to establish that $119 \in \mathcal{E}_{12}$. Lemma 3.1 establishes $\{121, 123, 127, 129\} \subset \mathcal{E}_{12}$. Thus no exceptions remain for side 12.

For side 13, use Theorem 4.1 with one point at infinity on a $\{4\} - GDD$ of type 3^4. This yields $x \in \mathcal{E}_{13}$ for all $4 \leq x \leq 146$, and $x \in \{2, 148, 149, 152\}$. Lemma 2.3 gives $x = 3$. Lemma 2.7 gives values 150 and 154. Now consider the $(13^2 - 5)$-orthogonal latin squares of side 13 given by:

3	1	h	7	9	8	x	z	y	4	6	5	2
8	h	1	y	7	9	5	x	z	2	4	6	3
h	5	3	9	2	7	z	8	x	6	y	4	1
x	z	y	6	4	h	1	3	2	7	9	8	5
5	x	z	2	h	4	8	1	3	y	7	9	6
z	8	x	h	y	6	3	5	1	9	2	7	4
4	6	5	x	z	y	9	7	h	1	3	2	8
2	4	6	5	x	z	y	h	7	8	1	3	9
6	y	4	z	8	x	h	2	9	3	5	1	7
7	9	8	1	3	2	4	6	5	z	x	h	y
y	7	9	8	1	3	2	4	6	5	h	x	z
9	2	7	3	5	1	6	y	4	h	8	z	x
1	3	2	4	6	5	7	9	8	x	z	y	h

3	h	1	x	z	y	**4**	**5**	**6**	7	8	9	2
h	3	2	z	y	x	**5**	**6**	**4**	8	9	7	1
2	1	h	y	x	z	**6**	**4**	**5**	9	7	8	3
7	8	9	6	h	4	x	y	z	1	2	3	5
8	9	7	h	6	5	y	z	x	2	3	1	4
9	7	8	5	4	h	z	x	y	3	1	2	6
x	y	z	1	2	3	9	h	7	4	5	6	8
y	z	x	2	3	1	h	9	8	5	6	4	7
z	x	y	3	1	2	8	7	h	6	4	5	9
4	5	6	7	8	9	**1**	**2**	**3**	z	h	x	y
5	6	4	8	9	7	**2**	**3**	**1**	h	z	y	x
6	4	5	9	7	8	**3**	**1**	**2**	y	x	h	z
1	2	3	4	5	6	7	8	9	x	y	z	h

This establishes $151 \in \mathcal{E}_{13}$. Performing type I switches on the two subsquares of the second square marked in bold yields $147 \in \mathcal{E}_{13}$. The only remaining value for side 13 is 153.

For side 14, Lemma 2.6 establishes that $x \in \mathcal{E}_{14}$ for $2 \leq x \leq 84$ except for $x \in \{81, 83\}$. Now start with an affine plane of order 4, which is a $\{4\} - GDD$ of type 4^4. Delete two points to form a $\{3,4\} - GDD$ of type $2^1 4^3$, and apply Theorem 4.1. Because the blocks of size three, and the group of size two, make no contribution to the sum, the exceptions are more prevalent in this application. Nonetheless, we obtain $x \in \mathcal{E}_{14}$ for $4 \leq x \leq 118$, and $x \in \{120, 121, 122, 124, 125, 128, 132\}$. Now Lemma 2.7 handles $x \in \{170, 172, 174, 175, 176, 178, 180\}$. There exist HMOLS of type $2^5 4^1$ [7]; filling the holes establishes that $\{160, 162, 164, 165, 168\} \subseteq \mathcal{E}_{14}$. Using HMOLS of type $2^5 3^1$ [12], apply Lemma 5.1 to obtain

$$\{142, 145, 150, 151, 154, 157, 163, 166, 169\} \subset \mathcal{E}_{14}.$$

Using a $\{4\} - GDD$ of type 2^7 [6], Theorem 4.1 establishes that $\{144, 156\} \subset \mathcal{E}_{14}$. In fact, we can do slightly better here. Suppose that B is a block of this GDD; it intersects each group in 0 or 1 elements. Fill the groups in such a way that on the element i intersecting B, that both squares contain i in position (i, i), and omit that entry from each square. This enables us to "fill" B using any entry from $\mathcal{E}_4$ instead of $\mathcal{I}_4$, and hence establishes that

$$\{134, 136, 137, 140, 146, 148, 149, 152, 158, 161\} \subseteq \mathcal{E}_{14}.$$

There exist HMOLS of type $3^4 1^1$, obtained from 3 MOLS(4). Applying Lemma 5.1 establishes that

$$\{138, 141, 143, 147, 153, 155, 159, 173\} \subset \mathcal{E}_{14}.$$

For some of the remaining values, we employ the single square defined by $\ell_{i+1,j} = \ell_{i,j-1}$, subscripts modulo 14. We present the permutations here:

$r - n$	Row Permutation	Column Permutation
119	3 0 6 13 12 7 9 11 2 4 8 10 1 5	7 4 5 1 9 2 6 13 12 8 0 11 3 10
123	9 10 13 7 0 12 6 5 3 4 1 8 2 11	6 10 8 12 3 4 0 7 13 5 9 2 1 11
126	2 1 11 7 12 5 3 8 10 0 13 6 4 9	13 6 1 5 9 8 11 12 10 0 4 3 7 2
127	3 10 13 6 9 1 12 2 4 5 11 0 8 7	2 8 9 7 13 1 5 0 12 10 6 11 3 4
129	5 1 13 3 8 10 9 4 12 0 6 2 11 7	12 1 8 13 7 2 5 6 3 4 9 11 0 10
130	2 3 5 8 0 4 11 12 6 7 1 13 10 9	3 11 2 1 13 4 0 8 6 9 7 10 5 12
131	13 11 4 5 9 3 1 0 7 10 2 8 6 12	3 13 7 0 10 1 11 8 5 9 4 12 6 2
133	2 4 9 11 13 3 12 5 7 8 10 6 0 1	0 6 13 2 9 3 12 4 8 1 7 11 5 10
135	4 7 0 3 10 13 6 9 12 11 2 5 8 1	3 4 8 5 7 9 6 13 10 12 0 11 1 2
139	13 12 8 6 10 5 1 3 11 7 2 9 0 4	13 5 6 12 4 1 7 8 0 11 3 9 10 2

Now consider the following two $(14^2 - 10)$-orthogonal squares:

3	1	h	7	9	8	x	z	y	6	4	k	2	5
8	h	1	y	7	9	5	x	z	2	k	4	3	6
h	5	3	9	2	7	z	8	x	k	y	6	1	4
x	z	y	6	4	h	3	1	k	7	9	8	5	2
5	x	z	2	h	4	8	k	1	y	7	9	6	3
z	8	x	h	y	6	k	5	3	9	2	7	4	1
4	6	5	z	x	k	9	7	h	1	3	2	8	y
2	4	6	5	k	x	y	h	7	8	1	3	9	z
6	y	4	k	8	x	h	2	9	3	5	1	7	x
9	7	k	1	3	2	4	6	5	z	x	h	y	8
y	k	7	8	1	3	2	4	6	5	h	x	z	9
k	2	9	3	5	1	6	y	4	h	8	z	x	7
1	3	2	4	6	5	7	9	8	x	z	y	h	k
7	9	8	x	z	y	1	3	2	4	6	5	k	h

3	h	1	x	y	z	6	4	5	9	k	7	2	8
h	3	2	y	z	x	4	5	6	k	9	8	1	7
2	1	h	z	x	y	5	6	4	8	7	k	3	9
9	7	8	6	h	4	z	k	x	**1**	**2**	**3**	5	y
7	8	9	h	6	5	k	z	y	**2**	**3**	**1**	4	x
8	9	7	5	4	h	y	x	k	**3**	**1**	**2**	6	z
x	y	z	3	k	1	9	h	7	4	5	6	8	2
y	z	x	k	3	2	h	9	8	5	6	4	7	1
z	x	y	2	1	k	8	7	h	6	4	5	9	3
6	k	4	7	8	9	1	2	3	z	h	x	y	5
k	6	5	8	9	7	2	3	1	h	z	y	x	4
5	4	k	9	7	8	3	1	2	y	x	h	z	6
1	2	3	4	5	6	7	8	9	x	y	z	h	k
4	5	6	1	2	3	x	y	z	7	8	9	k	h

Performing a type I or type III switch on the subsquare indicated in bold shows that 171, 167 $\in \mathcal{E}_{14}$. For 177 $\in \mathcal{E}_{11}$, one can find two IMOLS(14,2) one of which contains a 2×2 subsquare intersecting the hole in a single entry (we leave this as an easy exercise). Replacing the 2×2 subsquare then gives the desired result. For side 14, the remaining possible exception is 179.

For side 15, Belyavskaya [3] shows that

$$\mathcal{P}_{15} \setminus \{191, 193, 194, 205, 207\} \subseteq \mathcal{E}_{15}.$$

Lemma 3.1 establishes that $\{205, 207\} \subset \mathcal{E}_{15}$. Using HMOLS of type $2^5 4^1$ [7], apply Lemma 5.1 to obtain $\{191, 193, 194\} \subseteq \mathcal{E}_{15}$. Thus no exceptions remain for side 15.

For side 16, start with a $\{4\}-GDD$ of type 4^4 and add one point to the groups. Delete one point on a single block of size 5 to obtain a $\{4, 5\}-GDD$ of type $3^4 4^1$. Apply Theorem 4.1, as in the proof of Corollary 4.2 to obtain $\{17, \ldots, 229\} \subset \mathcal{E}_{16}$. Lemma 2.5 handles the remaining small values, and Lemma 2.7 handles

$$\{230, 231, 232, 233, 234, 236, 238\}.$$

For $235 \in \mathcal{E}_{16}$, proceed as in Lemma 3.1 to construct MOLS(15). We leave it as an exercise that one can border the two squares (replace the diagonal 3×3 subsquares by 4×4 subsquares) to obtain $(16^2 - 5)$-orthogonal latin squares of side 16. The only remaining possible exception is 237.

For side 17, Lemmas 2.5, 4.3 and 2.7 leave only the values 267 and 269. Again, $267 = 17^2 - 17 - 5$ can be handled by finding two IMOLS(17,2) one of which has a 2×2 subsquare intersecting the hole in a single position.

For side n, $18 \le n \le 26$, the only exception left by Lemmas 2.5, 4.3 and 2.7 is $n^2 - n - 3$. Lemma 3.1 handles this case for $n = 21, 24$.

Acknowledgments

Research of the first author is supported by NSERC Canada under grant number A0579. Both authors express their appreciation to NSERC Canada for a Foreign Researcher Award to support the visit of the second author to Waterloo.

REFERENCES

[1] G.B. Belyavskaya, "r-Orthogonal quasigroups I", *Math Issled.* 39 (1976) 32–39.

[2] G.B. Belyavskaya, "r-Orthogonal quasigroups II", *Math Issled.* 43 (1977) 39-49.

[3] G.B. Belyavskaya, "r-Orthogonal latin squares", Chapter 6 in *Latin Squares: New Developments* (J. Dénes and A.D. Keedwell, eds.) Elsevier North-Holland, 1992, pp. 169–202.

[4] F.E. Bennett, L. Wu and L. Zhu, "Further results on incomplete (3,2,1)-conjugate idempotent latin squares", *Discrete Math.* 84 (1990) 1–14.

[5] J.A. Bondy and U.S.R. Murty, *Graph Theory with Applications*, Macmillan, 1976.

[6] A.E. Brouwer, "The linear spaces on 15 points", *Ars Combinat.* 12 (1981) 3–35.

[7] J.H. Dinitz and D.R. Stinson, "MOLS with holes", *Discrete Math.* 44 (1983) 145–154.

[8] A.M. Hamel, W.H. Mills, R.C. Mullin, R.S. Rees, D.R. Stinson and J. Yin, "The spectrum of PBD($\{5, k^\star\}, v$) for $k = 9, 13$", *Ars Combinat.* 36 (1993) 7–26.

[9] H. Hanani, "Balanced incomplete block designs and related designs", *Discrete Math.* 11 (1975) 255–369.

[10] K. Heinrich, "Latin squares with and without subsquares of prescribed type", in: *Latin Squares: New Developments* (J. Dénes and A.D. Keedwell, eds.) Elsevier North-Holland, 1992, pp. 101–148.

[11] K. Heinrich and L. Zhu, "Existence of orthogonal latin squares with aligned subsquares", *Discrete Math.* 59 (1986) 69–78.

[12] D.R. Stinson and L. Zhu, "On the existence of MOLS with equal-sized holes", *Aequat. Math.* 33 (1987) 96–105.

[13] G. Tarry, "Le problème des 36 officiers", *C.R. Acad. Franc. Avanc. Sci. Nat.* 2 (1900) 170–203.

[14] Y. Xu, G. Ge and L. Zhu, "Existence of SOLS with holes of type $2^n u^1$", *J. Stat. Plan. Infer.*, to appear.

[15] L. Zhu, "Existence of self-orthogonal latin squares ISOLS(6m+2,2m)", *Ars Combinat.*, to appear.

GENERAL THEORY OF TRANSLATION INVARIANT SYSTEMS

Amir Daneshgar

Department of Mathematical Sciences
Sharif University of Technology,
P.O. Box: 11365-9415
Tehran, Iran
danshgar@irearn.bitnet

ABSTRACT

The basic goal of this article is to present an abstract system–theoretic approach to morphological filtering and the theory of translation invariant systems which is mainly based on residuated semigroups. Some new results as well as a number of basic questions are also introduced.

1 INTRODUCTION

Since morphological mathematics was introduced and developed in 1960's as a new alternative to traditional Fourier analysis and conventional methods of linear system theory, it has been widely used in signal processing with great influence on image analysis as one of its starting points [10,16,17,19,20]. Of course, a comprehensive survey of the entire field of mathematical morphology would necessarily be very superficial even if much more space were available, however, one should refer to set theory, integral geometry, convex analysis, stereology and geometrical probabilities, as a number of its mathematical origins. Our basic goal in this article is to introduce a modern formulation, from an abstract system–theoretic point of view. This approach is actually a continuation of the outstanding contribution to the theory which led Maragos and Schafer [15] to the concept of a translation invariant system. General attention to nonlinearity, development of fuzzy set theory as a modern approach to ordered structures and general interest in the theory of neural networks- especially those which are based on comparison and scaling- are some of the motivations for this new approach to obtain a sophisticated theory of translation invariant

C. J. Colbourn and E. S. Mahmoodian (eds.), Combinatorics Advances, 77–89.

(TI) systems. From this point of view, the theory of TI systems is a generalization of the theory of linear and shift invariant (LSI) systems which is based on Boolean algebras rather than vector spaces. Therefore, algebraically, one passes from the operational structure of a vector space to the comparative structure of a lattice. Intuitively, in both cases, one may compare the roles of zero (0) and minus infinity ($-\infty$, the least element in the lattice of valuation domain) as the candidates for "null element" or "zero degree of membership" to clarify this point. This also shows that techniques of Boolean algebra as well as its related fields, such as threshold logic, can be applied naturally [22].

A mathematical description of an Input-Output (I/O) system theory is the study of structure preserving maps of a suitably chosen function space which obey some invariance constraints. These maps are usually called operators (or systems) and the algebraic and topological structure of the function space as well as the invariance constraints should be strong enough to guarantee a reconstruction property. Strictly speaking, this reconstruction means that one should be able to compute the output for any input signal if the outputs for a certain specified set of signals are known. For example, in LSI systems the reconstruction property is usually expressed in the so called Convolution Theorem; and one can explicitly compute the output if the impulse response of the system is available. Therefore, in this case, that specific set of signals contains only one element and that element is the Dirac impulse function. Of course, in TI systems this set is called "the base" of the system and usually contains more than one element; which is quite natural since, intuitively, linearity is weakened to "DC gain one" (i.e. translation invariance on the valuation domain). We will see that the case of "a finite base" is of great importance, and one can think of the space of TI operators with a finite base as the direct image of the Boolean algebra of crisp sets under an extension which is known as "thresholding" (or level-cuts in fuzzy set theory). Considering a categorical point of view, it can be seen that this method is more or less a method of extension by sheaves. This categorical viewpoint will provide the necessary formalism to express known results as well as open problems in an abstract and well-defined form clearly. Along this direction, it will be shown that the theory of "topological residuated semigroups" is a very suitable framework for the whole theory of TI systems. Needless to say, this completely fits in the outgrowing pattern of new applications of category theory in mathematics such as theoretical computer science and modern logic.

The article is organized as follows. Section 2 is an introduction to basic definitions and concepts. Section 3 describes the categorical point of view and the relationship to residuated semigroups; and in section 4 we will introduce some basic open problems. Also it should be noted that it has not been our intention to add a complete bibliography to this article, however, the interested reader is encouraged to refer to the extensive list of [14].

2 THE MODEL

As it was mentioned in the previous section, the most important things to be considered are the signal space and the invariance constraints. Of course, the space of signals should be a function space (a set of maps); and since we are interested in the notion of translation, both domain and range (valuation domain) of these maps should have the structure of a group. Also in order to have a nice comparative structure, the range is chosen to be a complete lattice ordered group (L-group). Therefore, the signal space is more or less the set of maps from an abelian group (say $(G,+,-)$) to a complete L-group (say $(\Omega,*,\div)$). Now note that we have a function space (Ω^G) with suitable algebraic and topological properties; but since we are mainly interested in the ordered structure of this space, we would like to have a set theoretic description which asks for a least element (note that categorically, this is a weaker form of the topos of sheaves). With this aim, we add virtual universal bounds $-\infty$ and $+\infty$ to Ω. Also, the extension of group operations to $\Omega\cup\{-\infty,+\infty\}$ is defined as follows.

$$\div(-\infty)=+\infty \ , \quad \div(+\infty)=-\infty \ ,$$

$$(-\infty)*(+\infty)=(+\infty)*(-\infty)=(+\infty)*(+\infty)=(+\infty),\ (-\infty)*(-\infty)=(-\infty),$$

$$\forall\, p\in\Omega \ \ (+\infty)*p=p*(+\infty)=+\infty \ , \quad (-\infty)*p=p*(-\infty)=-\infty \ .$$

Moreover, hereafter, $\Phi \overset{\text{def}}{=} \{\ A : G \longrightarrow \Omega\cup\{-\infty\}\}$ is called the space of LG-fuzzy sets. Now, an operator on Φ is said to be translation invariant if

$$\forall\, g\in G,\ \forall\, p\in\Omega\cup\{-\infty\} \ ; \ Y(A[g,p])=Y(A)[g,p] \ ,$$

in which $A[g,p]$ is A translated by the vector (g,p); i.e.

$$A[g,p]=\{(x,A(x)) \ : \ x\in G\}[g,p]=\{(x+g,A(x)*p) \ : \ x\in G\}.$$

Needless to say, one can also define this invariance property in terms of group actions. Considering these definitions, the basic question is the classification of structure preserving TI operators on Φ; and the first case is the case of isotone TI operators, for which one can prove a nontrivial reconstruction theorem (Strong Reconstruction Theorem (SRT) [6,15]). The precise statement of this theorem is given in the next section, which shows the way that one can compute the systems output when one knows the answer for a specific set of signals, namely the kernel (or the base). For a TI operator Y, the kernel of Y, $\mathbf{K}(Y)$, is defined as follows.

$$\mathbf{K}(Y) \stackrel{\text{def}}{=} \{A \in \Phi \ : \ Y(A)_{(0)} \geq 0\}.$$

The most important property of $\mathbf{K}(Y)$ is that if there are two elements such as $D_1 \leq D_2 \in \mathbf{K}(Y)$ then D_2 can be excluded when one is going to use SRT. Therefore, for technical reasons, a TI operator is called well-defined if its kernel can be reduced to a set of bounded elements. Thinking in this direction, on should be very interested in the case for which $\mathbf{K}(Y)$ has minimal elements, since in this case one can use SRT only on this set which may happen to be considerably smaller than $\mathbf{K}(Y)$. Consequently, it is natural to define the base of Y, $\mathbf{B}(Y)$, as the set of minimal elements of $\mathbf{K}(Y)$ (if they exist).

SRT is the most general form of a reconstruction theorem for TI systems, however, a very interesting and important question concerns the structure of Ω when it has a finite representation. This is mainly because in real applications one usually should deal with a finite number of parameters. In [7] it is proved that if a complete L-group has a finite representation, then it is a product space such that each of its components is either isomorphic to $\mathbf{R}$ or isomorphic to $\mathbf{Z}$ (as totally ordered groups). This, of course, reduces most questions to the continuous case ($\Omega = \mathbf{R}$) or the discrete case ($\Omega = \mathbf{Z}$).

Another important notion to be considered is "thresholding". If A is an LG-fuzzy set, then its threshold set at level p is defined as

$$A_{[p]}(x) = \begin{cases} 0 & A(x) \geq p \\ -\infty & \text{otherwise} \end{cases}$$

Note that $A_{[p]}$ is a crisp set, when (0) means "existence" and $-\infty$ means "absence". Also it is helpful to consider the existence of Ω^+ and $-\infty$ as a trade-off to obtain an L-group with virtual universal bounds. From another point of view, thresholding can be considered as a method of extension by sheaves which helps to extend the Boolean algebra of crisp sets to the Kleene algebra of fuzzy sets [2,3]. As an advantage of this viewpoint, one can immediately see that TI operators which arise from isotone (positive) Boolean functions (stack filters) always commute with thresholding, i.e. one can reconstruct the systems output from the answers to the threshold signals. To be more precise, one can show that a positive TI operator commutes with thresholding if and only if it has a crisp base. Clearly, this property is of great importance since real digital systems are usually built according to crisp notions of sets.

At the end, it is constructive to look at a classical example. Let Ω be totally ordered and let G be a cyclic group of order 3, i.e.

$$G \stackrel{\text{def}}{=} < a \mid 3a = 0 > \quad .$$

Now let Y be the median operator on three points which corresponds to the Boolean function $x_1x_2+x_2x_3+x_3x_1$. It is easy to check that Y has the following base.

$$\mathbf{B}(Y) = \{\{(0,0),(a,-\infty),(2a,-\infty)\},\{(0,-\infty),(a,0),(2a,-\infty)\},$$
$$\{(0,-\infty),(a,-\infty),(2a,0)\}\}.$$

Moreover, using SRT, for a signal $A = \{(0,x_1),(a,x_2,(2a,x_3\}$ we have

$$Y(A)_{(0)} = Y(A)_{(a)} = Y(A)_{(2a)} = \sup(\inf(x_1,x_2),\inf(x_2,x_3),\inf(x_3,x_1)).$$

Note the fuzzy replacement of products and sums by (inf) and (sup) respectively. Also note that Y is a stack filter with a crisp base which commutes with thresholding. Another important aspect of this example is that it can be considered as the starting point for applications of threshold logic in TI systems. Delving into the details of this subject is evidently beyond the scope of this article, while the interested reader may find [22] a good reference.

3 A RESIDUATED SEMIGROUP

Again consider the signal space Φ. First note that Φ can be equipped with the natural pointwise structures as a product space. Consequently, if we assume $A = \{(x, A(x)) \; : \; x \in G\}$ then we may define [6,21]

Symmetric set	$A^s = \{(-x, A(x)) \; : \; x \in G\}$
Complement of a set	$A^c = \{(x, \div A(x)) \; : \; x \in G\}$
Restricted complement	$\overline{A}(x) = \begin{cases} \div A(x) & A(x) \neq -\infty \\ -\infty & \text{otherwise} \end{cases}$

Also, one can define the following binary operator $\oplus$ on Φ which is usually known as Minkowski addition.

$$A \oplus B = \sup_g A[g, B(g)].$$

Now, it is easy to check that $(\Phi, \oplus)$ is an ordered semigroup with the following identity

$$P_0(x) = \begin{cases} 0 & x = 0 \\ -\infty & \text{otherwise} \end{cases}$$

and the involution (A^s). Fortunately, this semigroup is residuated [4,5] which, intuitively, means that $\oplus$ has a best inverse approximation with respect to the order. Speaking in categorical terms, this means that for any $B \in \Phi$, translation by B has an adjoint, when one considers the poset Φ as a category in the usual sense. This adjoint map is commonly called the erosion (or morphological convolution) of A by B, however, for historical reasons - which are mainly based on duality- Minkowski subtraction is defined as

$$A \ominus B = \inf_g A[g, \div B(g)],$$

and one can easily check that

$$Er(A, B) = A \ominus B^s \ .$$

$$Di(A, B) = A \oplus B^s \ .$$

It is important to note that erosion is more natural than $\ominus$, since erosion is one side of the intrinsic Galois connection on Φ for $\oplus$. This becomes more evident even if one has a look at SRT which has the following statement.

Theorem 3.1 *(SRT) [6] If Y is an isotone TI operator on Φ then*

$$Y(A) = \sup_{D \in K(Y)} Er(A, D)$$

and if the base of Y exists then we also have

$$Y(A) = \sup_{D \in B(Y)} Er(A, D)$$

Now the Galois connection $(\oplus, Er)$ on Φ gives rise to a natural closure operator $C(A, B) = Er(A \oplus B, B)$ and an opening $O(A, B) = Er(A, B) \oplus B$ as its dual. It is quite interesting to note that almost all important morphological identities are immediate consequences of well-known results in category theory and the theory of residuated semigroups (e.g. note the applications of Adjoint Functor Theorem and the characterization of B-closed elements). Lists of properties for these basic morphological operators are given at the end of this section.

As it may have been clear from our discussion so far, duality plays an important role in the whole theory. Clearly, the most natural definition for the dual of a TI operator Y is $Y(A^c)^c$, however, one should be careful about the neighborhood of $-\infty$; and this will turn out to be closely related to the concept of continuity. This is mainly because of asymmetric behavior of $-\infty$ and $+\infty$ and the absence of the neighborhood of $+\infty$ in Φ. To clarify this point let's consider the following example. Again let G be the same group of our last example in section 2, and define $Y : \Phi \to \Phi$ as

$$Y(A)_{(g)} = \begin{cases} -\infty & A(g) = -\infty \\ A(g+a) & A(g) \neq -\infty \end{cases}$$

Then

$$\begin{array}{ll} K(Y) = & \{D : D = \{(0,s),(a,t),(2a,\nu\}, s \neq -\infty, t \geq 0\} . \\ A \stackrel{\text{def}}{=} & \{(0,x_1),(a,x_2),(2a,x_3)\}. \\ Y(A) = & \sup\limits_{D \in K(Y)} Er(A,D) \\ \Rightarrow Y(A) = & {}_{e \to -\infty}\{(0,\inf(x_2, x_1 \div e)),(a,\inf(x_3,x_2 \div e)),(2a,\inf(x_1,x_3 \div e))\}. \end{array}$$

To see that Y is not continuous, let $\lambda_\gamma \to -\infty$ and define

$$A_\gamma \stackrel{\text{def}}{=} \{(0,\lambda_\gamma),(a,M),(2a,M)\} ,$$

in which $M \in \Omega$ is a constant. Then

$$\begin{array}{ll} Y(A_\gamma) & = \{(0,M),(a,M),(2a,\lambda_\gamma)\} \\ \Rightarrow \lim Y(A_\gamma) & = \{(0,M),(a,M),(2a,-\infty)\} . \end{array}$$

But

$$\begin{array}{ll} \lim A_\gamma & = \{(0,-\infty),(a,M),(2a,M)\} \\ \Rightarrow Y(\lim A_\gamma) & = \{(0,-\infty),(a,M),(2a,-\infty)\} \neq \lim Y(A_\gamma) . \end{array}$$

In order to solve this problem, we confine ourselves to the space of continuous isotone TI operators. This is not a great limitation since it can be shown that an isotone TI operator is always continuous on Ω^G. Now one can define the dual of an operator Y as the operator Y^d which is the continuous extension of Y_r^d , the dual of the restriction of Y to Ω^G. Evidently, for such an operator one can also have the following dual reconstruction.

$$Y(A) = \sup_{D \in K(Y)} Er(A,D) = \inf_{D \in K(Y^d)} Di(A,D)$$

Note that this equality is not true for the discontinuous example which was discussed earlier. This is especially important when complements appear in the final expression, since in this case the exact value of the expression may be different from the limit. For instance, $P_0 \ominus P_0 = P_0$ while $((P_0)^c \oplus P_0)^c = One$. However, if $P_i \to P_0 \;\; (P_i \in \Omega^G)$ then

$$lim((P_i)^c \oplus P_0)^c = P_0 .$$

Hereafter, we use the following notation to show this equality of limits.

$$((P_0)^c \oplus P_0)^c \simeq P_0 \ .$$

Another interesting question is the characterization of continuous isotone TI operators in terms of some properties of their kernel. To do this, we should introduce the concept of a critical point.

Definition *The reduced kernel of Y, $(K_R(Y))$, is a subset of $K(Y)$ obtained by the process of exclusion for which if $\{D_\alpha\}$ is a nontrivial decreasing net in $K_R(Y)$ then*

$$\lim_\alpha D_\alpha = \inf D_\alpha \notin K_R(Y) \quad .$$

Definition *A point $x \in G$ is called a critical point of Y if and only if one can find a nontrivial net $\{D_\alpha\}$ in $K_R(y)$ such that $D_\alpha(x) \longrightarrow -\infty$. If for a critical point x, and for any net $\{D_\alpha\}$ such that $D_\alpha(x) \longrightarrow -\infty$, there exists $y \in G$ such that $D_\alpha(y) \longrightarrow +\infty$, then x is said to be a critical point of first type; and otherwise x is said to be of second type.*

As an example of a critical point of second type, one can consider 0 in our previous discontinuous example. Actually one can prove that an isotone TI operator is continuous if and only if it has no critical point of second type [7]. For a critical point of first type consider the two point mean value filter on $\underline{\mathbf{R}}^{\mathbf{Z}}$; i.e.

$$Y : \underline{\mathbf{R}}^{\mathbf{Z}} \rightarrow \underline{\mathbf{R}}^{\mathbf{Z}} \quad ,$$

$$Y(A)_{(g)} = 1/2(A(g) + A(g+1)) \ .$$

Then $K(Y) = \{\{(0, r_1), (1, r_2)\} : 1/2(r_1 + r_2) \geq 0\}$, which shows that both (0) and (1) are critical points of first type.

Now let's consider the semigroup $(\Phi, \oplus)$ once again. An element $A \in \Phi$ is called B-open if $O(A, B) = A$. A well-known result in the theory of residuated semigroups [4,5,8] states that an element is B-open if and only if there exists $E \in \Phi$ such that $A = E \oplus B$. With this characterization one can check that, for any $B \in \Phi$, the set of B-open elements is a sup-ideal of $(\Phi, \oplus)$ which is closed under translation. Actually, this is true for the fixpoint set of any opening operator on Φ. Note that it can be verified that no TI operator can induce a topological closure on Φ; therefore, the study of fuzzy (pointless) topologies [12] related to a TI operator should be an interesting subject which will be discussed in the next section.

On the other hand, for a given opening operator Y , one can show that Y has a representation in terms of morphological opening $(O(A, B))$ as

$$Y(A) = \sup_{D \in \mathbf{F}(Y)} O(A, D) ,$$

in which $\mathbf{F}(Y)$ is the fixpoint set of Y [8,10,16]. Of course, a dual statement holds for closure operators; and even granulometries with the usual classical representation theorems can be defined in this general framework [8,10,16].

According to our discussion so far, it seems that a lot is known about the topological residuated semigroup $(\Phi, \oplus)$, however, there are also some weak points too. For instance, Molinaro equivalences can not be used efficiently in this case since the negative cone is not rich enough; although, an spectral representation in terms of openings is still available, using the same techniques of [11]. In the next section this discussion about $(\Phi, \oplus)$ will be continued, with a number of possible directions to be considered.

In the following tables Φ is the space of LG-fuzzy sets and $\Phi^* = \Phi - \{zero\}$.

Table 3.1. $A, B, C, A_1, A_2 \in \Phi$

1. $A \oplus B = B \oplus A$
2. $(A \oplus B) \oplus C = A \oplus (B \oplus C)$
3. $(P_0[a, p]) \oplus A = A[a, p]$
4. $(A \oplus B)^{*} = A^{*} \oplus B^{*}$
5. $(A[a, p]) \oplus B = A \oplus (B[a, p]) = (A \oplus B)[a, p]$
6. $A \oplus B \simeq (A^c \ominus B)^c$
7. $A \oplus B \geq A \ominus \overline{B}$
8. $A_1 \leq A_2 \;\; \Rightarrow \;\; A_1 \oplus B \leq A_2 \oplus B$
9. $A \oplus (\sup(B, C)) = \sup(A \oplus B, A \oplus C)$
10. $A \oplus (\inf(B, C)) \leq \inf(A \oplus B, A \oplus C)$

Table 3.2. $A, C, A_1, A_2, \in \Phi$ and $B, B_1, B_2 \in \Phi^*$

1. $A \ominus B \simeq (A^c \oplus B)^c \simeq B^c \ominus A^c$
2. $(A \ominus B) \ominus C = A \ominus (B \oplus C)$
3. $(A \ominus B)^s = A^s \ominus B^s$
4. $(A[a,p]) \ominus B = (A \ominus B)[a,p]$
5. $A \ominus (B[a,p]) = (A \ominus B)[a, \div p]$
6. $A \ominus B \leq A \oplus \overline{B}$
7. $A_1 \leq A_2 \quad \Rightarrow \quad A_1 \ominus B \leq A_2 \ominus B$
8. $B_1 \leq B_2 \quad \Rightarrow \quad A \ominus B_1 \geq A \ominus B_2$
9. $A \ominus (\sup(B,C)) = \inf(A \ominus B, A \ominus C)$
10. $(\inf(B,C)) \ominus A = \inf(B \ominus A, C \ominus A)$
11. $A \ominus (\inf(B,C)) \geq \sup(A \ominus B, A \ominus C)$
12. $(\sup(B,C)) \ominus A \geq \sup(B \ominus A, C \ominus A)$
13. $A \oplus (B \ominus C) \leq (A \oplus B) \ominus C$

Table 3.3. $A, A_1, A_2 \in \Phi$ and $B, B_1, B_2 \in \Phi^*$

1. $O(A,B) \simeq (C(A^c, B))^c$
2. $O(A,B) \leq A$
3. $(O(A,B)) \ominus B^s = C(A \ominus B^s, B^s) = Er(A,B)$
4. $O(O(A,B),B) = O(A,B)$
5. $O(A[a,p],B) = (O(A,B))[a,p]$
6. $O(A, B[a,p]) = O(A,B)$
7. $A_1 \leq A_2 \quad \Rightarrow \quad O(A_1,B) \leq O(A_2,B)$
8. $O(\inf(A_1,A_2),B) \leq \inf\ (O(A_1,B), O(A_2,B))$
9. $O(\sup(A_1,A_2),B) \geq \sup\ (O(A_1,B), O(A_2,B))$
10. $O(A, \sup(B_1,B_2)) \leq \sup\ (O(A,B_1), O(A,B_2))$
11. $C(A,B) \geq A$
12. $(C(A,B)) \oplus B^s = O(A \oplus B^s, B^s) = Di(A,B)$
13. $C(C(A,B),B) = C(A,B)$
14. $C(A[a,p],B) = (C(A,B))[a,p]$
15. $C(A, B[a,p]) = C(A,B)$
16. $A_1 \leq A_2 \quad \Rightarrow \quad C(A_1,B) \leq C(A_2,B)$
17. $C(\inf(A_1,A_2),B) \leq \inf\ (C(A_1,B), C(A_2,B))$
18. $C(\sup(A_1,A_2),B) \geq \sup\ (C(A_1,B), C(A_2,B))$
19. $C(A, \inf(B_1,B_2)) \geq \inf\ (C(A,B_1), C(A,B_2))$
20. $C(A, B^c) \simeq Er(B, Er(B,A))$

4 SOME BASIC QUESTIONS

Our basic objective in this section is to introduce a number of questions which are believed to be of crucial importance. To begin, we would like to look at $(\Phi, \oplus)$ as a topological residuated semigroup. With this point of view, one could get interested in the structure of this kind of semigroups in general; and in that case the first fundamental question is "Does there exist a nice representation theorem under suitable finiteness conditions?" The answer to this question is very likely to be "yes" when one notes that the structure is something between topological semigroups and topological groups and that such a representation theorem already exist for Hungarian semigroups [18].

The next thing to be considered is the applications of fuzzy (pointless) topology [12]. In this direction, the first thing to be considered is the fixpoint set of an opening operator which has a very rich algebraic, topological and ordered structure as a sup-ideal of $(\Phi, \oplus)$. Clearly, this set induces a fuzzy topology on Φ and one may ask about the relationship between the properties of this topology and the corresponding operator. Of course, the same technique can be used in the space of isotone TI operators $\Gamma(\Phi)$, using the representation theorem of opening operators, however, in this case, the natural question should be about the classification of compact operators in the induced fuzzy topology.

Another important question concerns the idempotents of $(\Phi, \oplus)$, which is important in the algebraic study of this semigroup. This will turn out to be very closely related to the Cauchy functional equation [1], but this problem is somehow more general since one should deal with an inequality. Characterization of associates are also a very important algebraic problem, however, this equivalence relation does not have an adaptive behavior as far as the ordered structure is concerned.

A powerful operational calculus can be of great help, when one is going to use an I/O system theory, while the lack of such a thing is one of the weak points in the theory of TI systems. Now, considering the Hadamard-Walsh transformation as one of the most natural transformations on the Boolean algebra of crisp sets, one may ask "Is there a suitable extension of Hadamard-Walsh transform to the space of TI operators?" Also, it should be noted that presentation of a competent operational calculus for TI systems,undoubtedly, will help to have a better feeling about TI systems; and will be a step forward to solve the "Design Problem" for these systems.

To sum up, the author believes that the theory of TI systems is in harmony with the developments in mathematics as well as modern science and presents a vast arena for the continuing research.

REFERENCES

[1] J.Aczel, J.Dhombres, *Functional Equations in Several Variables.* in Encyclopedia of Mathematics and its Applications Vol. 31, Cambridge University Press, 1989.

[2] M. Barr, Fuzzy Sets and Topos Theory. *Canad. Math. Bull. 24 (1986c) 501-508.*

[3] M. Barr, C.Wells, *Category Theory for Computing Science.* International Series in Computer Science, Prentice-Hall 1990.

[4] G. Birkhoff, *Lattice Theory.* AMS Colloquium Publications, Vol. XXV, 1967.

[5] T.S.Blyth, M.F.Janowitz, *Residuation Theory.* Pergamon Press, 1972.

[6] A. Daneshgar, System Theory and Fuzzy Sets Part I: Reconstruction Properties of Translation Invariant Systems. *(Submitted)*

[7] A. Daneshgar, System Theory and Fuzzy Sets Part II: Topology and Duality in Translation Invariant Systems. *Institute for Studies in Theoretical Physics and Mathematics (IPM), Tech. Rep. 93-021*

[8] A. Daneshgar, System Theory and Fuzzy Sets Part III: Basic Morphological Concepts in Translation Invariant Systems. *Institute for Studies in Theoretical Physics and Mathematics (IPM), Tech. Rep. 93-039*

[9] A. Daneshgar, System Theory and Fuzzy Sets Part IV: Residuated Semigroups and Translation Invariant Systems. *(To Appear)*

[10] E.R. Dougherty, C.R. Giardina, *Morphological Methods in Image and Signal Processing.* Prentice-Hall, 1988.

[11] J. Goutsias, D.Schonfeld, Morphological Representation of Discrete and Binary Images, *IEEE Trans. Signal Processing Vol.39 No.6 (June 1991).*

[12] W. Guo-Jun, Theory of Topological Molecular Lattices, *Fuzzy Sets and Systems 47 (1992) 351-376.*

[13] M. Husek, Categories and Mathematical Morphology, *in Lecture Notes in Computer Science No.393, Springer-Verlag 1989.*

[14] P. Maragos, R.W. Schafer, Morphological Systems for Multidimensional Signal Processing, *Proc. IEEE 78 (April 1990) 690-710.*

[15] P. Maragos, R.W. Schafer, Morphological Filters- Parts I,II. *IEEE Trans. Acoust. Speech, Signal Processing, ASSP-35 (Aug. 1987) 1153-1184*
Also, "Corrections," ASSP-37 (April 1989) 597.

[16] G. Matheron, *Random Sets and Integral Geometry.* New York: John Wiley & Sons, 1975.

[17] G. Matheron, *Elements pour une Theorie des Milieux Poreux.* Paris: Masson, 1967.

[18] I.Z. Ruzsa, G.J. Szekely, *Algebraic Probability Theory.* John Wiley & Sons (1988).

[19] J. Serra, *Image Analysis and Mathematical Morphology.* New York: Academic Press, 1982.

[20] J. Serra, Ed. *Image Analysis and Mathematical Morphology Vol. 2: Theoretical Advances.* New York: Academic Press, 1988.

[21] D. Sinha, A General Theory of Fuzzy Arithmetic. *Fuzzy Sets and Systems 36 (1990) 339-363.*

[22] O. Yli-Harja, J. Astola, Y. Neuvo, Analysis of the Properties of Median and Weighted Median Filters Using Threshold Logic and Stack Filter Representation. *IEEE Trans. Signal Processing, Vol.39 No.2 (Feb. 1991).*

SOME MATHEMATICAL PROBLEMS ARISING IN MOLECULAR BIOINFORMATICS: THE CONCEPT OF BIOINFORMATICS

Andreas Dress

Universität Bielefeld
Germany

ABSTRACT

The advent and the upkeep of life is intimately related to the generation, processing and propagation of information on the molecular level. To understand and to analyse the basic processes of life in a scientifically satisfying and pertinent way, the concept of *information* will have to be added, therefore, to the list of basic concepts like *energy, matter, entropy*, etc., developed and clarified in the natural sciences during the last centuries [1]. Information, even though it is encoded in terms of specific spatial and/or temporal forms, patterns, or structures of matter, cannot be identified with its physical carrier or host—just as a river or even a simple wave cannot be identified with the water it carries along.

1 INTRODUCTION

To deal with information, Shannon's Information Theory [2] provides many useful tools which are based on measuring information in terms of *bits* or – more generally—in terms of (the minimal amount of) the complexity of structures needed to encode a given piece of information. Consequently, Shannon's theory can be applied to measure e.g. the danger of obliteration of information or, equivalently, the amount of work necessary to preserve or to transmit information properly through noisy channels.

The creation of this theory has surely been one of the great scientific events in this century. In particular, it established firmly that the concept of information has to be accepted as a fundamental, logically sound concept, amenable to

C. J. Colbourn and E. S. Mahmoodian (eds.), Combinatorics Advances, 91–109.

scientific scrutiny; it cannot be viewed as a solely anthropomorphic concept, useful in science on a metaphoric level only.

Still, Shannon's theory has never been intended and is not suitable to adequately describe the dynamics of generation, processing and propagation of information in living matter and its validation (or dis-validation) through the encounter of a living entity—be it a cell or an eco-system—with its environment. An understanding of these dynamics, though, is crucial in biology on many different levels: it is needed to understand information processing on the cellular level—the processing of genetic information as well as the orchestration of communication inside and between cells through the modulation of membranes—as well as on the organismic level, where information is processed through networks like the neural or the immune network; it is needed on the ecological and on the evolutionary level, where information is created and processed according to Darwinian theory through the interplay of replication, mutation and selection; and it is needed, of course, to understand the interaction between all these levels.

Bioinformatics, therefore, is neither just a biologically oriented application of Shannon's theory, nor is it the collection of computer science techniques necessary to build and to continuously update biological data banks. Rather, including these two aspects on its far ends, it aims at developing conceptual as well as practical tools to understand and to deal with the generation, processing and propagation of information on all levels of life. Of course, this is a rather long-range goal, the pursuit of which has just began about two decades ago, and to achieve this goal, much more work needs to be done.

2 THE CONCEPT OF SEQUENCE SPACE

Still, some fundamental insights emerged already which appear to have far-reaching consequences and some of these, which are connected with the concept of *sequence space* and which have been developed by Manfred Eigen and his group and co-workers, I would like to mention in some more detail.

To encode information in terms of spatiotemporal forms or patterns, the widely preferred method of choice is to combine simple building blocks according to well specified rules which allow the generation of sufficiently complex structures to incorporate all the information one wants to encode. If the building rules allow to make use of the full intricacy of 3-space, using all three degrees of freedom physical space provides, the amount of complexity which can be generated already with only a few simple building blocks will be enormous, as is clearly demonstrated by the universe of molecules, which can be built from, say, just carbon, oxygen, and hydrogen atoms using the most basic quantum mechanical bonding rules only. So, encoding information by combining even the simplest building blocks according to rules which make use of all three spatial degrees of freedom will definitely provoke extremely severe problems for the inverse process of decoding,—as is set in evidence for instance by the continuously ongoing debate about stereochemical nomenclature [3]. Such encoding procedures can therefore be useful only in highly specific contexts with highly developed recognition structures, that is, structures which recognize an encoded message the way a keyhole 'recognizes' its key.

In contrast, for general purpose procedures of encoding and decoding information, the spatial degrees of freedom for arranging the simple building blocks in admissible patterns have to be reduced drastically and—what is well known and equally important—they also can be reduced without loss of potential complexity to just one degree of freedom. That is, all sorts of information can indeed be encoded by aligning simple building blocks sequentially in the form of one-dimensional strings, shortly called *sequences*.

A formal model for the simplest universe capable of hosting all sorts of information, therefore, is what mathematicians call *sequence space*, that is, the set of all sequences of a given (and, if necessary, rather large) finite length, say ν, which can be constructed from a finite set $\mathbf{A}$ of 'symbols', called *alphabet* (which may be rather small), e.g. the set $\mathbf{A} = \{0, 1\}$, consisting of just two elements, '0' and '1'. According to standard terminology, this space is then denoted by $\mathbf{A}^\nu$.

The *distance* between any two 'points' in sequence space, that is, between any two sequences $a_1 a_2 \ldots a_\nu$ and $b_1 b_2 \ldots b_\nu$ in $\mathbf{A}^\nu$ is then defined according to R.W. Hamming [4] as the minimal number of point mutations necessary to transform $a_1 a_2 \ldots a_\nu$ into $b_1 b_2 \ldots b_\nu$, that is, the number of positions i between 1 and ν, at which a_i differs from b_i.

Obviously, genetic information is encoded this way, using the four-letter alphabet of nucleotides, and for two *aligned** genes their Hamming distance measures the *observed mutational distance* between these two genes and corresponds com-

*Of course, for non-aligned genes, a more complicated model is required, allowing e.g. sequences of varying length. Since the arising complications are of technical nature only, we will not enter into discussing such more realistic models here.

plementarily to their *homology*, that is, the number of positions $i \in \{1, \ldots, \nu\}$ with $a_i = b_i$.

3 THE GEOMETRY OF SEQUENCE SPACE

Just as insight into the abstract geometric properties of euclidean and/or non-euclidean 3-space is a necessary prerequisite for unraveling the dynamics which form solar systems or galaxies alike, insight into the abstract geometric properties of sequence space is a necessary prerequisite for unraveling the dynamics according to which genetic information evolves.

Such insight indeed helped to estimate the time necessary to develop the genetic code during primordial evolution [5] from an analysis of the geometric invariants [6] associated with the 'configuration of points in sequence space' representing present tRNA-sequences, or the age of the AIDS virus from corresponding sequence data [7].

To be specific, it is necessary to become a bit technical:

First, one has to observe that the most important distinction between the familiar geometry of euclidean space and the much less familiar geometry of sequence space is related to the concept of congruence, that is, the equivalence of figures, or *point configurations*, which can be transformed into each other by a congruence transformation, that is, a distance preserving map, defined on total space. Given two point configurations, say $P_1P_2\ldots P_n$ and $Q_1Q_2\ldots Q_n$, in euclidean space it is an almost too familiar fact that they are congruent if and only if the distance between any two points P_i and P_j from the first configuration coincides with the distance between the corresponding points Q_i and Q_j from the second configuration.

In arbitrary space, and so in particular in sequence space, this condition is obviously still a necessary condition for congruence, but in general it will not be sufficient. To give the most simple counterexample in sequence space, consider the two 'triangles', given by the sequences $P_1 := A\,B\,B$, $P_2 := A\,A\,C$, $P_3 := C\,B\,C$ and $Q_1 := P_1 = A\,B\,B$, $Q_2 := P_2 = A\,A\,C$, $Q_3 := A\,C\,A$ (where $\mathbf{A} := \{A, B, C\}$;—so this example lives in $\{A, B, C\}^\nu$ for $\nu := 3$). Even though both triangles are equilateral and have the same edge length 2, they cannot be transformed into each other by a congruence transformation, since such a congruence transformation would transform the sequence $P := ABC$, which has distance 1 to P_1, P_2 and P_3, into a corresponding sequence Q having distance 1 to Q_1, Q_2, and Q_3 which just does not exist in $\{A, B, C\}^{3}$ *)[†] To clarify the situation, we restrict ourselves to *four*-point configurations $P_1 = a_1 a_2 \ldots a_\nu$, $P_2 = b_1 b_2 \ldots b_\nu$, $P_3 = c_1 c_2 \ldots c_\nu$, $P_4 = d_1 d_2 \ldots d_\nu$ (cf. [8] for the general case of n-point configurations): For every position i between 1 and ν altogether 15 different *patterns of coincidence* need to be distinguished: all four letters a_i, b_i, c_i and d_i may coincide (the 4:0 pattern), three letters may coincide and differ from the fourth one (a 3:1 pattern of which there are four, depending on whether a_i, b_i, c_i or d_i is different from the rest), there may be two distinct pairs of coinciding letters (a 2:2 pattern of which there are three, depending on whether a_i coincides with b_i, c_i or d_i), there may be just one pair of coinciding letters (a 2:1:1 pattern of which there are six and which are possible only if the alphabet **A** contains at least three distinct letters), and finally, all four letters may be different (the 1:1:1:1 case which is possible only if **A** contains at least four different letters).

Using these distinctions, one can show [8] that two four-point configurations are congruent in sequence space, if and only if for each of the fifteen patterns of coincidence both configurations possess the same number of positions exhibiting this particular pattern (cf. Fig. 1).

In other words, these fifteen cases lead to fifteen numerical invariants associated with any given four-point configuration in sequence space which together characterize the geometry of such configurations in this space in just the same way distances characterize configurations according to Felix Klein's fundamental 'Erlanger Programm' [9] in euclidean space, once one is willing to adapt this program to the study of the geometry of sequence space. Moreover, while distances are related to each other by the fundamental equations and inequalities of euclidean distance geometry [10], these fifteen invariants are related only by the fact that their sum must coincide with the total length ν of the sequences considered.

Having this in mind one can now analyze sequence families like e.g. the family of tRNA-molecules or the family of the aids-virus sequences by considering for,

---

[†]Similar counterexamples exist even for a two-letter alphabet, the smallest one being two four-point configurations built from sequences of length 4, e.g. $P_1 := AAAB$, $P_2 := AABA$, $P_3 := ABAA$, $P_4 := BAAA$ and $Q_1 := P_1 = AAAB$, $Q_2 := P_2 = AABA$, $Q_3 := P_3 = ABAA$, $Q_4 := P_4 = ABBB$. Again, all six distances between distinct points in both configurations have the value 2, but there is no sequence Q in $\{A, B\}^4$ which—corresponding to the sequence $P := AAAA$, having distance 1 to P_1, P_2, P_3, and P_4—has distance 1 to Q_1, Q_2, Q_3, and Q_4.

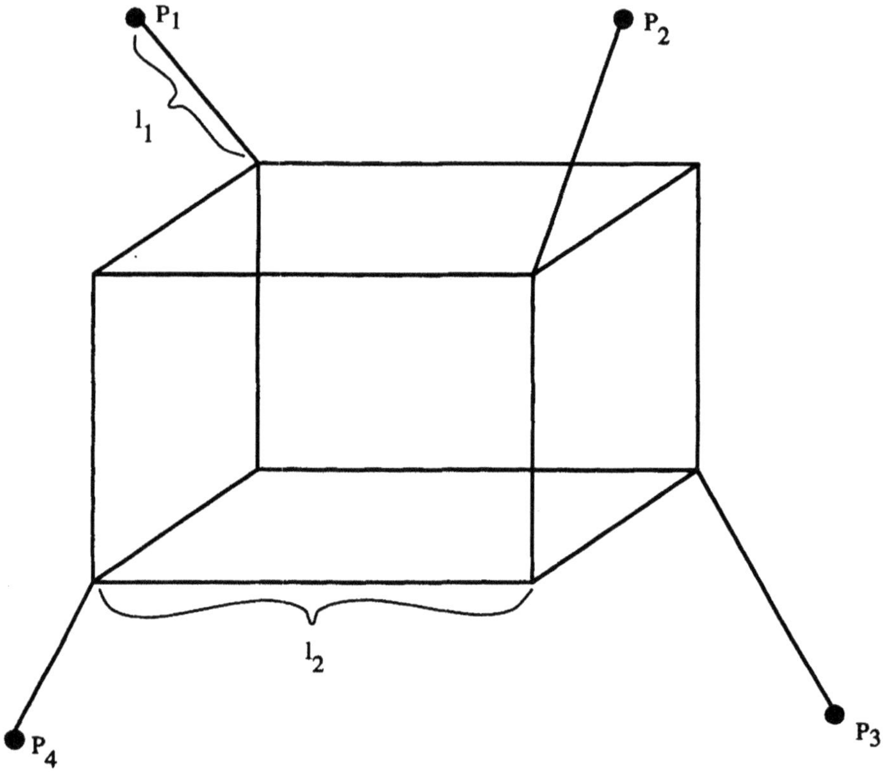

For a two-letter alphabet **A** there are 8 possible patterns of coincidence at every position of a four-point configuration. The associated numerical invariants add up to the total length ν, so only seven of them can be chosen freely. They can be visualized by the edge lengths of the box graph associated with such a four-point configuration according to the theory of median graphs: in the above figure, the eight vertices of the box represent eight sequences from $\mathbf{A}^\nu$ which are uniquely determined by the conditions that e.g. the distance l_1 represents the number of positions where P_1 differs from the remaining sequences P_2, P_3, P_4, the distance l_2 represents the number of positions, where P_1 and P_4 coincide and differ from P_2 and P_3, and so on.

Figure 1

say, any quadruple of them the associated numerical invariants and their overall statistical properties and by comparing these with the corresponding values from other sequence families. These may come from other biological data sets, they may represent sequence families at early nodes in the phylogenetic tree derived by appropriate reconstruction methods, or they may be generated by all sorts of computer simulation programs.

This way it became possible [5] to estimate the relative time spans of chemical versus biological evolution, resulting in a 1:3 ratio compatible with the hypothesis that the origin of life took place on earth (cf. Fig. 2), or to study the age of the aids-virus, leading to an estimate of 600 ± 300 years [7].

4 CLUSTER ANALYSIS

Geometric observations of this sort are also fundamental for all procedures which aim at

- • unraveling kinship relations between present day species from corresponding sequence data to reconstruct (this or that part of) the phylogenetic tree (used e.g. in the analysis presented in Fig. 2),
- • at understanding the intraspecies evolution of various classes of proteins,
- • at classifying the underlying diffusion and branching processes,
- • and at discriminating between convergent, parallel, and divergent evolution.

Obviously, such insights are fundamental for evolutionary biology and they can be very helpful for many tasks in biotechnology as well—from genetic engineering to drug design.

Hence, many different procedures from mathematical *cluster theory* [11] are available for such tasks. This is not surprising since rather similar problems arise in many other branches of the sciences as well, from astronomy (classification of stars, star clusters, galaxies etc.), geology (classification of strata), psychology (classification of conceptualizations), linguistics (classification of languages), and archaeology (classification of e.g. pottery in terms of the chemical composition of pot-shards etc.) to library science and artificial intelligence (e.g. the classification of form elements in various pattern recognition techniques).

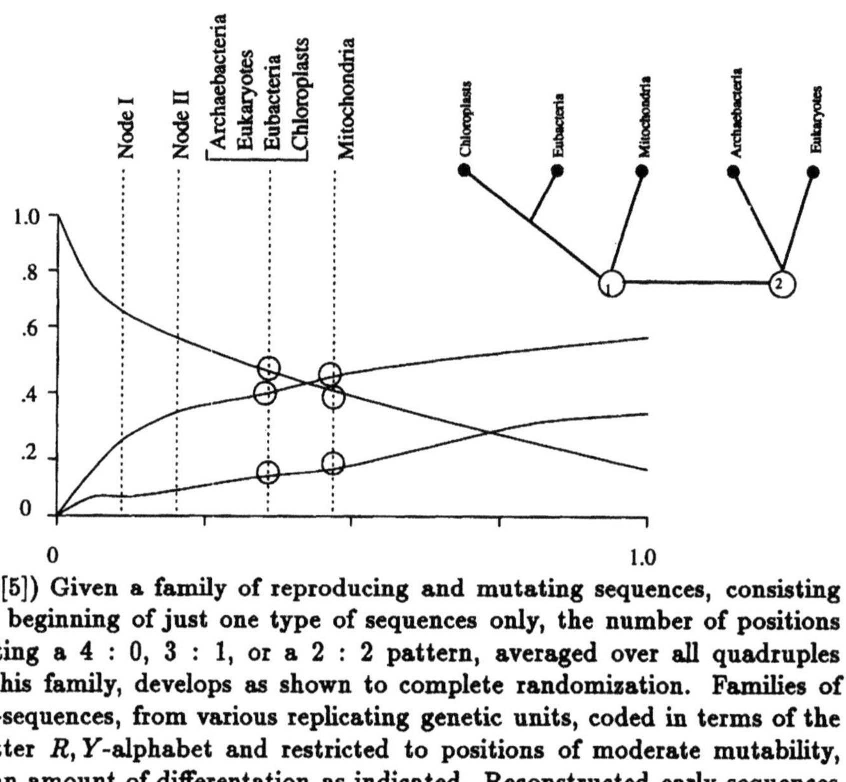

(from [5]) Given a family of reproducing and mutating sequences, consisting at the beginning of just one type of sequences only, the number of positions exhibiting a 4 : 0, 3 : 1, or a 2 : 2 pattern, averaged over all quadruples from this family, develops as shown to complete randomization. Families of tRNA-sequences, from various replicating genetic units, coded in terms of the two-letter R, Y-alphabet and restricted to positions of moderate mutability, show an amount of differentation as indicated. Reconstructed early sequences from the two nodes I and II in the inserted phylogenetic tree show about one fourth of that, II showing significantly more than I, indicating that I is closest to the primordial cell and that differentiation of tRNA-sequences during *chemical* evolution needed (at most) one third of the time used up so far in *biological* evolution.

Figure 2

Sometimes, the goal of cluster theory is 'nothing but' an efficient book keeping strategy (those who have worked in this area know very well that even this often constitutes a highly non-trivial problem), sometimes the goal is to establish basic conceptual categories on which all further understanding of the subject in question may depend, sometimes—and in particular in biology— classification is inseparably intertwined with the analysis of evolutionary dynamical processes which are supposed to have produced the variety of objects to be analyzed.

Cluster theory aims at structuring a set X by specifying a system $\mathcal{C}(\mathcal{X})$ of subsets of X, often called *clusters* (and representing, if you wish, the *extent* of *concepts*, relevant for X), subject to the following conditions:

- • The clusters should collect *similar* objects, that is, objects in a given cluster $C \in \mathcal{C}(\mathcal{X})$ should somehow be more similar to each other than to objects outside C.
- • The specifying procedure should be reasonably *stable*, that is, additions, eliminations, or small changes of a few aspects (e.g. positions of sequences) or even small changes of X should not result in a drastically different system of clusters.
- • $\mathcal{C}(\mathcal{X})$ should be *informative*, that is, it should be reasonably small (e.g. it should not grow more than polynomially with the size of X) and, simultaneously, it should be reasonably large (not just $\{\emptyset, X\}$, for instance) and, if possible, it should contain reasonably sized subsets (not only very small or very large subsets).
- • $\mathcal{C}(\mathcal{X})$ should be *computable*: without a reasonably fast algorithm to compute the clusters in $\mathcal{C}(\mathcal{X})$, even the very best theory could not be used in practice.
- • Sometimes, e.g. in evolutionary biology, the clusters should not be overlapping and thus form a *hierarchy*, that is, a system of subsets of X such that the intersection $C \cap C'$ should either be empty or equal to C or to C' for all clusters C, C' in that system (which automatically reduces the cardinality $\#\mathcal{C}(\mathcal{X})$ of $\mathcal{C}(\mathcal{X})$ to at most $2 \cdot \#X - 2$!).

All this could be achieved easily if it were not for the notorious *intransitivity* of similarity, the crux of cluster theory. Many attempts have been made and many clever schemes have been designed to overcome this problem. While some of the most popular classification procedures aim directly at constructing hierarchical classification schemes (or treelike structures) which approximate as accurately as may be possible a given *scheme of diversity* (mostly a *metric space* according to mathematical terminology), others are less restrictive and allow to detect parallel and convergent evolution events as well as hybridization effects due to gene exchange in addition to phylogenetic kinship relations, leading to trees only if the data support unambiguously a unique tree. The less restrictive methods include the *spectral analysis* of phylogenetic data, introduced by M.D. Hendy and D. Penny [12], the analysis of *weak hierarchies* associated with distance data [13] and the *split-decomposition* method [14].

5 SPLIT DECOMPOSITION

We experimented in particular with the latter and we found it amazingly suitable to analyze, say, viral evolution [15] or controversial bifurcation points in the phylogenetic tree, e.g. the man-chimpanzee-gorilla bifurcation, the plant-animal-fungus bifurcation or the phylogenetic status of the archaebacteria.

It all starts with the basic and well known observation [16] that if a phyletic line separates two pairs of taxa $1, 2$ and $3, 4$, then—with respect to the phyletic distance p—the sum $p_{12} + p_{34}$ (of distances between 1 and 2 and between 3 and 4) is smaller than $p_{13} + p_{24} = p_{14} + p_{23}$ ("treelikeness" alias "additivity" or "4-point condition", cf. Fig. 3). Sequence distances d, representing only an estimate of true phyletic relationships, normally do not even match the ordering $d_{12}+d_{34} < d_{13}+d_{24}$ and $d_{12}+d_{34} < d_{14}+d_{23}$; but one could hope that $d_{12}+d_{34}$ is at least not the largest of the three sums.

Given this as a working hypothesis, we would then expect such a relationship to be observed whenever the two taxa 1 and 2 are chosen from a group $\mathcal{J}$ which is separated from its complementary group $\mathcal{K}$ by a phyletic line, while taxa 3 and 4 are chosen from the complement $\mathcal{K}$. Consequently, any complementary pair $\mathcal{J}, \mathcal{K}$ satisfying this (comparatively weak) condition will be called a *d-split*.

To any such d-split one can, moreover, associate a positive weight, the *isolation index*, which is defined to be the largest positive number α such that $d_{12} + d_{34} + 2\alpha$ is not larger than both $d_{13} + d_{24}$ and $d_{14} + d_{23}$, for all taxa $1, 2$ from $\mathcal{J}$ and $3, 4$ from $\mathcal{K}$ and which in the case of treelike distances would yield the length of the corresponding branch in the representing tree (cf. Fig. 3 again). However, there may be more d-splits than those supported by true phyletic relationships. These, typically exhibiting a low isolation index, often reflect traits of penetrating parallelism.

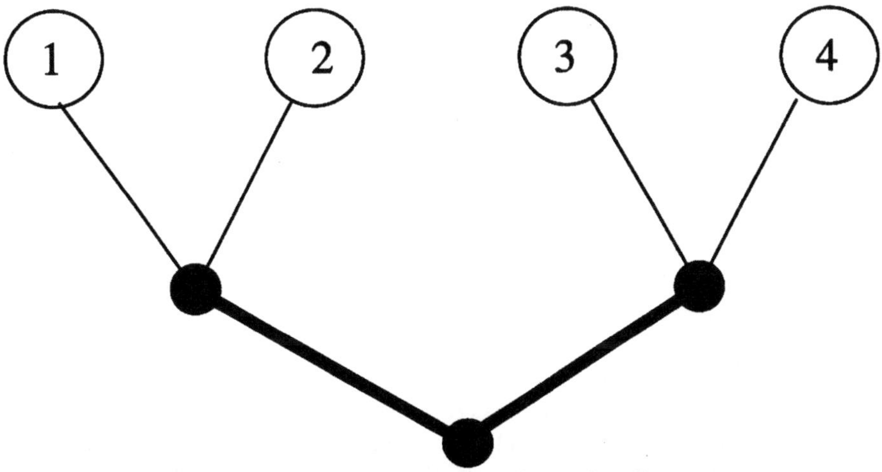

In a perfect treelike distance relationship as above the distance sums $p_{13} + p_{24}$ and $p_{14} + p_{23}$ both coincide with the sum of the length of all thin branches and twice the length of the two thick branches; so they coincide and are larger than the third sum $p_{12} + p_{34}$, representing the sum of the length of the thin branches only.

Figure 3

If—for example—an observed distance d is the sum $d = p + e$ of a linearly scaled phyletic distance p plus an error term e such that e itself happens to be realized by some tree different from the one representing p, then the d-splits would consist exactly of all splits which are either p-splits or e-splits or both (with isolation indices of p and e adding up to the indices for d). Which of the splits belong to p and which to e, though, cannot be decided unambiguously. If the error term e has considerably smaller entries than p, then the d-splits with larger isolation indices would belong to p rather than e.

The theory of split decomposition [14] predicts at most $\binom{n}{2}$ d-splits for n sequences. This bound is considerably larger than $2n - 3$, the maximum number of splits in a tree connecting n sequences, yet it is small enough to have all d-splits computed efficiently. Re-analysis of numerous distance data [17] derived from sets of 10 to 25 aligned ribosomal RNA sequences confirms that biologically relevant data typically bring about twice as many splits, a large portion of which fitting together on a single tree, and leave a small residue (cf. Figures 4–7). In contrast, randomly generated distances tend to have a rather large residue and to produce mostly *trivial* splits separating one taxon from all the remaining ones, and only very few others, generally separating no more than two or at most three taxa from the rest.

I described this particular method in some detail because it provides a good example for the specific mathematical methods being used in bioinformatics not at all related to standard calculus, and because of its potential to also disentangle kinship relations between other sequence families, say, families of weakly homologous proteins in one species, e.g. man,—an obviously rather important task for many applied areas of molecular biology.

6 CONCLUDING REMARK

Sequence space can also be used to model biological information processing through dynamic optimization processes, defined on populations 'living' in sequence space [19]. Many important phenomena like e.g. the *error-threshold* phenomenon, relating rates of mutation with the amount of complexity of evolving information to be processed by replication and mutation and validated by selection [20], or the *quasi-species* phenomenon [21], neutral drift [22] (related to the degeneracy of the sub-optimal states), and the statistical characteristics of the so-called *ruggedness of the fitness landscape* [23] can be studied and clarified this way.

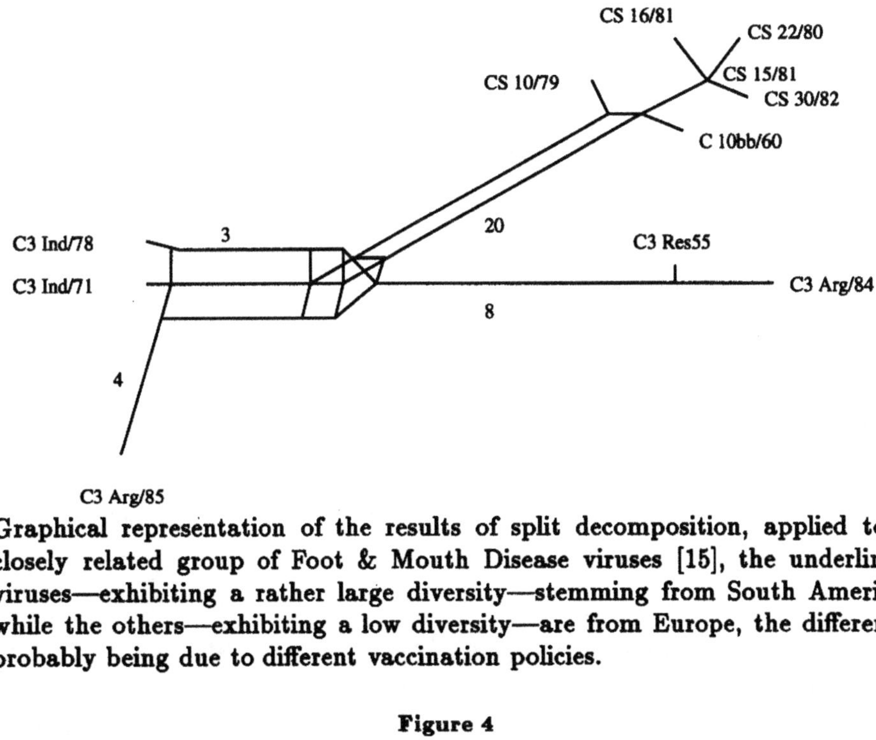

Graphical representation of the results of split decomposition, applied to a closely related group of Foot & Mouth Disease viruses [15], the underlined viruses—exhibiting a rather large diversity—stemming from South America, while the others—exhibiting a low diversity—are from Europe, the difference probably being due to different vaccination policies.

Figure 4

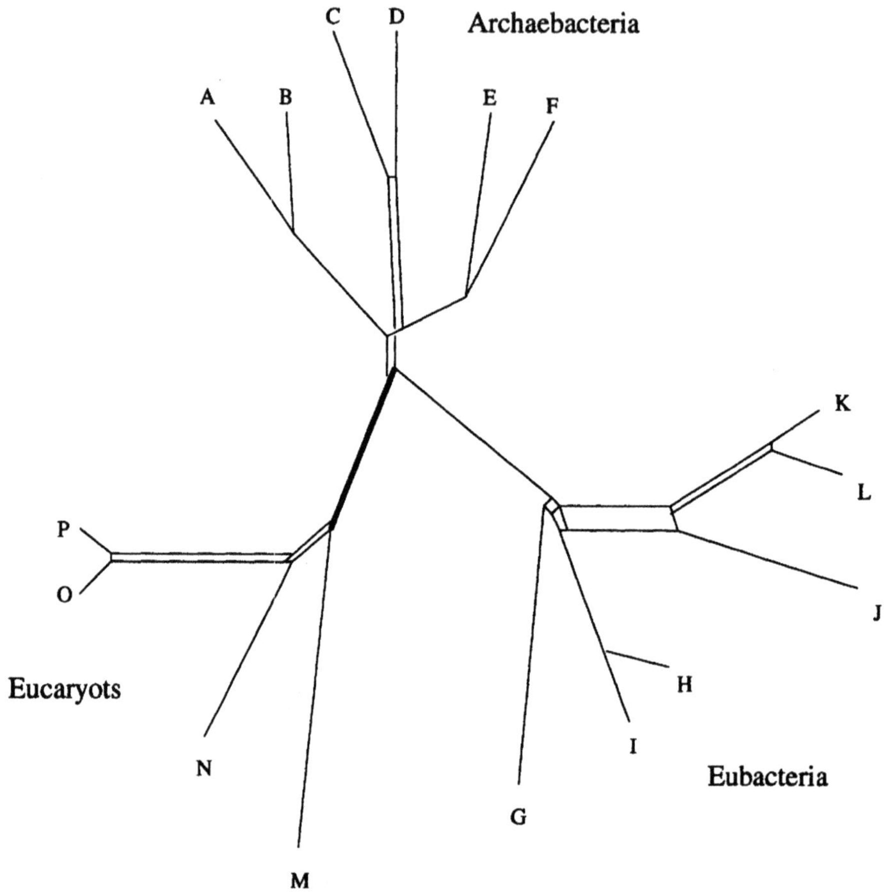

Graphical representation of the results of split decomposition, applied to 23S ribosomal RNA sequences of 6 archaebacteria, 6 eubacteria (including 2 chloroplasts), and 4 eukaryotes, studied by H. Leffers et. al. [18] (taxon symbols being as follows: (*A*) *Desulfurococcus mobilis*; (*B*) *Thermoproteus tenax*; (*C*) *Halococcus morrhuae*; (*D*) *Halobacterium halobium*; (*E*) *Methanococcus vanillii*; (*F*) *Methanobacterium thermoautrophicum*; (*G*) *Escherichia coli*; (*H*) *Bacillus stearothermophilus*; (*I*) *Bacillus subtilis*; (*J*) *Anacystis nidulans*; (*K*) *Zea mays* chloroplast; (*L*) Tobacco chloroplast; (*M*) *Physarum polycephalum*; (*N*) *Saccharomyces cerevisiae*; (*O*) *Xenopus laevis*; (*P*) Mouse).

Figure 5

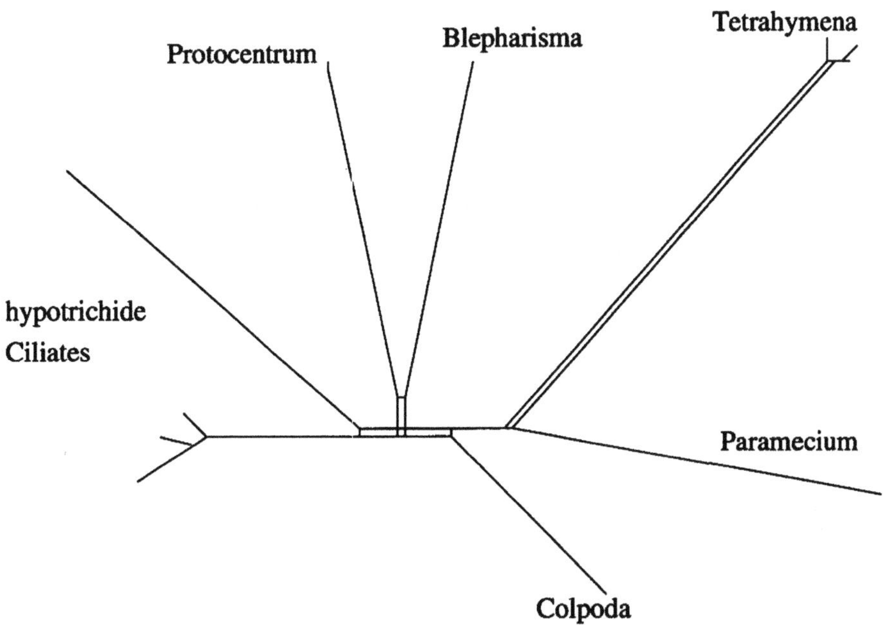

Graphical representation of the results of split decomposition, applied to a group of ciliates.

Figure 6

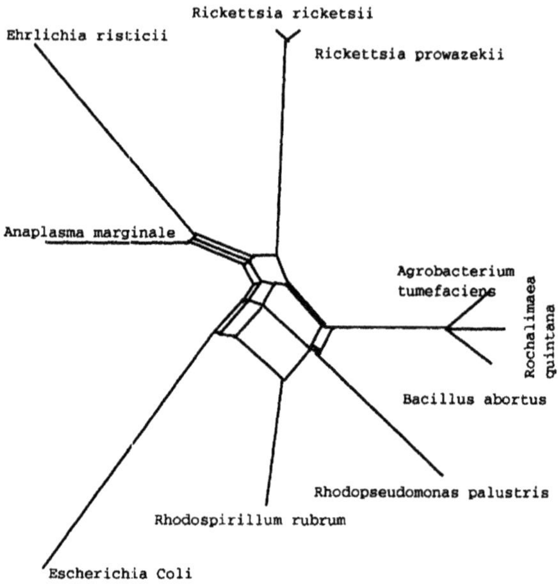

Graphical representation of the results of split decomposition, applied to a group of bacteria.

Figure 7

Equally important, but much less understood are the recognition processes regulating the life cycle of cells, the immune system or the neural network (not to mention the relations between information processing and social interaction among higher developed animals, e.g. bees, ants, or man). To study these processes on the molecular level, it will be unavoidable to study more complicated procedures of encoding information by using the full degree of freedom provided by 3-space or, in other words, to clarify on a rather fundamental level the relation between *structure* and *function*. To this end, one needs of course almost everything the natural sciences, mathematics, and computer science have to offer, from data bank management or methods to identify structure by X-ray crystallography and/or spectroscopy through molecular dynamics, cluster chemistry and membrane analysis to dynamical systems theory, distance geometry and combinatorics, to name only a few relevant topics.

So, progress in this rather comprehensive field can be achieved only as a result of more or less independent progress in many different individual fields. Still, I hope that the vision of bioinformatics as presented here will lead to many unexpected and fruitful interactions between these fields which will contribute to understand in rather general terms the process of information generation, validation, and propagation in all living matter.

REFERENCES

[1] Cf. e.g. L. Boltzmann, Wiss. Abh., ed. F. Haenoehrl (Chelsea, New York, 1968).

[2] C. Shannon and W. Weaver, The Mathematical Theory of Communication (U. Of Ill. Press, Urbana-Ill., 1969), see also L. Brillouin, Science and Information Theory (Acad. Press, New York, 1971).

[3] Cf. e.g. UPAC, Nomenclature of Organic/Inorganic Chemistry (Pergamon Press, Oxford, 1979/1981) or How to Name an Inorganic Substance (Pergamon Press, Oxford, 1981).

[4] R.W. Hamming, Bell Syst. Tech. J. **29**, 147 (1950).

[5] M. Eigen et al., Science **244**, 673 (1989).

[6] M. Eigen et al., Proc. Natl. Acad. Sci. U.S.A. **85**, 5913 (1988).

[7] M. Eigen and K. Nieselt-Struwe, Aids **4** (suppl. 1), 885 (1990).

[8] A.W.M. Dress, Statistische Geometrie von Konfigurationen und deren Evolution in Sequenzräumen—Definition und Probleme; ein Programmvorschlag, Preprint, Bielefeld (1986).

[9] F. Klein, Das Erlanger Programm—vergleichende Betrachtungen über neuere geometrische Forschungen, ed. H. Wussing (Akad. Verl.-Ges. Geest & Portig, Leipzig, 1974).

[10] Cf. e.g. L. Blumenthal, Theory and Applications of Distance Geometry (Chelsea Publ. Comp., Bronx-N.Y., 1970).

[11] H.H. Bock, Automatische Klassifikation (Vandenhoeck and Ruprecht, Göttingen, 1974). M. Nei, Molecular Evolutionary Genetics (Columbia University Press, New York, 1987), A. Gordan, J.R. Statist. Soc. **A 150**, 119 (1987).

[12] M.D. Hendy and D. Penny, Spectral Analysis of Phylogenetic Data, J. Classificatin **10**, 5–24 (1993).

[13] H.-J. Bandelt and A.W.M. Dress, Bull. Math. Biology **51**, 133 (1989).

[14] H.-J. Bandelt and A.W.M. Dress, A Canonical Decomposition Theory for Metrics on a Finite Set, Advances in Mathematics **92**, 47–105 (1992).

[15] J. Dopazo et al., Split Decomposition: a New Technique to Analyse Viral Evolution, PNAS **90** (1993).

[16] J.M.S. Simões-Pereira, J. Combin. Theory Ser. B **6** 303 (1969), P. Bunemann, J. Combin. Theory Ser. B **17**, 48 (1974).

[17] H.-J. Bandelt and A.W.M. Dress, Split Decomposition: A New and Useful Approach to Phylogenetic Analysis of Distance Data, Molecular Phylogenetics and Evolution, Vol. 1, No. 3, September, pp. 242–252, 1992.

[18] H. Leffers et al., J. Mol. Biol. **195**, 43 (1987).

[19] A.W.M. Dress, in Ordnung aus dem Chaos, ed. B.O. Küppers (Piper, München Zürich, 1987), 103.

[20] M. Eigen and P. Schuster, Naturwissenschaften **64**, 541 (1977), **65**, 341 (1978).

[21] M. Eigen et al., in Adv. Chem. Physics **75**, 149, eds. I. Prigogine & S.A. Rice (John Wiley & Son, 1989); M. Eigen et al., J. Phys. Chem. **92**, 6881 (1988).

[22] M. Kimura, The Neutral Theory of Molecular Evolution (Cambridge Univ. Press, Cambridge, 1984).

[23] S. Wright, in Proceedings 6th Congress on Genetics **1**, 356 (1932), S. Kaufmann and S. Levin, J. Theor. Biol. **128**, 11 (1987).

AN ALGORITHMIC APPROACH TO TILINGS

A. Dress
O. Delgado Friedrichs
D. Huson

Universität Bielefeld
Germany

The obvious *structuredness* inherent to regular periodic tilings has been so self-evident that it took almost two and a half thousand years after the first complete classification and enumeration results had been obtained before people started to try to explicitly address this particular structuredness in terms of appropriate formal concepts capturing its *essence*.

The first essential breakthrough occurred when, in the last century, people observed that the symmetries of each of those regular tilings form what became to be named a *group*, and that those groups can be studied, classified and listed on a rather abstract level, quite independently from the particular geometric entity whose symmetry group they are.

Still, the classification of the crystallographic groups is, by far, not enough to classify geometric objects like tilings up to structural equivalence.

Consequently, geometers started to develop particular 'symbolisms' to capture this essence and to rigorously get done with the classification task (cf. [5], [19], [18]).

From one of these many efforts (cf. [13], [6], [14]) — most of which were begun, if not carried through quite unaware of the other people's parallel efforts — the following universal strategy for symbolically characterizing regular or, more precisely, equivariant tilings of simply connected manifolds emerged (cf. [15]):

For a given natural number n, let $\Sigma_n := \langle\, \sigma_0, \ldots, \sigma_n \mid \sigma_i^2 = 1 \,\rangle$ denote the group which is freely generated by a family $\sigma_0, \ldots, \sigma_n$ of $n+1$ involutions and

C. J. Colbourn and E. S. Mahmoodian (eds.), Combinatorics Advances, 111–119.

consider pairs $(\mathcal{D}, M)$ consisting of a (right) transitive Σ_n-set $\mathcal{D}$ and a map

$$M: \mathcal{D} \to (\mathbf{N})_{(n+1)\times(n+1)}: D \mapsto (m_{ij}(D))_{i,j=0,\ldots,n}$$

from $\mathcal{D}$ into the set of integral $(n+1) \times (n+1)$ matrices such that

$$\begin{aligned} &D(\sigma_i\sigma_j)^{m_{ij}(D)} = D, \\ &m_{ij}(D) = m_{ji}(D) = m_{ij}(D\sigma_i), \\ &m_{ii}(D) = 1, \end{aligned}$$

and

$$m_{ij}(D) = 2 \text{ in case } |i-j| > 1$$

for all $D \in \mathcal{D}$ and $i, j \in \{0, \ldots, n\}$.

Obviously, a transitive Σ_n-set $\mathcal{D}$ can also be described as a set together with a family of $n+1$ permutations

$$\overline{\sigma}_i: \mathcal{D} \xrightarrow{\sim} \mathcal{D} \quad (i = 0, \ldots, n)$$

of order at most 2 such that the subgroup $\langle \overline{\sigma}_i \mid i = 0, \ldots, n \rangle$ of the full permutation group acts transitively on $\mathcal{D}$. Equivalently, it can also be described as a set $\mathcal{D}$ together with a subset $E \subseteq \mathcal{D} \times \mathcal{D} \times \{0, \ldots, n\}$ such that E coincides with

$$E^* := \{ (D_1, D_2; i) \mid (D_2, D_1; i) \in E \}$$

and such that for every $D \in \mathcal{D}$ and $i \in \{0, \ldots, n\}$ there is precisely one $D' \in \mathcal{D}$ with $(D, D'; i) \in E$ while the induced graph

$$\left(\mathcal{D}, \left\{ \{D, D'\} \in \binom{\mathcal{D}}{2} \mid \text{ there exists } i \in \{0, \ldots, n\} \text{ with } (D, D'; i) \in E \right\} \right)$$

is connected.

For any two such pairs $(\mathcal{D}_1, M_1)$ and $(\mathcal{D}_2, M_2)$ let $\mathrm{Hom}((\mathcal{D}_1, M_1), (\mathcal{D}_2, M_2))$ denote the set of maps $\varphi: \mathcal{D}_1 \to \mathcal{D}_2$ such that $M_1(D) = M_2(\varphi(D))$ and $\varphi(D\sigma_i) = \varphi(D)\sigma_i$ for all $D \in \mathcal{D}_1$ and $i \in \{0, \ldots, n\}$. Note that this way we get a category whose objects are those pairs and whose morphisms are the Hom-sets we have just defined. Note also that any morphism is necessarily an epimorphism, and it is an isomorphism if and only if it is bijective or — equivalently — if and only if it is injective; hence, if $\#\mathcal{D}_1 = \#\mathcal{D}_2 < \infty$, then any morphism from $\mathcal{D}_1$ into $\mathcal{D}_2$ is an isomorphism.

The basic observation on which all further developments are based is the following one: given

- a connected n-manifold X without boundary,
- the i-skeleta $X_i \subseteq X$ $(i = 0, \ldots, n)$ of a regular CW-structure T on X, and
- a discrete group Γ of automorphisms of X respecting this CW-structure (that is, satisfying $\gamma X_i = X_i$ for all $i = 0, \ldots, n$ and $\gamma \in \Gamma$),

one can associate with these data a pair $(\mathcal{D}, M) = (\mathcal{D}_{(X,T,\Gamma)}, M_{(X,T,\Gamma)})$ as above as follows:

First put $X_{-1} := \emptyset$; then consider

$$\mathcal{F} := \{ F = (F_0, \ldots, F_n) \in \prod_{i=0}^{n} \pi_0(X_i - X_{i-1}) \mid F_{i-1} \subseteq \overline{F_i} \text{ for } i = 1, \ldots, n \},$$

the set of (maximal) flags of X and observe that

(a) for each flag $F = (F_0, \ldots, F_n)$ and each $i \in \{0, \ldots, n\}$, there exists precisely one flag $F' = (F'_0, \ldots, F'_n)$ with $F_i \neq F'_i$ and $F_j = F'_j$ for all $j \neq i$, so there is a well-defined natural Σ_n-action on $\mathcal{F}$ defined by $F\sigma_i := F'$ whenever F, F' and i are as above;

(b) $\mathcal{F}$ becomes a transitive Σ_n-set this way;

(c) $m_{ij}(F) := \min(m \in \mathbf{N} \mid \mathbf{F}(\sigma_i\sigma_j)^m = \mathbf{F})$ always exists (that is, for every $F \in \mathcal{F}$ and $i, j \in \{0, \ldots, n\}$ there exists some $m \in \mathbf{N}$ with $F(\sigma_i\sigma_j)^m = F$), and one has $m_{ij}(F) = m_{ji}(F) = m_{ij}(F\sigma_i)$ as well as $m_{ii}(F) = 1$ and $m_{ij}(F) = 2$ for $|i - j| > 1$ for all $F \in \mathcal{F}$ and $i, j \in \{0, \ldots, n\}$;

(d) the action of Γ on X induces a free action of Γ on $\mathcal{F}$ which satisfies $(\gamma F)(\sigma_i) = \gamma(F\sigma_i)$ and $m_{ij}(\gamma F) = m_{ij}(F)$ for all $\gamma \in \Gamma$, $F \in \mathcal{F}$ and $i, j \in \{0, \ldots, n\}$.

Finally conclude that, in consequence, one can associate to the triple (X, T, Γ) a pair $(\mathcal{D}, M) = (\mathcal{D}_{(X,T,\Gamma)}, M_{(X,T,\Gamma)})$ as above by

- putting $\mathcal{D} := \Gamma \backslash \mathcal{F}$, the set of orbits $\Gamma \cdot F := \{ \gamma F \mid \gamma \in \Gamma \}$ of Γ on $\mathcal{F}$,

and

- observing that the relations $(\gamma F)(\sigma_i) = \gamma(F\sigma_i)$ and $m_{ij}(\gamma F) = m_{ij}(F)$ imply that there is an induced transitive (right) Σ_n-action on $\mathcal{D}$ and that

$$M(\Gamma \cdot F) := (m_{ij}(F))_{i,j=0,\ldots,n}$$

is well-defined and satisfies the relations mentioned above.

More importantly, one can then invoke standard facts about how to compute the fundamental group of a CW-complex to conclude that given two triples (X, T, Γ) and (X', T', Γ') consisting of *simply* connected manifolds X, X' with regular CW-structures T, T' and discrete automorphism groups Γ, Γ', the resulting symbols

$$(\mathcal{D}, M) = (\mathcal{D}_{(X,T,\Gamma)}, M_{(X,T,\Gamma)})$$

and

$$(\mathcal{D}', M') = (\mathcal{D}_{(X',T',\Gamma')}, M_{(X',T',\Gamma')})$$

are isomorphic if and only if there exists a homeomorphism $\varphi: X \xrightarrow{\sim} X'$ with $\Phi(X_i) = X_i'$ for $i = 0, \ldots, n$ and $\Gamma' = \{ \Phi\gamma\Phi^{-1} \mid \gamma \in \Gamma \}$, while there exists a morphism $\varphi: (\mathcal{D}, M) \to (\mathcal{D}', M')$ if and only if there exists a homeomorphism $\Phi: X \xrightarrow{\sim} X'$ with $\Phi(X_i) = X_i'$ for all $i = 0, \ldots, n$ and

$$\Gamma' \supseteq \{ \Phi\gamma\Phi^{-1} \mid \gamma \in \Gamma \}.$$

The transformation from the geometry of tilings to the combinatorics of (M-ornamented) equivariant Σ_n-sets which can be based on this observation has turned out to be as fundamental a starting point for the development of a rigorous mathematical (and even *computerizable*) theory of tilings as the transformation from classical geometry to algebra and analysis based on Descartes' coordinate systems was for the development of a rigorous mathematical theory of space and spatial structures in general. It also allows to manipulate tilings on the symbolic level and, hence, to construct algorithms which — performing on the symbolic level — can handle all sorts of questions related to the comparative analysis, classification and enumeration of tilings conforming with essentially any preassigned list of geometric specifications (cf. [16], [17], [9], [4] and more references given below). And finally, by reversing this transformation (whenever possible), one can actually generate *geometric realizations* as well as visual displays of all such (classes of) tilings on the computer screen, and one can thoroughly manipulate these visual displays by translating mouse-clicks and -motions on the screen into corresponding actions and transformations on the symbolic level (cf. [10]).

In the past 10 years, the symbolic approach as described above has led to an elegant and consistent classification theory for two-dimensional tilings, based on the following three main observations:

First, if $(\mathcal{D}, M)$ is a finite ornamented Σ_2-set as above, then the *curvature* of $(\mathcal{D}, M)$ is defined as

$$K(\mathcal{D}, M) := \sum_{D \in \mathcal{D}} \frac{1}{m_{01}(D)} + \frac{1}{m_{12}(D)} - \frac{1}{2}.$$

One can show that $(\mathcal{D}, M)$ is the symbol of a tiling of the plane, the hyperbolic plane or the sphere if and only if its curvature is zero, negative or positive, respectively, and if, in the spherical (i.e. positive curvature) case, some additional integrality conditions hold which are closely related to the well-known characterization of 'bad' orbifolds in dimension 2.

Second, if Γ is a discrete group of automorphisms of some connected 2-manifold X with compact orbit space X/Γ, then there is some number $t = t_\Gamma \in \mathbf{N}$ such that for any number $k \in \mathbf{N}$ and for any *tile-k-transitive* tiling (X, T, Γ), the size of the corresponding Σ_2-set $\mathcal{D}_{(X,T,\Gamma)}$ is bounded by tk. Here, a tiling (X, T, Γ) is called tile-k-transitive if the number of Γ-orbits in the set of tiles (i.e. cells of maximal dimension) of T is exactly k.

Third, any tiling of the euclidean or hyperbolic plane or of the sphere can be derived systematically from some so-called *fundamental* tiling by applying a finite number of simple geometric operations. A fundamental or, more precisely, 1-fundamental tiling is a tile-(1-)transitive one with the additional property that none of its symmetries leave any tile invariant.

The first two of these observations lead to algorithms for the classification of tile-k-transitive tilings, where k is a small integer. For example, in the euclidean case, the number t_Γ is at most 12. In [12], all the tile-2-transitive tilings of the euclidean plane have been classified by means of a computer by first enumerating all the possible symbols of size at most 24 and then extracting those having zero curvature. This method, however, is rather inefficient, since it generates lots of symbols which do not correspond to tilings of the desired type.

The third observation, which was translated to the symbolic level and implemented as a computer program in [20], leads to algorithms which produce tilings of prescribed symmetry, which makes it possible to classify all the tile-1-, 2- and 3-transitive tilings of the euclidean plane and the sphere. Due to the fact that there are infinitely many discrete co-compact groups of isometries of the hyperbolic plane and therefore infinitely many equivalence classes of hyperbolic two-dimensional tilings, to achieve analogous classification results for tilings of the hyperbolic plane, it was necessary to choose a finite collection of such groups as the symmetry groups of those tilings to be classified.

A variant of the third observation also is the basis for efficient methods to construct geometric realizations of given two-dimensional tiling symbols. A general method to construct realizations of symbols corresponding to fundamental tilings is described in [24] and [3]. An arbitrary symbol is realized by first reducing it to a fundamental one, then realizing the fundamental symbol, and finally reversing the sequence of reductions on the level of geometric realizations (c.f. [24], [7]).

The combination of our symbolic approach with geometric construction methods has also proved to be fruitful for certain more specific classification problems, like the classification of the so-called 4-colourable tilings of the euclidean plane, i.e. those tile-4-transitive tilings with the property that no two tiles in the same Γ-orbit share an edge (c.f. [22]), or of the so-called net-like partial tilings of the euclidean plane (c.f. [21]). A net-like partial tiling can be considered as one allowing a partition of the tiles into two classes in such a way that the first class consists of one Γ-orbit while the second class consists of all the other Γ-orbits, and no two tiles of the second class may share an edge. The interiors of the tiles in the second class may then be considered as *holes* cut out of the plane, while the tiles in the first class may be considered to be forming a tiling of the rest of the plâne.

As mentioned before, there are a number of alternative ways to describe the symbol corresponding to a tiling. Although the advantages of the chosen algebraic formulation are evident, most of the early computer implementations do not make use of this algebraic background. This is probably due to the fact that the algebraic structure of two-dimensional symbols is comparatively simple; so, for the necessary operations on these symbols, specialized ad-hoc implementations have been used instead of sophisticated general algorithms from computational group theory or related fields.

This situation changes significantly when one considers tilings of higher dimension and their symbols. In [8], the author discusses practical algorithmic methods to decide whether a given pair $(\mathcal{D}, M)$, where $\mathcal{D}$ is a finite Σ_3-set and M is a map as described above, is the symbol of a tiling of the usual euclidean space. This question can be reduced, by non-trivial means, to a purely topological one, which is related to the problem of deciding whether two given three-dimensional triangulated manifolds are homeomorphic. No practical general solution for the latter problem is currently available, and, in effect, the same is true of our original problem. There are, however, practical criteria based on the algebraic structure of the possible symmetry groups of such tilings, namely the so-called three-dimensional crystallographic space groups, which can be used to significantly reduce the number of candidate symbols.

Here, by a tiling of the euclidean space, we mean a tiling (X, T, Γ), where X is the three-dimensional real vector space $\mathbf{R}^3$ and Γ is a discrete co-compact group of isometries of X w.r.t. the usual euclidean metric. Bieberbach, in his classical works [1] and [2], showed that each such group contains a normal subgroup of finite index which is generated by three linearly independent translations (and thus is abstractly isomorphic to the direct sum of three copies of the infinite cyclic group). He also showed that (in any given dimension) there is only a finite number of such groups. For dimension 3, these had been completely classified by Fedorov and by Schönflies in 1890. One can use these facts to derive a necessary algebraic condition for the euclidicity of a three-dimensional tiling symbol, which can be implemented by constructing a finite presentation of the group Γ from the data $(\mathcal{D}_{(X,T,\Gamma)}, M_{(X,T,\Gamma)})$ and then testing — applying several standard algorithms from computational group theory — whether there is an appropriate normal subgroup $N \trianglelefteq \Gamma$, which satisfies $N/[N, N] \cong \mathbf{Z}^3$.

Since efficient implementations of the above-mentioned algorithms are far from being straightforward, the computer algebra system GAP ([23]) was chosen as a platform for the implementation of the euclidicity criterion. The GAP-system contains a number of routines for finitely presented groups, which are very useful for our purposes, as well as a high-level programming language, which, among other things, contains permutations and extensible arrays as basic data types. On this basis it was fairly easy to re-implement many of the standard operations on tiling symbols, now making explicit use of the algebraic structures involved and using a number of the higher level group-theoretic routines available in GAP.

More advanced euclidicity criteria have also been implemented in GAP, as well as routines for the systematic generation of symbols. The application of these programs establishes the fact that there are exactly 923 symbols of size up to 10 of *non-degenerate* (c.f. [8]) euclidean space tilings. These have been found by investigating a list of about 33000 'locally euclidean' candidates. As a second application, a result from [11] about face-transitive euclidean space tilings, which had been found by using the traditional computational techniques together with a large amount of case by case analysis done 'by hand', has been reproduced completely automatically.

REFERENCES

[1] L. Bieberbach, *Über die Bewegungsgruppen der Euklidischen Räume I*, Math. Ann. **70** (1911), 297.

[2] L. Bieberbach, *Über die Bewegungsgruppen der Euklidischen Räume II*, Math. Ann. **72** (1912), 400.

[3] J. Bornhöft, *Zum einfachen Zusammenhang von Kantenkoronen und einem sich daraus ergebenden iterativen Algorithmus zur (Re-)Konstruktion planarer Pflasterungen*, Master's thesis, University of Bielefeld, 1994.

[4] O. Delgado Friedrichs, A.W.M. Dress, A. Müller, and M.T. Pope, *Polyoxometalates: A class of compounds with remarkable topology*, Molecular Engineering **3** (1993), 9–28.

[5] B.N. Delone, *Teoriya planigonov*, Ivz. Akad. Nauk SSSR, ser. Matem. **23** (1961), 365–386.

[6] M.S. Delaney, *Quasisymmetries of space group orbits*, Match **9** (1980), 73–80.

[7] O. Delgado Friedrichs, *Die automatische Konstruktion periodischer Pflasterungen*, Master's thesis, University of Bielefeld, 1990.

[8] O. Delgado Friedrichs, *Euclidicity criteria for three-dimensional branched triangulations*, Ph.D. thesis, University of Bielefeld, 1994.

[9] A.W.M. Dress and D.H. Huson, *Heaven and hell tilings*, Revue Topologie Structurale **17** (1991), 25–42.

[10] O. Delgado Friedrichs and D.H. Huson, REPTILES, University of Bielefeld, 1992, Shareware macintosh-program.

[11] A.W.M. Dress, D.H. Huson, and E. Molnár, *The classification of face-transitive 3d-tilings*, Acta Crystallographica **A49** (1993), 806–817.

[12] O. Delgado Friedrichs, D.H. Huson, and E. Zamorzaeva, *The classification of 2-isohedral tilings of the plane*, Geometriae Dedicata **42** (1992), 43–117.

[13] A.W.M. Dress, *On the classification and generation of two- and higher-dimensional regular patterns*, Match **9** (1980), 81–100.

[14] A.W.M. Dress, *Regular polytopes and equivariant tessellations from a combinatorial point of view*, Algebraic Topology (Göttingen), SLN 1172, Göttingen, 1984, pp. 56–72.

[15] A.W.M. Dress, *Presentations of discrete groups, acting on simply connected manifolds*, Adv. in Math. **63** (1987), 196–212.

[16] A.W.M. Dress and R. Scharlau, *Zur Klassifikation äquivarianter Pflasterungen*, Mitteilungen aus dem Math. Seminar Giessen **164** (1984), 83–136.

[17] A.W.M. Dress and R. Scharlau, *The 37 combinatorial types of minimal, non-transitive, equivariant tilings of the euclidean plane*, Discrete Math. **60** (1986), 121–138.

[18] B. Grünbaum and G.C. Shephard, *Tilings and patterns*, W.H. Freeman and Company, New York, 1987.

[19] H. Heesch, *Reguläres Parkettierungsproblem*, Westdeutscher Verlag, Köln-Opladen, 1968.

[20] D.H. Huson, *The generation and classification of tile-k-transitive tilings of the euclidean plane, the sphere and the hyperbolic plane*, Geometriae Dedicata **47** (1993), 269–296.

[21] D.H. Huson, *Tile-transitive partial tilings of the plane*, Contributions to Geometry and Algebra **34** (1993), no. 1, 87–118.

[22] D.H. Huson, *A four-color theorem for periodic tilings*, Geometriae Dedicata **51** (1994), 47–61.

[23] Martin Schönert et al., *GAP - Groups, Algorithms, and Programming*, Lehrstuhl D für Mathematik, Rheinisch Westfälische Technische Hochschule, Aachen, Germany, third ed., 1993.

[24] K. Westphal, *Zur Konstruktion zweidimensionaler Pflasterungen in allen drei Geometrien*, Master's thesis, University of Bielefeld, 1991.

A NEW CONNECTION BETWEEN CONVEX GEOMETRY AND THRESHOLD LOGIC

M. R. Emamy-K.

University of Puerto Rico at Rio Piedras
Rio Piedras, P.R., 00931
and
Institute for Studies in Theoretical
Physics and Mathematics
Tehran, Iran

ABSTRACT

A threshold Boolean function is a Boolean function defined on $\{0,1\}^n$ whose On-vertices and Off-vertices are strictly separable by a hyperplane in $\Re^n$. Threshold logic is the main source for study of threshold Boolean functions, while Boolean algebraic methods have been the classical tools to study these objects.

Recently, we have claimed there exists a purely geometric approach to these linearly separable Boolean functions. The principal motivation to this claim is the fact that these functions are just linearly separable cubical complexes and their place is convex geometry and polytopes. Here, we present a brief overview of a few results justifying this new connection.

1 INTRODUCTION

Threshold logic, an extension of switching theory, was formally initiated by J. Von Neumann [14, 15] and E. Goto [9, 10] in Fifty's, solely for its applications in electronic circuits. In fact, innovation of advanced electronic computers was the main motivation for progress in these efforts. Later, this research was accelerated by C. K. Chow [2, 3], C. C. Elgot [4], S. Muroga [13], S. T. Hu [12] and R. O. Winder [16]. The central elements of study in threshold logic are threshold Boolean functions and the dominant strategy in this study has been restricted to Boolean algebraic methods .

C. J. Colbourn and E. S. Mahmoodian (eds.), Combinatorics Advances, 121–128.

On the other hand, a threshold Boolean function of n variables is a Boolean function, i.e. a $0-1$ valued function on $\{0,1\}^n$, whose On-vertices are separable from its Off-vertices by a hyperplane of $\Re^n$. Thus, a threshold Boolean function is (roughly speaking) just a separable cubical complex, called a cut-complex here. The geometric and combinatorial study of cubical complexes belongs to convex geometry and polytopes. Hence, naturally, polytopes should play an essential role in study of these complexes. Surprisingly, this has not happened and no trace of polytopes are found in the literature of threshold logic, see [13].

In this paper, we present a purely geometric approach to these Boolean functions which is Boolean free, in the sense that our method does not depend on Boolean expression of Boolean functions. This latter restriction is merely to stress more on the geometric nature of this study. Any nontrivial well-known property of threshold functions, rediscovered only by geometric means is interesting and useful to constitute a base for this geometric threshold logic. However, we present a few results here which have not been addressed in conventional threshold logic, merely for the geometric nature of their concepts.

We begin with a simple enumerating algorithm for cut-complexes which interacts with a generation of convex polytopes through a natural manner. Since enumeration of distinct threshold Boolean functions becomes an exhaustive task in higher dimensions, then normally different equivalent classes of these functions are introduced and the enumeration for these classes are carried out, see [12, 13, 16].

Here, a class of nonisomorphic cut-complexes is introduced which is defined independent of the Boolean expression for the corresponding Boolean functions. Then, the enumerating algorithm can enumerate these nonisomorphic complexes together with the different nonisomorphic convex polytopes generated by the intersection of the n-cube with the separating hyperplanes corresponding to the cut-complexes. Section 2 includes this enumeration and the subject will be continued in section 3 by geometric study of isomorphic cut-complexes.

2 ALGORITHMIC ENUMERATION OF NONISOMORPHIC CUT-COMPLEXES AND A GENERATION OF CONVEX POLYTOPES

2.1 Basic Terminology

Let $B^n = \{0,1\}^n$ denote the n-dimensional Boolean cube. A k-dimensional face of B^n, or a k-Boolean subcube, is obtained by fixing any n - k coordinates of the n-vectors in B^n. In the next section $c^n = [0,1]^n$ will stand for the n-cube in $\Re^n$. In fact, $B^n = V(c^n)$ is the set of vertices of c^n. Faces of c^n are also subcubes of the n-cube, so vertices of a k-face of c^n are exactly elements of a k-face of the Boolean n-cube. Edges of B^n (or c^n) are its 1-dimensional faces. From now on, by the n-cube we mean the Boolean n-cube B^n, unless otherwise is stated.

A cubical complex is a nonempty set C of faces of the n-cube such that for any two faces F_1 and F_2 of B^n with $F_1 \subset F_2$, $F_2 \in C$, then $F_1 \in C$ also holds. What we call faces of C are exactly those faces of B^n lying in C. Two cubical complex C_1 and C_2 are isomorphic if there is a bijection $\varphi : C_1 \to C_2$ which preserves incidence; i.e. for any two faces F_1 and F_2 of C_1 ; $F_1 \subset F_2 \Rightarrow \varphi(F_1) \subset \varphi(F_2)$. An exterior edge of a cubical complex C is an edge of the n-cube which has exactly one vertex in C. For more about cubical and general complexes the reader is referred to Grünbaum [11]. One of the most important class of cubical complexes is the one of cut-complexes. A cut-complex of the n-cube is a cubical complex whose vertices are strictly separated from the rest of the vertices of the n-cube by a hyperplane of $\Re^n$. We are now ready for the simple algorithm.

2.2 A General Algorithm

Here is the enumeration algorithm which can list all the nonisomorphic cut-complexes of the n-cube.

- Step 1) For a given $v \in B^n$ set $C_1 = \{v\}$ as the initial cut-complex.
- Step 2) Generate C_{k+1} from C_k as follows :

Choose an exterior edge uw of C_k $(u \in V(C_k), w \notin V(C_k))$.

If $C_{k+1} = (C_k, , w)$, the complex spanned by C_k and w , is a cut-complex , then choose one from its isomorphic class.

- Step 3) If $k < 2^n$ continue, otherwise stop.

The implementation of the algorithm above and its computational results can be found in [5, 6, 7]. The algorithm, basically has two major tasks

1. To check if a cubical complex is a cut-complex.
2. To check if a cubical complex is isomorphic to those already chosen.

Task 1 is standard in computational threshold logic and is equivalent to threshold recognition of Boolean functions. However, in higher dimensions this is equivalent to solve linear programs which are very large in size.

Task 2 is computationally related to isomorphic graph algorithms and has not been treated in threshold logic.

The computational results of [6] shows that the number of nonisomorphic cut-complexes is exactly equal to the number of symmetry type classes of threshold functions for dimensions ≤ 6, see winder [16]. It seems that for higher dimensions the number of symmetry type grows much faster than the one of nonisomorphic cut-complexes, and an interesting question is to find the minimum dimension when the two numbers are not equal.

2.3 A Generation of Simple Convex Polytopes

A convex polytope is convex hull of a finite set in $\Re^n$. By a polytope here, we always mean a convex one. For a positive integer $2 \leq d \leq n$, a d-dimensional convex polytope is a said to be simple if its graph is a regular graph of degree d. A k-simplex is just a convex hull of $k+1$ affinely independent points of $\Re^n$. A face of a convex polytope P is its intersection with a supporting hyperplane of $\Re^n$. It is well-known that faces of a simple polytope are simple polytopes. In the sequel, Q is always a simple d-polytope, for some integer $2 \leq d \leq n$, which has a p-simplex face for $0 \leq p \leq d-1$. We define a simplex deformation map φ

on the set of all simple d-polytopes possessing a p-simplex face, which satisfies the following:

1. $\varphi(Q)$ is a simple d-polytope for any simple d-polytope Q with a p-simplex face F.
2. The simplex face F of Q is deformed to a new q-simplex face of $\varphi(Q)$ with $p+q=d-1$.
3. All exterior edges of F in Q will remain to be exterior edges of the new q-simplex $\varphi(F)$ in $\varphi(Q)$.
4. Those faces of Q, disjoint from F, remain faces combinatorially unchanged in $\varphi(Q)$.

To have a better intuition about φ, let us consider the case when $d = 3$; i.e. when Q is a simple 3-polytope and check all of the different possibilities for F.

- Case 1. Let F be a vertex of the simple 3-polytope Q, then $\varphi(F)$ will be a triangle of $\varphi(Q)$. The three edges incident to F are exactly the exterior edges of F. In this case $\varphi(Q)$ simply can be obtained from Q by cutting the vertex F off and form a new triangle $\varphi(F)$ in $\varphi(Q)$. In fact, comparing to Q, its image $\varphi(Q)$ has three new vertices instead of F, three new edges of $\varphi(F)$ and a 2-dimensional triangular face $\varphi(F)$.
- Case 2. Let F be an edge vw of Q.

  Then $\varphi(F)$ will also be an edge, say xy, of $\varphi(Q)$. Thus, there are two 2-faces, say F_1 and F_2, containing $F = vw$ and two more faces say G_1 and G_2 containing v and w respectively. Note that neither F_1 nor F_2 can be a triangle here, since the deformation can not be acceptable. The edge, $\varphi(F) = xy$ will be shared by the two 2-faces $\varphi(G_1)$ and $\varphi(G_2)$ and the 2-faces $\varphi(F_1)$, $\varphi(F_2)$ contain x and y respectively. The simplex deformation map φ in a dual formulation has been extensively studied in Ewald and Shephard [8]. The significance of this deformation comes from the following theorem of [8].

Theorem 2.1 *Every simple convex d-polytope Q can be obtained from a d-simplex by successive applications of simplex deformations.*

A natural question related to Theorem 2.1 is that how Q can be constructed from a given d-simplex, or in other words how to find a sequence of simplex deformations by which a d-polytope Q can be constructed. We will show that the algorithm of section 2.2 is useful to answer this question for polytopal sections of the n-cube, in small dimensions.

2.4 Generation of Simple Polytopal Sections

From now on, by the n-cube we mean the usual convex polytope c^n and not the Boolean-cube. Here, the real interaction of simple polytopes and the enumeration algorithm of 2.2 comes to the surface. Let H_k be the hyperplane strictly separating $V(C_k)$ from the rest of the vertices of the n-cube. In a typical step of this algorithm a cut-complex $C_{k+1} = C_k \cup \{w\}$, $w \in B^n - V(C_k)$ is produced. So, the separating hyperplane H_k corresponding to C_k is transformed to a new separating hyperplane H_{k+1} corresponding to C_{k+1}. Similarly, the simple $(n-1)$-polytope $Q_k = H_k \cap c^n$ is deformed to a new simple $(n-1)$-polytope $Q_{k+1} = H_{k+1} \cap c^n$. A natural question which is raised is that if the combinatorial type of Q_{k+1} can be obtained from that of Q_k. As a matter of fact, simplex deformations are convenient tools to obtain a positive answer to this question. Recall that w is the appropriate vertex which generates the cut-complex C_{k+1}.

Assume that $p+1$ edges of w are intersecting the hyperplane H_k, and so the other $n-(p+1)$ edges intersect H_{k+1}. The intersection of these $p+1$ edges and H_k are vertices of a p-simplex F which is a face of $Q_k = H_k \cap c^n$. In order to obtain C_{k+1}, we move H_k toward w such that this becomes a new separating hyperplane for C_{k+1} with $V(C_{k+1}) = V(C_k) \cup \{w\}$.

In this smooth deformation, Q_k is transformed to Q_{k+1} exactly by a simplex deformation, where the p-simplex F is transformed to a new q-simplex with $q = (n-2) - p$. Those vertices of the q-simplex are intersection of H_{k+1} and the other $n-(p+1)$ edges incident to w. Summarizing, the enumeration algorithm of section 2.2 not only enumerates the nonisomorphic cut-complexes, but simultaneously, can also generate simple polytopal sections of the n-cube joint with their Ewald-Shephard sequence of simplex deformations.

More details about existence of the vertex w at each step of the algorithm and various computational acceleration methods of the algorithm can be found in [6, 7]. For computer generation of simple polytopal sections of the 5-cube see [1]. In the next section, we will describe geometric results which might be very helpful in acceleration of these algorithms in high dimensions.

3 GEOMETRIC PROPERTIES

Let $S \subset B^n$, a map $\varphi : S \to B^n$ is called an isometry if it preserves the Euclidean distance. That is to say,$d(\varphi(x), \varphi(y)) = d(x, y)$ for all $x, y \in S$, where $d(x, y)$ denotes Euclidean distance between x and y . In general, isomorphic cubical complexes may not be isometric, but it can be shown that isomorphic cut-complexes are isometric.

For a given $S \subset B^n$, conv(S) denotes the smallest cube containing S. In a cubical complex C with a vertex $v \in C$, by $deg_C(v)$ we mean the degree of v in the graph formed by the vertices and the edges of C. A vertex v of C is said to be an interior vertex if $deg_C(v) = n$, i.e. V has no neighbor in $B^n - C$. A leaf of a cubical complex is an edge uv of C such that $deg_C(v) = 1$. The vertex u is called the root of the leaf.

Now, we can state a basic theorem related to geometric properties of cut-complexes. In this theorem $C - v$ is just the complex induced by $V(C) - \{v\}$.

Theorem 3.1 *Let C be a cut-complex of the n-cube, $n \geq 2$. Then the root u of any leaf uv of C is an interior vertex of the complex $C' = C - v$ in cub(C'), provided C' has at least one interior vertex in cub(C').*

Assume that k is the smallest dimension for which a cut-complex C is embedded in the k-cube. Intuitively, k seems to be the same as the maximum degree of vertices in C. This turns out to be true and will be stated as follows.

Theorem 3.2 *For a cut-complex C of the n-cube, $n \geq 2$ if $\max_{v \in C} deg_C(v) = m$, then dimcub($C$) $= m$.*

Finally, the theorems above joint with a special labeling of vertices in the n-cube, leads to the following:

Theorem 3.3 *Isomorphic cut-complexes of the n-cube are isometric cubical complexes.*

The proofs of the theorems above appear elsewhere.

REFERENCES

[1] C. Caiseda and L. González, "A Computer Generation of Polytopes that are Sections of the 5-cube", to appear in *Congressus Numerantium.*

[2] C. K. Chow, "On the Characterization of Threshold Functions", *SCTLD* (1961), 34-38.

[3] C. K. Chow, "Boolean Functions Realizable with Single Threshold Devices", *Proc. IRE* 49 (1961), 370-371.

[4] C. C. Elgot, "Truth Functions Realizable by Single Threshold Organs", *SCTLD* (1961), 225-245.

[5] M. R. Emamy-K. and L. Lazarte, "On the Cut-complexes of the 5-cube", *Discrete Math.* 78 (1989), 239-256.

[6] M. R. Emamy-K., C. Caiseda, P. Pei, "Enumeration of Threshold Pseudo-Boolean Functions for 5 and 6-cubes", to appear.

[7] M. R. Emamy-K. and C. Caiseda, "An Efficient Algorithm for Characterization of Cut-complexes", *Congressus Numerantium* 84 (1991).

[8] G. Ewald and G. C. Shephard, "Stellar Subdivisions of boundary Complexes of Convex polytopes", *Math. Ann.* 210 (1974), 7-16.

[9] E. Goto, Japanese Patent, 1954.

[10] E. Goto, *Threshold Logic*, Lecture Notes, MIT, 1963.

[11] B. Grünbaum, *Convex Polytopes*, Inter-Science, London, 1967.

[12] S. T. Hu, *Threshold Logic*, University of California Press, Berkeley, 1965.

[13] S. Muroga, *Threshold Logic and its Applications*, Wiley Interscience, Toronto, 1971.

[14] J. Von Neumann, U. S. Patent (1954).

[15] J. Von Neumann, "Probabilistic Logics and the Synthesis of Reliable Organism from Unreliable Components", Automata Studies, Princeton University Press, 1956, p.p. 43-98.

[16] R. O. Winder, "Threshold Logic", Ph.D. Thesis, Mathematics Department, Princeton University, May 1962.

THE UNITY OF COMBINATORICS

Richard K. Guy

University of Calgary
Calgary, Alberta, Canada

1 INTRODUCTION

One reason why Combinatorics has been slow to become accepted as part of mainstream Mathematics is the common belief that it consists of a bag of isolated tricks, a number of areas:

combinatorial number theory (partitions, integer sequences), combinatorial set theory, Ramsey theory, partially ordered sets, lattices (in both the poset and number-theoretic senses), error-correcting codes, combinatorial designs (latin squares and rectangles, projective and affine geometries, Steiner systems, Kirkman's schoolgirls problem), combinatorial games, enumerative combinatorics (recurrence relations, generating functions), 0-1 matrices, graph theory (including tournaments, topological properties, coloring problems), combinatorial geometry, packing & covering (in number-theoretic, set-theoretic, graph-theoretic or geometric contexts),

with little or no connexion between them. We shall see that they have numerous threads weaving them together into a beautifully patterned tapestry.

2 LANGFORD'S PROBLEM

We will start with Dudley Langford's small son playing with his colored building blocks. He arranged them with one block between the red ones,

C. J. Colbourn and E. S. Mahmoodian (eds.), Combinatorics Advances, 129–159.

R		R

two between the blue ones,

R	B	R		B

and three between the green;

G	R	B	R	G	B

to which we can add two white ones with zero blocks between:

G	R	B	R	G	B	W	W

Replace the colors by the numbers of blocks between

3	1	2	1	3	2	0	0

and then add one to each number:

4	2	3	2	4	3	1	1

Bang showed that there is such an arrangement of n pairs just if $n \equiv 0$ or 1 mod 4.

The problem generalizes in an obvious way: observe the spacing of the triples $i \ldots i \ldots i$ in

0 0 0 1 9 1 2 1 8 2 4 6 2 7 9 4 5 8 6 3 4 7 5 3 9 6 8 3 5 7

We return to this when we discuss the $x + y = 2z$ problem.

3 SKOLEM'S PROBLEM

If we label the blocks from 1 to 8:

4	2	3	2	4	3	1	1
1	2	3	4	5	6	7	8

we find that we have solved a problem of Skolem (in those cases, $n \equiv 0$ or 1 mod 4, where there is a solution): partition the numbers from 1 to $2n$ into n pairs whose differences are the numbers 1 to n:

$$8 - 7 = 1, \qquad 4 - 2 = 2, \qquad 6 - 3 = 3, \qquad 5 - 1 = 4$$

4 BEATTY SEQUENCES

If we don't restrict ourselves to a particular value of n, we can partition the positive integers into two sequences with the differences 1, 2, 3, ..., between corresponding terms:

	1	3	4	6	8	9	11	12	14	16	17	19	...
	2	5	7	10	13	15	18	20	23	26	28	31	...
difference	1	2	3	4	5	6	7	8	9	10	11	12	...

These are called **Beatty sequences**. Such a pair can be generated by $\lfloor n\alpha \rfloor$ and $\lfloor n\beta \rfloor$ where α and β are irrational numbers satisfying $1/\alpha + 1/\beta = 1$.

5 PENROSE PIECES

This particularly famous pair of Beatty sequences is $\lfloor n\tau \rfloor$, $\lfloor n\tau^2 \rfloor$, generated by taking α as the golden number $\tau = (1+\sqrt{5})/2$. It occurs in many places, since it is associated with the **Fibonacci numbers and Lucas numbers.**

rank	⋯	−4	−3	−2	−1	0	1	2	3	4	5	6	7	8	9	⋯
Fibonacci	⋯	−3	2	−1	1	0	1	1	2	3	5	8	13	21	34	⋯
Lucas	⋯	7	−4	3	−1	2	1	3	4	7	11	18	29	47	76	⋯

A striking example occurs in the arrangement of short and long "bow-ties" in the "Conway worms" in a **Penrose aperiodic tiling** of the plane with "kites" and "darts". Kites and darts are made by joining a point of golden section of the long diagonal of a 72° rhombus to the 108° vertices. Note that the rules demand that every vertex in a tiling must have only heads or only tails of the kites and darts. In each worm the short bow-ties (comprising 3 kites & 2 darts) are separated by just one or two long ones (which each comprise 5 kites & 3 darts). If you successively label the bow-ties in a worm, then the labels of the long and short bow-ties form sequences like the pair of Beatty sequences we gave above.

6 WYTHOFF'S GAME

is another example. It is played with two heaps of beans. The two players alternately take either any number of beans from one heap or equal numbers from both heaps, the last player being the winner. If you play so that the numbers in the two heaps are corresponding members of the two sequences, you will always win.

Rufus Isaacs unwittingly invented an isomorph of this game; played with a queen on a chess board. Players alternately move the queen in one of three directions taking it nearer to a corner; the player who moves the queen to the corner wins. In our version, Wyt Queens, any number of queens may occupy the same square; i.e., we play a **disjunctive compound** or sum of copies of Wythoff's game, which itself contains *non-disjunctive* moves in which both heaps of beans are altered. See Figure 1.

Here is a table of nim-values for a queen on various squares. The **Sprague-Grundy theory** asserts that all (positions in) impartial games are equivalent

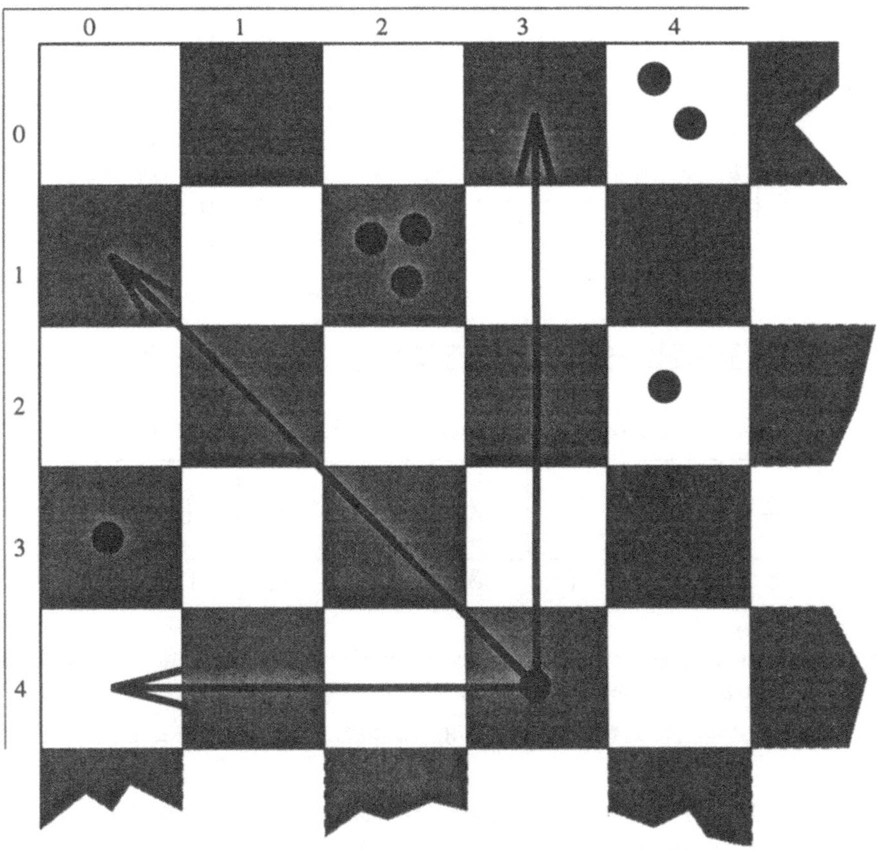

Figure 1 Wyt Queens, a disjunctive form of Isaacs's game.

to a heap of beans in the game of Nim. The number of beans in the heap is the nim-value of the position. Note that the $\mathcal{P}$-positions (previous-player-winning positions, nim-value zero) have coordinates taken from a column in the table of Beatty sequences, which should, perhaps, have begun with a column of zeros.

	0	1	2	3	4	5	6	7	8	9	10	11	12
0	0	1	2	3	4	5	6	7	8	9	10	11	12
1	1	2	0	4	5	3	7	8	6	10	11	9	13
2	2	0	1	5	3	4	8	6	7	11	9	10	14
3	3	4	5	6	2	0	1	9	10	12	8	7	15
4	4	5	3	2	7	6	9	0	1	8	13	12	11
5	5	3	4	0	6	8	10	1	2	7	12	14	9
6	6	7	8	1	9	10	3	4	5	13	0	2	16
7	7	8	6	9	0	1	4	5	3	14	15	13	17

7 TRIPLES SATISFYING $X + Y = Z$

Let's go back to Langford's son's blocks and add n to the labels:

4	2	3	2	4	3	1	1
5	6	7	8	9	10	11	12

We have solved another problem: partition the numbers from 1 to $3n$ into n triples which are solutions of $x + y = z$:

$$1 + 11 = 12, \quad 2 + 6 = 8, \quad 3 + 7 = 10, \quad 4 + 5 = 9$$

Since we added 1 to the numbers on the blocks, we see that the $x + y = z$ problem can be solved just if $n \equiv 0$ or $1 \bmod 4$. We will use such solutions to find packings and coverings of the complete graph with triangles, and to find cyclic Steiner triple systems.

8 TRIPLES SATISFYING $X + Y = 2Z$

The corresponding problem where $x + y = z$ is replaced by $x + y = 2z$ does have solutions for all values of n. For example for $n = 10$ we have

$1+3=2\cdot 2$	$7+13=2\cdot 10$	$11+21=2\cdot 16$	$12+26=2\cdot 19$	$17+29=2\cdot 23$
$4+8=2\cdot 6$	$5+25=2\cdot 15$	$9+27=2\cdot 18$	$14+30=2\cdot 22$	$20+28=2\cdot 24$

In other words, ten arithmetic progressions with differences 1, 2, ..., 10, which give us the positions in which to place the ten triples in the generalization of Langford's problem at the end of §2.

9 COIL DIAGRAMS

Although we cannot always partition the numbers from 1 to $3n$ into n triples satisfying $x + y = z$, we can partition those from 1 to $3n + 1$, omitting one number, for example $\lfloor(5n + 3)/2\rfloor$. Represent the numbers $[n + 1, 3n + 1]$ by the points in Figure 2, which are labeled slightly differently in case n is odd or even. E.g., for $n = 4$ (or 5), the points are 5 to 13 (or 6 to 16), 11 (or 14) is unused, and the two points near the middle are 8 & 9 (or 11 & 12). A triple is a pair connected by a semicircle together with its diameter. This use of coil diagrams is very similar to that of Ringel & Youngs when they confirmed Heawood's conjecture for the genus of the complete graph (the simplest case is Heawood's map below). Indeed, the connexion is less surprising when you realize that their main tool was a trivalent "current graph" on which Kirchhoff's current law requires $x + y = z$ at each node.

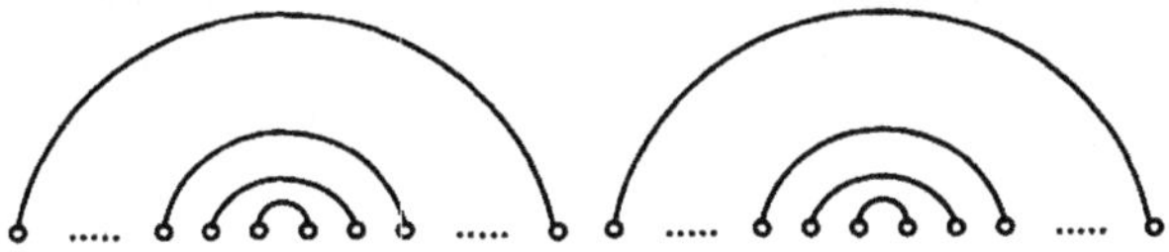

Figure 2 A coil diagram displaying n triples satisfying $x + y = z$.

10 SQUARING THE SQUARE

Current graphs and Kirchhoff's laws are also the basis of the Brooks-Smith-Stone-Tutte method of solving the problem of "squaring the square". Duijvestijn has found the lowest order possible, 21. Figure 3 shows his square, with the corresponding current diagram. Figure 4 shows the associated dual networks.

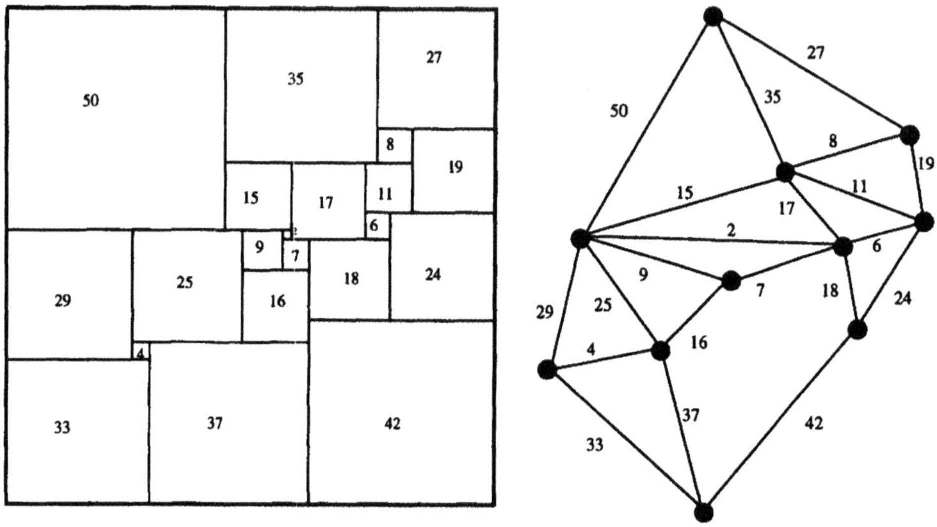

Figure 3 Duijvestijn's squared square and the related current diagram.

Notice that the lengths of the sides of the squares may be regarded as the magnitudes of the currents flowing in the "wires" of the circuit, or, if these wires are thought of as having unit resistance, then as potential differences between the various levels in the square. An alternative current diagram can be drawn from left to right instead of from top to bottom. This gives an example of a pair of **dual graphs** imbedded in the plane with each vertex of one graph corresponding to a region of the other and corresponding edges intersecting each other.

Duijvestijn has also recently discovered pairs of squares containing the same elements but differently arranged and has given examples of squares whose largest element is not on the boundary.

11 PACKING OR COVERING THE COMPLETE GRAPH

We can use the partition of the numbers [1,...,12] into four triples that satisfy $x + y = z$ as the "sides" of four different triangles, covering the 12 different "lengths" of chord in a circle of 25 equally spaced points. If we rotate these four triangles into each of their 25 possible different positions, we will have used up

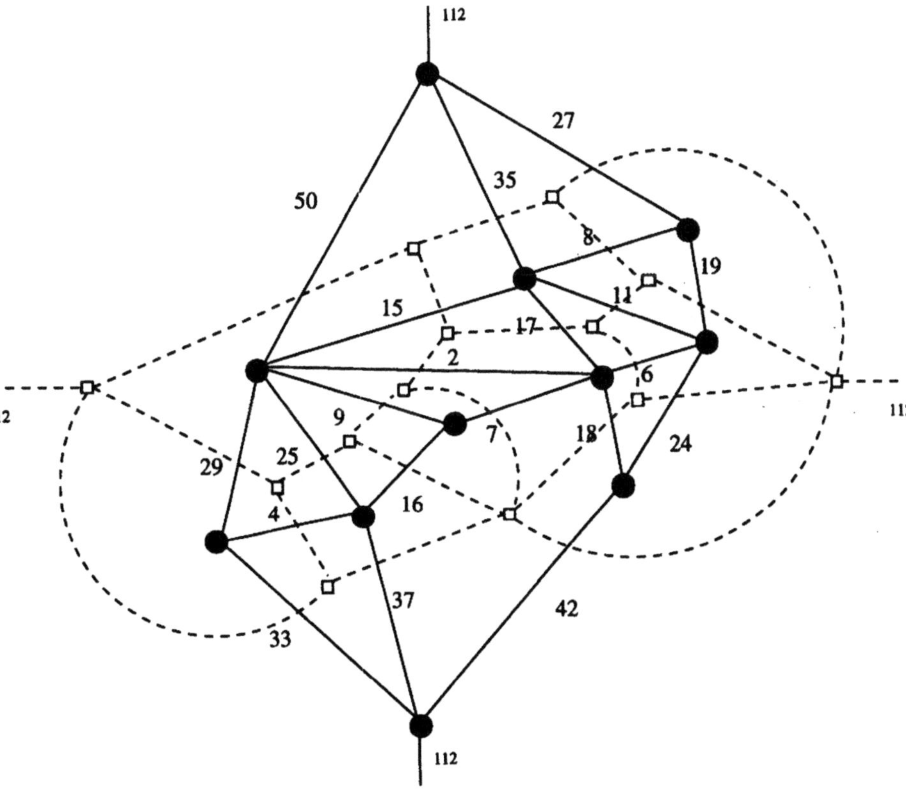

Figure 4 Dual networks associated with Duijvestijn's square.

all 25 chords of each of the 12 different lengths. This gives an exact packing or covering of K_{25}, the complete graph on 25 points, with 100 copies of K_3. See Figure 5.

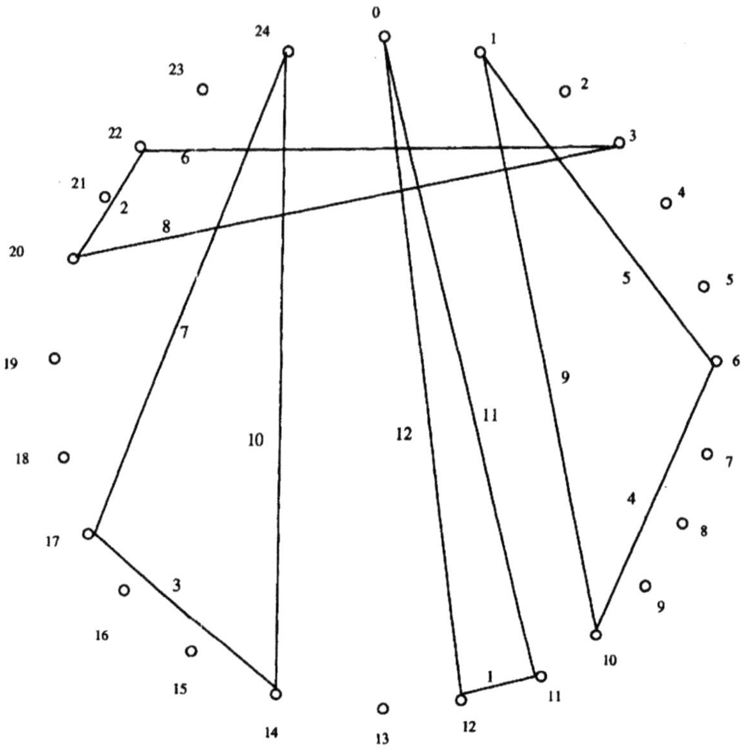

Figure 5 Four triangles whose sides cover 12 different chord lengths.

12 HANANI'S CYCLIC STEINER SYSTEMS

This is Hanani's construction for **cyclic Steiner triple systems**, e.g.,

{0,11,12}	{0,2,8}	{0,3,10}	{0,5,9}
{1,12,13}	{1,3,9}	{1,4,11}	**{1,6,10}**
{2,13,14}	{2,4,10}	{2,5,12}	{2,7,11}
{3,14,15}	{3,5,11}	{3,6,13}	{3,8,12}
...	...	...	...
...	...	**{14,17,24}**	...
...	...	...	...
...	**{20,22,3}**	...	...
...	...	...	...
{23,9,10}	{23,0,6}	{23,1,8}	{23,3,7}
{24,10,11}	{24,1,7}	{24,2,9}	{24,4,8}

which is a set of 100 triples with the property that each of the $\binom{25}{2}$ pairs chosen from 0 through 24 occurs in just one triple. The boldface entries correspond to the four triangles depicted in Figure 5.

Steiner triples on n objects exist just if $n \equiv 1$ or 3 mod 6.

13 PERFECT DIFFERENCE SETS

Ringel used these triples to find the toroidal thickness of the complete graph. Before we look at what that means, and how he did it, let's construct an easier example by taking a **perfect difference set**. The differences of the numbers

$$1, \quad 2, \quad 4$$

are $\pm 1, \pm 2, \pm 3$, that is all the nonzero residue classes mod 7, and we can rotate the 'triangle' 124 round a circle of seven equally spaced points (Figure 6).

to produce the simplest nontrivial Steiner triple system

$$\{1,2,4\} \quad \{2,3,5\} \quad \{3,4,6\} \quad \{4,5,0\} \quad \{5,6,1\} \quad \{6,0,2\} \quad \{0,1,3\}$$

There are perfect difference sets with $n+1$ members, giving all the nonzero residues modulo n^2+n+1, whenever n is a prime power. For example,

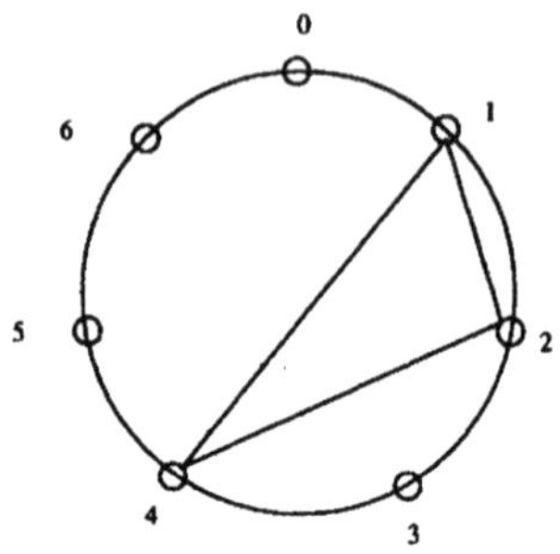

Figure 6 The difference set {1,2,4} generates a Steiner triple system.

n	n^2+n+1	difference set
3	13	{0, 1, 3, 9}
2^2	21	{0, 2, 7, 8, 11}
5	31	{0, 1, 3, 8, 12, 18}
7	57	{0, 4, 11, 20, 25, 26, 28, 38}
2^3	73	{0, 9, 19, 20, 22, 26, 34, 50, 55}
3^2	91	{0, 7, 12, 28, 32, 42, 43, 45, 51, 69}

It is conjectured that there are no perfect difference sets if n is not a prime power.

14 PROJECTIVE PLANES

This Steiner system is also the simplest nontrivial example of a **projective plane**, the projective plane of order 2 or **Fano configuration** (although, as Charles Colbourn remarked in his lecture, it goes back to Kirkman and Woolhouse). The triples are "lines" and the numbers are "points" — every pair of lines meets in a point, every pair of points determine a line. See Figure 7.

Moreover, such a geometry is **self-dual** in the sense that points and lines are interchangeable. For example, label the sides of the triangle opposite 0, 1, 4 in Figure 7 with the same symbols, 0, 1, 4, respectively, the corresponding medians 2, 5, 3, and the curved line as 6, then the same incidences occur as before: just as the points 0,4,5 are collinear (on line 1), so are the lines 0,4,5 concurrent (in the point 1).

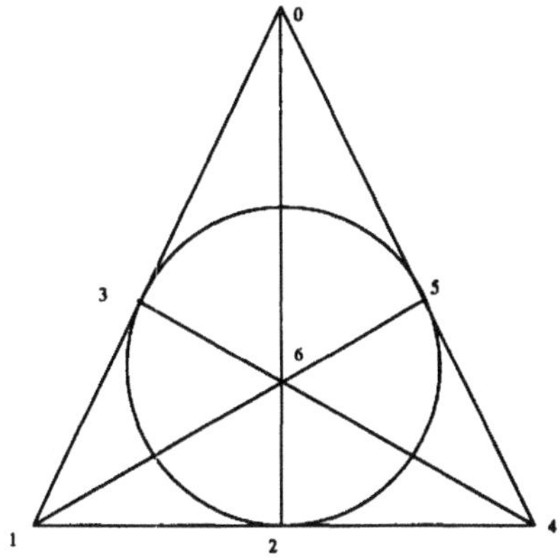

Figure 7 The Fano configuration; the projective plane of order 2.

We can use the perfect difference set $\{0, 1, 3, 9\}$ to form a projective plane with 13 points and 13 lines:

(0,1,3,9) (1,2,4,A) (2,3,5,B) (3,4,6,C) (4,5,7,0) (5,6,8,1) (6,7,9,2)
(7,8,A,3) (8,9,B,4) (9,A,C,5) (A,B,0,6) (B,C,1,7) (C,0,2,8)

Here is a different but fairly obvious construction:

(0,1,2,3)	(0,4,5,6)	(0,7,8,9)	
(A,1,4,7)	(A,2,5,8)	(A,3,6,9)	
(B,1,5,9)	(B,2,6,7)	(B,3,4,8)	
(C,1,6,8)	(C,2,4,9)	(C,3,5,7)	(0,A,B,C)

15 AFFINE GEOMETRIES

If we regard the points of one line of a projective geometry as being at infinity, the remaining points lie on families of parallel lines. For example, delete the line 0ABC from the geometry just constructed:

(1,2,3)	(4,5,6)	(7,8,9)
(1,4,7)	(2,5,8)	(3,6,9)
(1,5,9)	(2,6,7)	(3,4,8)
(1,6,8)	(2,4,9)	(3,5,7)

and we have 9 points on 12 lines, the 4 rows being sets of 3 parallel lines.

There's a close connexion with

16 MAGIC SQUARES

We can't arrange that *all* of the lines of an affine geometry are 'magic' in the sense that if the points are numbers then the sum of the points on each line is the same, but we can manage quite a few. An n by n **latin square** is an arrangement of n copies of n symbols in a square with just one of each symbol in each row or column. It is also possible to find *diagonal* latin squares in which the diagonals (including the broken diagonals) also contain just one of each symbol. Two latin squares are said to be **orthogonal** if, when they are superposed, all n^2 ordered pairs of symbols occur (once each). Here are two 5 by 5 orthogonal diagonal latin squares, next shown superposed, and finally read off as two digit numbers in base 5 to yield the numbers 0, 1, ... 24 arranged in **a pandiagonal magic square.**

0	2	4	1	3	0	1	2	3	4		00 21 42 13 34
1	3	0	2	4	2	3	4	0	1		12 33 04 20 41
2	4	1	3	0	4	0	1	2	3		24 40 11 32 03
3	0	2	4	1	1	2	3	4	0		31 02 23 44 10
4	1	3	0	2	3	4	0	1	2		43 14 30 01 22

0	11	22	8	19
7	18	4	10	21
14	20	6	17	3
16	2	13	24	5
23	9	15	1	12

The rows, columns and diagonals, including the broken diagonals such as $14 + 2 + 15 + 8 + 21$ or $16 + 20 + 4 + 8 + 12$, all have magic sum 60. These can be thought of as 20 lines of an affine geometry on 25 points; the other 10 lines are obtained by making chess knight's moves on a torus (draw a square round the array and identify opposite sides):

(0,20,15,10,5), (11,6,1,21,16), (22,17,12,7,2), (8,3,23,18,13), (19,14,9,4,24)
(0,4,3,2,1), (11,10,14,13,12), (22,21,20,24,23), (8,7,6,5,9), (19,18,17,16,15)

Only the third member of the first set and the second member of the second set have magic sum 60.

17 KIRKMAN'S SCHOOLGIRLS PROBLEM

Kirkman asked for a construction of a Steiner triple system of 15 schoolgirls in 7 parallel sets of 5 triples, each set containing all 15 girls. For 3^n girls we can use the n-dimensional affine geometry with three points on a line. We've just seen the case $n = 2$ with 9 girls walking out in 3 rows of 3 on four different days. We can solve Kirkman's original problem with 5 triples for each day of the week by putting the girls in a circle. Rotate the equilateral triangle (0,5,10) for Sunday (\$) and the triangles (0,2,8) and (1,2,5) (moved round two clicks at a time to reveal the pattern better) will yield 2×15 triples which split into 6 parallel sets for the week-days (MTWΘFS) (Figure 8).

0	5	10	\$
1	6	11	\$
2	7	12	\$
3	8	13	\$
4	9	14	\$

	0	2	8	M
	1	3	9	M
	2	4	10	T
	3	5	11	T
	4	6	12	W
	5	7	13	W
	6	8	14	T
	7	9	0	T
	8	10	1	W
	9	11	2	W
F	10	12	3	
	11	13	4	M
	12	14	5	M
Θ	13	0	6	
S	14	1	7	

	1	2	5	Θ
	3	4	7	Θ
	5	6	9	F
	7	8	11	F
	9	10	13	S
	11	12	0	S
	13	14	2	F
	0	1	4	F
	2	3	6	S
	4	5	8	S
M	6	7	10	
	8	9	12	Θ
	10	11	14	Θ
T	12	13	1	
W	14	0	3	

Figure 8 Solution of Kirkman's schoolgirls problem.

18 HEAWOOD'S MAP ON THE TORUS

We can use the 6 differences modulo 7 from the perfect difference set 0, 1, 3 to label the nodes of a triangular lattice along its 6 different directions (Figure 9).

If we identify all nodes having the same label we obtain a torus with seven nodes, each of them adjacent to the other six; the complete graph on 7 points,

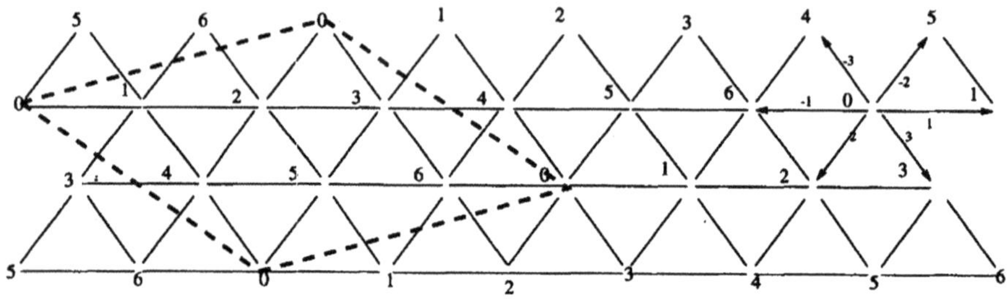

Figure 9 Finding 7 nodes and $\binom{7}{2}$ adjacencies on a torus.

K_7. Think of them as capitals of seven countries, with roads between each pair, crossing the frontiers of every pair of contiguous countries. This dual map is Heawood's famous map on the torus, showing that at least 7 colors are needed to color maps on the torus (Figure 10). We also see that the genus of K_7 is one, i.e. the least genus of a surface in which we can imbed it; clearly we can't imbed K_7 in the sphere or plane.

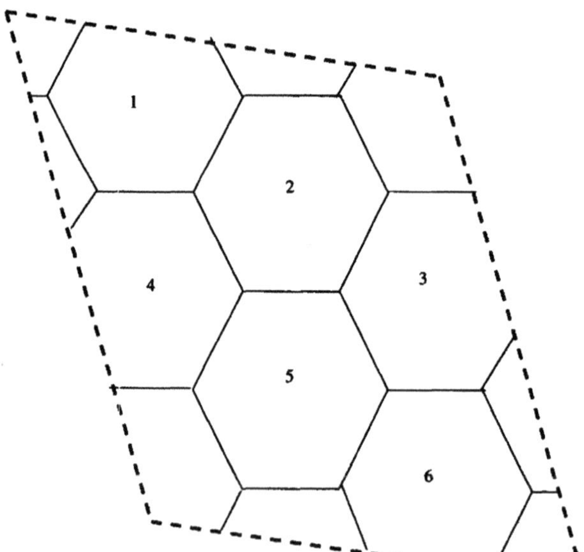

Figure 10 Heawood's map on the torus.

19 THE TOROIDAL THICKNESS OF THE COMPLETE GRAPH

Equally clearly, we can't imbed K_{25} in the torus. But we can ask the question: how few parts can we partition its edges into with each part forming a graph that's imbeddable in the torus? This is called the **toroidal thickness** of the graph. And in this case the answer's four, because we can take our four equations from §7

$$1 + 11 = 12, \quad 2 + 6 = 8, \quad 3 + 7 = 10, \quad 4 + 5 = 9$$

and use them in place of $1 + 2 = 3$ to make four maps like Heawood's, but working mod 25 instead of mod 7.

In the same way you can use the triangles with "sides" 1, 3, 4 and 2, 5, 7 in a circle of 13 points to show that the toroidal thickness of K_{13} is two and three triangles, 1, 7, 8; 2, 3, 5; 4, 6, 9 in a circle of 19 points to give $\theta_1(K_{19}) = 3$. See also Figure 11. Ringel has proved that in general

$$\theta_1(K_n) = \left\lceil \frac{n-1}{6} \right\rceil .$$

20 NIM ADDITION

There's another way of labelling the Fano configuration, depicted in Figure 12.

At first sight it looks as though I've just added one to each of the labels, but it's much more subtle than that and I don't know of any simple rule for getting from one labelling to the other. The motivation for this labelling is seen if you write the labels in binary, as in Figure 13.

Then the nim-sum of 3 labels on a line is zero. What is nim-sum? Vector or coordinate-wise addition over GF(2); addition without carry in base 2; exclusive or, XOR. The automorphism group for the projective plane of order 2 can be represented by the nonsingular 3×3 0-1 matrices. The first row can be written down in $2^3 - 2^0 = 7$ ways, the second in $2^3 - 2^1 = 6$ ways and the third in $2^3 - 2^2 = 4$ and the order of the group is $7 \times 6 \times 4 = 168$ (We select a **triangle of reference** by choosing one point, then a second, then the third not on the line joining the first two). This is also the automorphism group for the Heawood map.

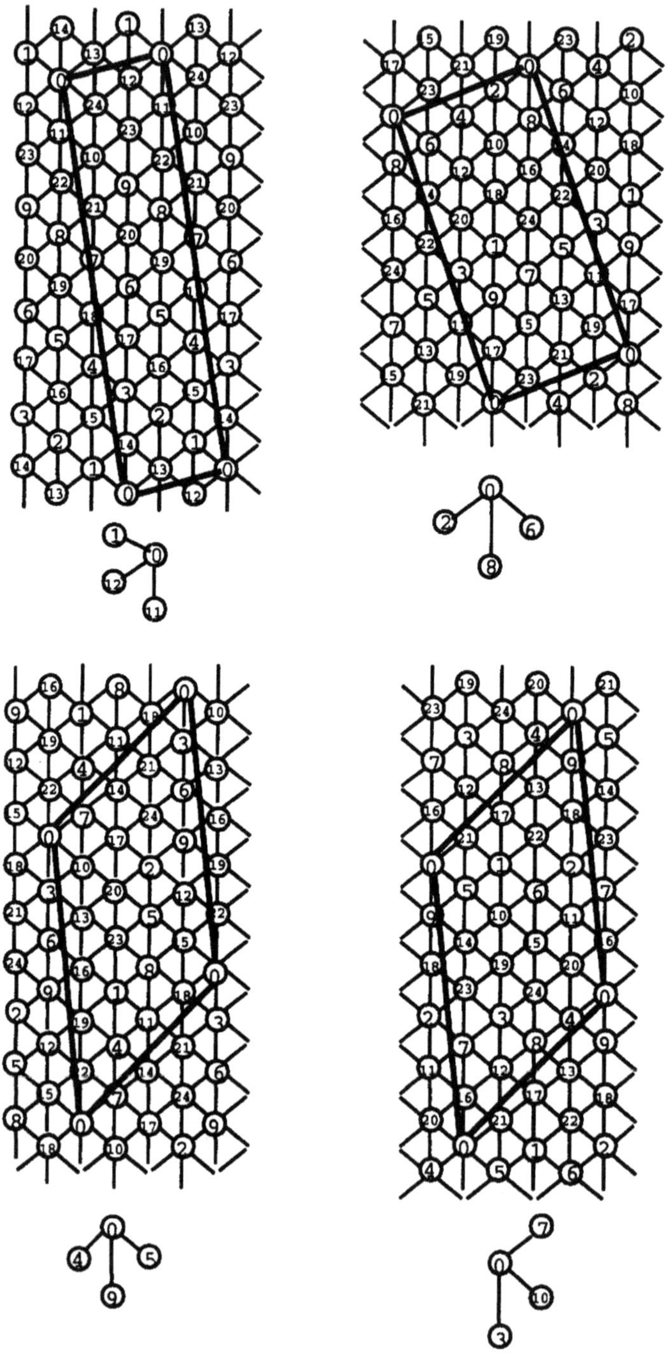

Figure 11 The toroidal thickness of K_{25} is four.

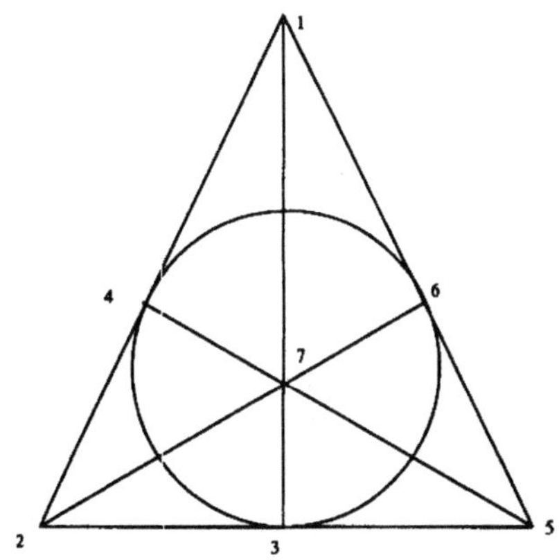

Figure 12 The Fano configuration relabelled.

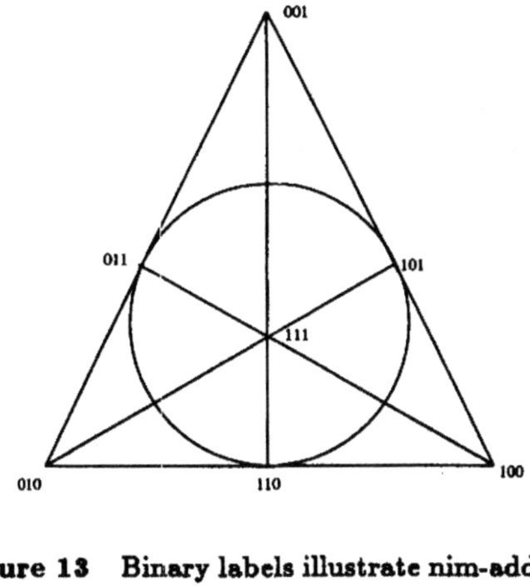

Figure 13 Binary labels illustrate nim-addition.

[Query: why is $PSL_2(7)$ isomorphic to $GL_3(2)$?]

21 INCIDENCE MATRICES

Incidence matrices of projective geometries, e.g., that of the plane of order 2:

	1	2	3	4	5	6	7
1	1	1	1	0	0	0	0
2	1	0	0	1	1	0	0
3	1	0	0	0	0	1	1
4	0	1	0	1	0	1	0
5	0	1	0	0	1	0	1
6	0	0	1	1	0	0	1
7	0	0	1	0	1	1	0

where the rows are lines and the columns points (or vice versa, since we have seen that a projective plane is self-dual) can be used to solve special cases of

22 ZARANKIEWICZ'S PROBLEM

What is the maximum number $Z(m, n; i, j)$ of ones that can occur in an $m \times n$ 0-1 matrix which doesn't contain an $i \times j$ submatrix consisting entirely of ones? From the incidence matrix above we see that $Z(7, 7; 2, 2) = 21$.

But incidence matrices have many uses and we'll next see how they are connected to

23 ERROR-CORRECTING CODES

Choose any two rows from the matrix and compare them. They differ in exactly four places. If we used them as code-words to send a message and one of the bits got changed en route, the recipient would still be able to tell which word was meant. If two bits got changed then she'd know there was an error, but wouldn't know which of two words was intended. We can amplify this code by adding the unit word (the first row in Figure 14), and then the complements of these eight words, and finally a **parity check bit** (the first column) making the sum of each row even (one word of weight 8, 14 of weight 4, and one of weight 0). This is the **extended Hamming code** $\mathcal{H}_8$, as in Figure 14.

1	1	1	1	1	1	1	1
1	1	1	1	0	0	0	0
1	1	0	0	1	1	0	0
1	1	0	0	0	0	1	1
1	0	1	0	1	0	1	0
1	0	1	0	0	1	0	1
1	0	0	1	1	0	0	1
1	0	0	1	0	1	1	0
0	1	1	0	1	0	0	1
0	1	1	0	0	1	1	0
0	1	0	1	1	0	1	0
0	1	0	1	0	1	0	1
0	0	1	1	1	1	0	0
0	0	1	1	0	0	1	1
0	0	0	0	1	1	1	1
0	0	0	0	0	0	0	0

Figure 14 The extended Hamming code $\mathcal{H}_8$.

24 HADAMARD MATRICES

If, in the first 8 rows of this array, we replace zeros by minus ones (represented by minus signs) and represent ones by plus signs, we obtain a **Hadamard matrix**,

$$\begin{bmatrix} + & + & + & + & + & + & + & + \\ + & + & + & + & - & - & - & - \\ + & + & - & - & + & + & - & - \\ + & + & - & - & - & - & + & + \\ + & - & + & - & + & - & + & - \\ + & - & + & - & - & + & - & + \\ + & - & - & + & + & - & - & + \\ + & - & - & + & - & + & + & - \end{bmatrix}$$

i.e. an $n \times n$ matrix of plus or minus ones in which any pair of rows (or columns) is orthogonal, $HH^{tr} = nI$ (here n=8). n must be 1, 2 or a multiple of 4. It is believed that they exist for all multiples of 4, but many remain to be constructed.

25 CYCLIC HADAMARD MATRICES

The Hadamard matrix we've just constructed came from the Hamming code, which came from the incidence matrix of the projective geometry. We could also construct a cyclic Hadamard matrix from the *other* labelling of the Fano configuration, which came from the difference set $\{1, 2, 4\}$:

	0	1	2	3	4	5	6	7
0	+	+	+	+	+	+	+	+
1	+	+	+	−	+	−	−	−
2	+	−	+	+	−	+	−	−
3	+	−	−	+	+	−	+	−
4	+	−	−	−	+	+	−	+
5	+	+	−	−	−	+	+	−
6	+	−	+	−	−	−	+	+
7	+	+	−	+	−	−	−	+

As a matter of fact, this is the same Hadamard matrix as before, but that is because $n = 8$ is a small number (arrange the rows in the order 01753426 and then the columns in order 01423765). As n gets larger the number of nonisomorphic Hadamard matrices of order n increases rapidly.

26 FACTORING WITH QUADRATIC FORMS

The same pattern occurred when the Lehmers were using their sieve to factor numbers via quadratic forms:

$x^2 - dy^2 : d =$	1	−1	2	−2	3	−3	6	−6
$24k + 1$	+	+	+	+	+	+	+	+
17	+	+	+	+	−	−	−	−
13	+	+	−	−	+	+	−	−
5	+	+	−	−	−	−	+	+
23	+	−	+	−	+	−	+	−
7	+	−	+	−	−	+	−	+
11	+	−	−	+	+	−	−	+
19	+	−	−	+	−	+	+	−

A row represents a residue class mod 24 of a number to be factored. A column represents a quadratic form. A plus sign means that the number is representable by the form, a minus that it is not. If two representations are found, a factorization easily follows. Suppose we wish to factor a number $\equiv 13 \bmod 24$. It must break as 1×13, 17×5, 23×11 or 7×19. Apart from column 1 (which contains all plus signs, corresponding to Fermat's method of difference of squares), row 13 has plus signs in columns −1, 3 and −3, corresponding to forms $x^2 + y^2$, $x^2 - 3y^2$ and $x^2 + 3y^2$, which are the ones to use on such a number. The plus signs in these three columns correspond to factorizations (1×13 and) 17×5, 23×11 and 7×19.

$x^2 - dy^2 : d =$	1	−1	2	−2	3	−3	6	−6	5	−5	10	−10	15	−15	30	−30
$120k + 1$, 49	+	+	+	+	+	+	+	+	+	+	+	+	+	+	+	+
73, 97	+	+	+	+	+	+	+	+	−	−	−	−	−	−	−	
41, 89	+	+	+	+	−	−	−	−	+	+	+	+	−	−	−	−
17, 113	+	+	+	+	−	−	−	−	−	−	−	−	+	+	+	+
61, 109	+	+	−	−	+	+	−	−	+	+	−	−	+	+	−	−
13, 37	+	+	−	−	+	+	−	−	−	−	+	+	−	−	+	+
29, 101	+	+	−	−	−	−	+	+	+	+	−	−	−	−	+	+
53, 77	+	+	−	−	−	−	+	+	−	−	+	+	+	+	−	−
71, 119	+	−	+	−	+	−	+	−	+	−	+	−	+	−	+	−
23, 47	+	−	+	−	+	−	+	−	−	+	−	+	−	+	−	+
31, 79	+	−	+	−	−	+	−	+	+	−	+	−	−	+	−	+
7, 103	+	−	+	−	−	+	−	+	−	+	−	+	+	−	+	−
11, 59	+	−	−	+	+	−	−	+	+	−	−	+	+	−	−	+
83, 107	+	−	−	+	+	−	−	+	−	+	+	−	−	+	+	−
19, 91	+	−	−	+	−	+	+	−	+	−	−	+	−	+	+	−
43, 67	+	−	−	+	−	+	+	−	−	+	+	−	+	−	−	+

For example, if we discover that $130813 = 347^2 + 102^2 = 357^2 + 58^2$, then

$$\gcd(\tfrac{1}{2}(357 + 347), \tfrac{1}{2}(102 - 58)) = 22 \ ,$$

$$\gcd(\tfrac{1}{2}(357 - 347), \tfrac{1}{2}(102 + 58)) = 5 \ ,$$

and $22^2 + 5^2 = 509$ is a factor of 130813.

In fact the Lehmers used the prime 5 as well as 2 and 3, giving the above scheme mod 120, which is a Hadamard matrix of order 16.

27 PROJECTIVE GEOMETRIES

This array corresponds to the 3-dimensional projective geometry of order 2, just as the previous array corresponded to the projective plane of order 2. Omit the first row and column and we are left with the incidence matrix for 15 points, lying 7 on each of 15 (Fano) planes.

The 15 points may be thought of as:

the 4 vertices, $(1, 0^3)$ (i.e., $1 = 0001_2$, $2 = 0010_2$, $4 = 0100_2$, $8 = 1000_2$) of a tetrahedron,

the 6 midpoints, $(1^2, 0^2)$ (i.e., 3, 5, 6, 9, 10, 12) of its edges,

the 4 centroids, $(1^3, 0)$ (7, 11, 13, 14) of the faces, and

the body centroid, (1^4) (i.e.,15).

Dually, the planes are:

the 4 faces, $x = 0$, $y = 0$, $z = 0$, $w = 0$, of the tetrahedron;

the 6 'medial' planes, such as $x + y = 0$, containing an edge and the midpoint of the opposite edge;

4 'cones' such as $x + y + z = 0$, joining a vertex to the incircle of the opposite face; and

a 'sphere' $x + y + z + w = 0$, through the midpoints of the edges together with its centre.

Each plane contains 7 lines and each line lies in 3 planes, so that there are 35 lines. These are the triples of distinct numbers chosen from $\{1, \ldots, 15\}$ whose nim-sum is zero [choose the first number in one of 15 ways, the second in 14 ways and the nim-sum rule determines the third: $15 \times 14/3! = 35$].

28 SPHERE PACKING

In the extended Hamming code $\mathcal{H}_8$, if we replace the zero vector by 8 vectors $(1, 0^7)$, each of the fourteen vectors of weight four by 16 vectors $((\pm\frac{1}{2})^4, 0^4)$ and the weight eight vector by 8 vectors $(-1, 0^7)$, we have the 240 minimal vectors of the lattice E_8, the best lattice packing of spheres in 8 dimensions, displaying the **kissing number** 240 in 8 dimensions, one of the few that is known. These vectors also correspond to the 240 unit (norm 1) **Cayley numbers** or **octaves.**

E_8 can also be identified with the ring of **icosians**, i.e. finite sums of members of the **icosian group**, the group consisting of 120 quaternions: 8 of shape $(\pm 1, 0^3)$, 16 of shape $(\pm\frac{1}{2}^4)$ and the 96 even permutations of $\frac{1}{2}(0, \pm 1, \pm\sigma, \pm\tau)$ where $\sigma, \tau = \frac{1}{2}(1 \mp \sqrt{5})$.

29 MOCK TURTLES

When you play the game of Mock Turtles with just eight coins (kinder than using turtles) you're allowed to turn one, two or three coins over, provided the rightmost one turned goes from heads to tails. The winner is the one who leaves all tails. The codewords of the Hamming code $\mathcal{H}_8$ will give you the $\mathcal{P}$-positions (previous-player-winning positions) of the heads (or tails). The nim-values of the turtles which are right-way-up are:

n	0	1	2	3	4	5	6	7	8	9	10	11	12	13	14	15	16	$\cdots$
$\mathcal{G}(n)$	1	2	4	7	8	11	13	14	16	19	21	22	25	26	28	31	32	$\cdots$

These numbers are the **odious numbers**, i.e. those whose binary representation contains an odd number of ones. In other words, write n in binary and append a check digit 0 or 1 to make the sum of the digits odd.

If each player is allowed to turn up to five coins at each move, then we have the game of Moebius, named because when played on a row of 18 coins, the $\mathcal{P}$-positions can be obtained from each other by Möbius transformations modulo

17:

$$x \longrightarrow \frac{ax+b}{cx+d} \quad \text{with} \quad ad-bc=1$$

acting on the following labels for the 18 coins:

$\infty \quad 1 \quad 4 \quad 0 \quad -4 \quad -1 \quad 5 \quad 6 \quad -8 \quad 2 \quad -3 \quad -5 \quad 8 \quad 3 \quad -7 \quad 7 \quad -6 \quad -2$

The $\mathcal{P}$-positions form the words of a **quadratic residue code.**

Correspondingly, if the move is to turn up to seven coins, we have the game of Mogul, whose $\mathcal{P}$-positions, when played on 24 coins, manifest the **extended Golay code**, obtained by appending a check digit to the quadratic residue code, mod 23. Its automorphism group is the Mathieu group M_{24} and it is intimately related to the **Leech lattice.**

Notice that the third letter in moCk turtles, moEbius & moGul is respectively the 3rd, 5th & 7th letter of the alphabet.

These remarkable relationships between combinatorial games and codes inspired Conway & Sloane to discover **lexicodes.**

[To avoid cluttering the text, reference numbers are not given there. Where the title doesn't indicate the relevance, an explanatory bracket is added.]

REFERENCES

[1] V. E. Alekseev, On the Skolem method of constructing Steiner triple systems (Russian), *Mat. Zametki*, **2**(1967) 145–156; *MR* **35** #5341. [$x+y=z$]

[2] W. W. Rouse Ball & H. S. M. Coxeter, *Mathematical Recreations & Essays*, 12th edition, Univ. of Toronto, 1974. [36–40 Nim & Wythoff's Game; 115–116 squaring the square; 149–152 sphere packing; 189–192 latin squares; 193–221 magic squares; 222–242 map coloring; 271–311 combinatorial designs, an excellent survey written by J. J. Seidel]

[3] Thøger Bang, On the sequence $[n\alpha]$, $n = 1, 2, \ldots$, supplementary note to the previous paper by Th. Skolem, *Math. Scand.*, **5**(1957) 69–76; *MR* **19**, 1159h. [Skolem problem]

[4] S. Beatty, Problem 3173, *Amer. Math. Monthly*, **33**(1926) 159; **34**(1927) 159. [Beatty sequences]

[5] Claude Berge, *The Theory of Graphs and its Applications*, Methuen, London, 1962, p. 38. [Isaacs's Game]

[6] E. R. Berlekamp, J. H. Conway & R. K. Guy, *Winning Ways for Your Mathematical Plays*, Academic Press, London, 1982. [sum of games, p. 32; nim addition, pp. 60–61; Wyt Queens, p. 61; nim-values, pp. 82–87; $\mathcal{P}$-positions, p. 83; Turning Turtles, p. 429]

[7] T. Beth, D. Jungnickel and H. Lenz, *Design Theory*, Cambridge University Press, Cambridge, 1986. [a comprehensive reference for Steiner systems and difference methods]

[8] Charles L. Bouton, Nim, a game with a complete mathematical theory, *Ann. of Math., Princeton*,(2) **3**(1901–02) 35–39.

[9] R. L. Brooks, C. A. B. Smith, A. H. Stone & W. T. Tutte, The dissection of rectangles into squares, *Duke Math. J.*, **7**(1940) 312–340; *MR* **2**, 153.

[10] R. H. Bruck & H. J. Ryser, The non-existence of certain finite projective planes, *Canad. J. Math.*, **1**(1949) 88–93; *MR* **10**, 319.

[11] J. H. Conway, *On Numbers and Games*, Academic Press, London, 1976. [impartial games & Nim, chap. 11]

[12] J. H. Conway & N. J. A. Sloane, *Sphere Packings, Lattices and Groups*, Springer Grundlehren der mathematischen Wissenschaften **290**, 1988. [Hamming code p. 80; Golay code, p. 84; Cayley numbers, p. 122; lexicodes, p. 327]

[13] J. H. Conway & N. J. A. Sloane, Lexicographic codes: error-correcting codes from game theory, *IEEE Trans. Info. Theory*, **IT-32**(1986) 337–348; *MR* **87f**:94049.

[14] H. S. M. Coxeter, The golden section, phyllotaxis and Wythoff's game, *Scripta Math.*, **19**(1953) 135–143; *MR* **15**, 2

[15] Hallard T. Croft, Kenneth J. Falconer & Richard K. Guy, *Unsolved Problems in Geometry*, Springer, 1991. [C2 squaring the square; D10 packing spheres; E9–E13 lattice points]

[16] R. O. Davies, On Langford's problem, II, *Math. Gaz.*, **43**(1959) 253–255; *MR* **22** #5581.

[17] J. Dénes & A. D. Keedwell, *Latin Squares and Their Applications*, Academic Press, 1974.

[18] J. Doyen, Recent developments in the theory of Steiner systems, *Atti dei Conv. Lincei*, **17**(1976) Tomo I, 277–285; *MR* **55** #10286. [excellent bibliography with many early items, previously overlooked]

[19] A. J. W. Duijvestijn, Simple perfect squared square of lowest order, *J. Combin. Theory Ser. B*, **25**(1978) 240–243; *MR* **80a**:05051.

[20] A. J. W. Duijvestijn, Simple perfect squared squares and 2×1 squared rectangles of order 25, *Math. Comput.*, **62**(1994) 325–332: *MR* **94c**:05023 (& see 05017). [Squares with the same elements differently arranged; squares whose largest element is not on the boundary.]

[21] G. Fano, Sui postulati fondamentali della geometria proiettiva, *Giorn. Mat.*, **30**(1892) 114–124. [the 'Fano' configuration was anticipated by Kirkman in 1850]

[22] Branko Grunbaum & G. C. Shephard, *Tilings and Patterns*, W. H. Freeman, New York, 1987, [76–81 squaring the square, 531–583 Penrose aperiodic tiles]

[23] P. M. Grundy, Mathematics and games, *Eureka*, **2**(1939) 6–8; [Sprague-Grundy theory of impartial games]

[24] Richard K. Guy, A many facetted problem of Zarankiewicz, in *The Many Facets of Graph Theory*, Springer, New York, 1969, 129–148; *MR* **41** #91. [adjacency matrices, incidence matrices, packing & covering, Turán & Ramsey problems, tournaments, steiner triples, affine & projective planes, difference sets]

[25] Richard K. Guy, Packing $[1, n]$ with solutions of $ax + by = cz$ — the unity of combinatorics, *Atti dei Conv. Lincei*, **17**(1976) Tomo II, 173–179; *MR* **57** #9565. [$x + y = z$, Langford-Skolem, Wythoff, Isaacs, Steiner, Hanani, Ringel, Zarankiewicz]

[26] Richard K. Guy, The Penrose pieces, *Bull. London Math. Soc.*, **8**(1976) 9–10.

[27] Richard K. Guy & Cedric A. B. Smith, The *G*-values for various games, *Proc. Cambridge Philos. Soc.*, **52**(1956) 514–526; *MR* **18**, 546.

[28] J. Hadamard, Résolution d'une question relative aux déterminants, *Bull. Sci. Math.*(2) **17**(1893) 240–248. [Hadamard matrices]

[29] R. W. Hamming, Error correcting and error detecting codes, *Bell Sys. Tech. J.*, **29**(1950) 147–160; *MR* **12**, 35c.

[30] H. Hanani, A note on Steiner triple systems, *Math. Scand.*, **8**(1960) 154–156; *MR* **23** #A2330.

[31] H. Hanani, On quadruple systems, *Canad. J. Math.*, **12**(1960) 145–157; *MR* **22** #2558.

[32] T. P. Kirkman, On a problem in combinations, *Cambridge & Dublin Math. J.*, **2**(1847) 191–204.

[33] T. P. Kirkman, Note on an unanswered prize question, *Cambridge & Dublin Math. J.*, **5**(1850) 255–262.

[34] T. P. Kirkman, On triads made with fifteen things, *London, Edinburgh & Dublin Phil. Mag.*, **37**(1850) 169–171.

[35] T. P. Kirkman, On the perfect r-partitions of $N = r^2 - r + 1$, *Trans. Historic Soc. Lancs. & Cheshire*, **9**(1856–57) 127–142.

[36] T. P. Kirkman, On the puzzle of the fifteen young ladies, *London, Edinburgh & Dublin Phil. Mag.*(4), **23**(1862) 198–204.

[37] Thomas Kövári, Vera Sós & Paul Turán, On a problem of K. Zarankiewicz, *Colloq. Math.*, **3**(1954) 50–57; *MR* **16**, 456.

[38] C. Dudley Langford, Problem, *Math. Gaz.*, **42**(1958) 228.

[39] John Leech, Some sphere packings in higher space, *Canad. Math. J.*, **5**(1964) 657–682; *MR* **29** #5166; **19**(1967) 251–267; *MR* **35** #878.

[40] Emma Lehmer, On residue difference sets, *Canad. J. Math.*, **5**(1953) 425–432; *MR* **15**, 10.

[41] Al. A. Markov, On a certain combinatorial problem (Russian), *Problemy Kibernetiki*, **15**(1965) 263–266; *MR* **35** #1497. [$x + y = z$]

[42] R. J. Nowakowski, *Generalizations of the Langford-Skolem problem*, M.Sc. thesis, The Univ. of Calgary, 1975.

[43] R. J. Nowakowski, *Zarankiewicz's problem*, Ph.D. thesis, The Univ. of Calgary, 1978.

[44] C. J. Priday, On Langford's problem, I, *Math. Gaz.* **43**(1959) 250–253; *MR* **22** #5580.

[45] D. K. Ray-Chaudhuri & R. M. Wilson, Solution of Kirkman's schoolgirl problem, *Proc. Symp. Pure Math.*, **19** Amer. Math. Soc. 1971, 187–203; *MR* **47** #3195.

[46] G. Ringel, Die toroidale Dicke des vollständigen Graphen, *Math. Zeit.*, **87**(1965) 19–26; *MR* **30** #2489.

[47] G. Ringel, *Map Color Theorem*, Grundlehren der math. Wissenschaften, **209**, Springer, 1974.

[48] G. Ringel & J. W. T. Youngs, Solution of the Heawood map-coloring problem, *Proc. Nat. Acad. Sci. U.S.A.*, **60**(1968) 438–445; *MR* **37** #3959.

[49] G. Ringel & J. W. T. Youngs, Solution of the Heawood map-coloring problem – case 11, *J. Combin. Theory*, **7**(1969) 71–93; *MR* **39** #1360; – case 2, 342–352; –case 8, 353–363; *MR* **41** #6723.

[50] A. Rosa, A note on Steiner triple systems (Slovak), *Mat.-Fyz. Časopis Sloven. Akad. Vied*, **16**(1966) 285–290; *MR* **35** #2759. [$x + y = z$]

[51] H. J. Ryser, *Combinatorial Mathematics*, Carus Math. Monograph **14**, Math. Assoc. Amer., 1963.

[52] J. Sedláček, On a set system, *Ann. New York Acad. Sci.*, **175**(1970) 329–330; *MR* **42** #117. [$x + y = z$]

[53] J. Singer, A theorem in finite projective geometry and some applications to number theory, *Trans. Amer. Math. Soc.*, **43**(1938) 377–385.

[54] Th. A. Skolem, On certain distributions of integers in pairs with given differences, *Math. Scand.*, **5**(1957) 57–68; *MR* **19**, 1159g.

[55] Th. A. Skolem, Some remarks on the triple systems of Steiner, *Math. Scand.*, **6**(1958) 273–280; *MR* **21** #5582.

[56] Th. A. Skolem, Über einige Eigenschaften der Zahlenmengen [$\alpha n + \beta$] bei irrationalem mit einleitenden Bemerkungen über einige kombinatorische Probleme, *Norske Vid. Selsk. Forh., Trondheim*, **30**(1957) 42–49; *MR* **19**, 1159i.

[57] R. P. Sprague, Über mathematische Kampfspiele, *Tôhoku Math. J.*, **41**(1935–36) 438–444; *Zbl.* **13**, 290. [Sprague-Grundy theory of impartial games]

[58] R. P. Sprague, Beispiel einer Zerlegung des Quadrats in lauter verschiedene Quadrate, *Math. Z.*, **45**(1939) 607–608. [squaring the square]

[59] Thomas Storer, *Cyclotomy and Difference Sets*, Markham, Chicago, 1967.

[60] C. M. Terry, L. R. Welch & J. W. T. Youngs, Solution of the Heawood map-coloring problem - case 4, *J. Combin. Theory*, **8**(1970) 170–174; *MR* **41** #3321.

[61] A. Tietäväinen, On the non-existence of perfect codes over finite fields, *SIAM J. Appl. Math.*, **24**(1973) 88–96; *MR* **48** #3609.

[62] Thomas M. Thompson, *From Error-correcting Codes through Sphere Packings to Simple Groups*, Carus Math. Monograph **21**, Math. Assoc. Amer., 1983.

[63] J. A. Todd, A combinatorial problem, *J. Math. and Phys.*, **12**(1933) 321–323.

[64] W. S. B. Woolhouse, Prize question 1733, *Lady's and Gentleman's Diary*, 1844.

[65] W. A. Wythoff, A modification of the game of Nim, *Nieuw Arch. voor Wisk.*(2), **7**(1905-07) 199–202.

[66] J. W. T. Youngs, Solution of the Heawood map-coloring problem - cases 3, 5, 6 and 9, *J. Combin. Theory*, **8**(1970) 175–219; - cases 1, 7, and 10, 220–231; *MR* **41** #3322-3.

[67] K. Zarankiewicz, Problem P101, *Colloq. Math.*, **2**(1951) 301.

[68] Š. Znám, On a combinatorical problem of K. Zarankiewicz, *Colloq. Math.*, **11**(1963) 81–84; *MR* **29** #37; **13**(1965) 255–258; *MR* **32** #7434.

UNSOLVED PROBLEMS IN COMBINATORIAL GAMES

Richard K. Guy

University of Calgary
Calgary, Alberta, Canada

We have retained the numbering from the list of unsolved problems given on pp. 183–189 of Amer. Math. Soc. *Proc. Sympos. Appl. Math.* **43**(1991) and added in some new material. For many more references than we list, see Fraenkel's Bibliography. WW refers to

Elwyn Berlekamp, John Conway & Richard Guy, *Winning Ways for your Mathematical Plays*, Academic Press, 1982.

1. Subtraction games are known to be periodic. Investigate the relationship between the subtraction set and the length and structure of the period. The same question can be asked about *partizan* subtraction games, in which each player is assigned an individual subtraction set. See Fraenkel & Kotzig (1987).

See also Subtraction Games in WW, 83–86, 487–498. A move in the game $S(s_1, s_2, s_3, \ldots)$ is to take a number of beans from a heap, provided that number is a member of the **subtraction-set**, $\{s_1, s_2, s_3, \ldots\}$. Analysis of such a game and of many other heap games is conveniently recorded by a **nim-sequence**,

$$n_0 n_1 n_2 n_3 \ldots,$$

meaning that the nim-value of a heap of h beans is n_h, $h = 0, 1, 2, \ldots$, i.e., that the value of a heap of h beans in this particular game is the **nimber** $*n_h$. To avoid having to print stars, we say that the nim-value of a position is n, meaning that its value is the nimber $*n$.

Table 1 shows some examples: the first is a manifestation of She-Loves-Me-She-Loves-Me-Not; the last is Nim. If the subtraction-set is finite, the nim-sequence

C. J. Colbourn and E. S. Mahmoodian (eds.), Combinatorics Advances, 161–179.

is (ultimately) periodic. But little is known about the length of the period *vis à vis* the membership of the subtraction set.

Subtraction game	Nim-sequence	(ultimate) Period	
$S(1)$	$\dot{0}\dot{1}01010101\ldots$		2
$S(2)$	$\dot{0}01\dot{1}0011001100\ldots$		4
$S(3)$	$\dot{0}0011\dot{1}0001110001110\ldots$		6
$S(1,2)$	$\dot{0}1\dot{2}0120120120\ldots$		3
$S(1,2,3)$	$\dot{0}12\dot{3}0123012301230\ldots$		4
$S(1,2,3,4)$	$\dot{0}123\dot{4}01234012340\ldots$		5
$S(2,4,7)$	$00112203\dot{1}0\dot{2}10210210\ldots$		3
$S(2,5,6)$	$\dot{0}011021302\dot{1}0011021302100\ldots$		11
$S(4,10,12)$	$\dot{0}00011110022113300221\dot{1}0000\ldots$		22
$S(1,2,3,4,\ldots)$	$0123456789\ldots$	(saltus 1 &)	1

Table 1 Nim-sequences and periods for subtraction games.

In subtraction games the nim-values 0 and 1 are remarkably related by **Ferguson's Pairing Property** [Ferguson (1974); WW, 86, 422]: if s_1 is the least member of the subtraction-set, then

$$\mathcal{G}(n) = 1 \qquad \text{just if} \qquad \mathcal{G}(n - s_1) = 0.$$

Here and later "$\mathcal{G}(n) = v$" means that the nim-value of a heap of n beans is v.

2. Are all finite octal games ultimately periodic? Resolve any number of outstanding particular cases, e.g., **·6** (Officers), **·06**, **·14**, **·36**, **·64**, **·74**, **·76**, **·004**, **·005**, **·006**, **·007** (One-dimensional tic–tac–toe, Treblecross), **·016**, **·104**, **·106**, **·114**, **·135**, **·136**, **·142**, **·143**, **·146**, **·162**, **·163**, **·172**, **·324**, **·336**, **·342**, **·362**, **·371**, **·374**, **·404**, **·414**, **·416**, **·444**, **·454**, **·564**, **·604**, **·606**, **·644**, **·744**, **·764**, **·774**, **·776** and Grundy's Game (split a heap into two unequal heaps). Find a game with a period longer than 149459. Explain the structure of the periods of games known to be periodic.

[If the binary expansion of the kth code digit in the game with code $\mathbf{d}_0{\cdot}\mathbf{d}_1\mathbf{d}_2\mathbf{d}_3\ldots$ is

$$\mathbf{d}_k = 2^{a_k} + 2^{b_k} + 2^{c_k} + \ldots,$$

where $0 \le a_k < b_k < c_k < \ldots$, then it is legal to remove k beans from a heap, provided that the rest of the heap is left in exactly a_k or b_k or c_k or $\ldots$ non-empty heaps. See WW, 81–115 and Gangolli & Plambeck (1989).]

Table 2 exhibits some specimen games. Anil Gangolli & Thane Plambeck established the ultimate periodicity of four octal games which were previously unknown:

·**16** has period 149459 (a prime!), the last exceptional value being $\mathcal{G}(105350) = 16$. The game ·**56** has period 144 and last exceptional value $\mathcal{G}(326639) = 26$. The games ·**127** and ·**376** each have period 4 (with cycles of values 4, 7, 2, 1 and 17, 33, 16, 32 respectively) and last exceptional values $\mathcal{G}(46577) = 11$ and $\mathcal{G}(2268247) = 42$.

3. Examine some hexadecimal games. Obtain conditions for arithmetic periodicity.

[**Hexadecimal games** are those with code digits $\mathbf{d}_k$ in the interval from **0** to **15**. See WW, 115–116.]

The last three games in Table 2 are hexadecimal games with $\mathbf{d}_k \leq$ **15**= **F**. Such games may be arithmetically periodic, but most of the known ones are fairly trivial: ·**8** is a first cousin of Triplicate Nim, ·**A**, ·**B**, ·**D** and ·**F** are essentially She-Loves-Me-She-Loves-Me-Not, while ·**1A** and ·**1B**, after a couple of exceptional values, display period 2 and saltus 1. Kenyon's Game, ·**3F** in Table 3, whose saltus of 3 countered a conjecture of Guy & Smith that it should always be a power of two, is a little more interesting, but there must be many others such, which are possibly accessible to present-day computers.

4. Extend the analysis of **Domineering** to larger boards. In an earlier version of this paper it was proposed to find the values of 4×4 and 4×5 boards. David Wolfe has found these to be respectively $\pm\{0, \{2|0, 2+_2 \,||\, 2|0, -_2\} \,||\, 0\}$ and 1. Berlekamp has shown than the value of a $4 \times n$ board, if n is odd, is twice that of the corresponding $2 \times n$ board plus a correction term of lower temperature.

[Left & Right take turns to place dominoes on a checker-board. Left orients her dominoes North-South and Right orients his East-West. Each domino exactly covers two squares of the board and no two dominoes overlap. A player unable to play loses. See Berlekamp (1988) and WW, 495–498.]

5. Analyze endgame positions in the game of **Go**.

[Compare Berlekamp (1988), and his book with Wolfe.]

Code	Game	Nim-sequence
·77	**Kayles.** Knock down one skittle, or two contiguous skittles, from a row. [Dudeney (1907), Loyd (1914)].	Ultimate period $p = 12$, $\dot{4}1281472182\dot{7}$ except $n = 0, 3, 6, 9, 11, 15, 18, 21,$ 22,28,34,39,57,70, nim-value is resp. 0,3,3,4,6,7,3,4,6,5,6,3,4,6.
·137	**Dawson's Chess.** $3 \times n$ board. White & Black pawns on ranks 1 & 3. Capturing obligatory. Looks partizan but isn't. [Dawson (1934, 1935)].	$\dot{8}112031103322445593301130211045 37\dot{4}$ except 0 for $n = 0, 14, 34$ & 2 for $n = 16, 17, 51$. $p = 34$.
·07	**Dawson's Kayles.** Knock down 2 skittles, but only if they're contiguous.	As for **·137**, but shifted one term: 0011203 ... in place of 011203 ...
·6	**Officers.** Take 1 counter from any *longer* row. [B. Descartes (1953)].	No period found. $\mathcal{G}(10342) = 256$.
·007	**Treblecross.** One-dimensional tic-tac-toe. [WW, 93–94].	No pattern yet found.
·077	**Duplicate Kayles.** Knock down 2 or 3 contiguous skittles [Guy & Smith (1956)].	$p = 24$. Kayles with each nim-value repeated, 00112233114433 ...
·7777	**Double Kayles.** Take up to 4 beans from a heap; leave rest in at most 2 heaps. [Guy & Smith (1956); WW, 98].	$p = 24$. Kayles with each nim-value g replaced by the pair $2g, 2g + 1$ or $2g + 1, 2g$ (according to a certain rule), 01234567328976543289 45 ...
·156	See Kenyon (1967).	$p = 349$.
·165	See Austin (1976).	$p = 1550$.
4·3	**Lasker's Nim**	0124356879 ... $p = s = 4$.
·8	(1st cousin of) **Triplicate Nim.** Take 1 from heap, rest left in exactly 3 non-empty heaps.	Arithmetically periodic, $p = 3$, saltus = 1. 0000111222333444 ...
·3F	(F=15) **Kenyon's Game.** Take 1 from heap or take 2 & leave rest in any number of heaps up to 3. [Kenyon (1967)].	$p = 6$, $s = 3$ 01201234534567 86789 ...
·E	(E =14) Take 1, leave rest in just 1, 2 or 3 heaps.	001234153215826514 ... $\mathcal{G}(246) = 128$. No known pattern.

Table 2 Some sample take-and-break games.

6. In an earlier edition of this paper I asked if **Go-Moku** (Five-in-a-row, Go-Bang, Pegotty) is a win for the first player. An affirmative answer for the unrestricted version and for the version where 6-in-a-row doesn't count as a win was given by Allis, van den Herik & Huntjens (1992).

7. Complete the analysis of impartial **Eatcakes** (WW, 269, 271, 276–277).

[Eatcakes is an example of a **join** or **selective compound** of games. Each player plays in **all** the component games. It is played with a number of rectangles, $m_i \times n_i$; a move is to remove a strip $1 \times n_i$ or $m_i \times 1$ from each rectangle, either splitting it into two rectangles, or reducing the length or breadth by one. Winner removes the last strip.]

For fixed breadth the remoteness becomes constant when the length is sufficiently large. But 'sufficiently large' seems to be an increasing function of the breadth and doesn't, in the hand calculations already made, settle into any clear pattern. Perhaps computer calculations will reveal something.

8. Complete the analysis of **Hotcakes** (WW, 279–282).

[Also played with integer-sided rectangles, but as a **union** or **selective compound** in which each player moves in **some** of the components. Left cuts as many rectangles vertically along an integer line as she wishes, and then rotates one from each pair of resulting rectangles through a right angle. Right cuts as many rectangles as he wishes, horizontally into pairs of integer-sided rectangles and rotates one rectangle from each pair through a right angle. The **tolls** for rectangles with one dimension small are understood, but much remains to be discovered.]

9. Develop a misère theory for unions of partizan games.

[In a union of two or more games, you move in as many component games as you wish. In misère play, the last player *loses*.]

10. Extend the analysis of **Squares Off** (WW, 299).

[Played with heaps of beans. A move is to take a perfect square (> 1) number of beans from any number of heaps. Heaps of 0, 1, 2 or 3 cannot be further reduced. A move leaving a heap of 0 is an overriding win for the player making it. A move leaving 1 is an overriding win for Right, and one leaving 2 is an overriding win for Left. A move leaving 3 doesn't end the game unless all other heaps are of size 3, in which case the last player wins.]

11. Extend the analysis of **Top Entails** (WW, 376–377).

[Played with stacks of coins. Either split a stack into two smaller ones, or remove the top coin from a stack. In the latter case your opponent's move must use the same stack. Last player wins. Don't leave a stack of 1 on the board, since your opponent must take it and win, since it's now your turn to move in an empty stack!]

Julian West wrote a program to check a student's work and calculated the first 38,000 values. He found loony positions at 2403 coins, 2505 coins, and 33,243 coins. The authors of *Winning Ways* did not know of a loony stack of more than 3 coins. These results are typical of the apparently quite unpredictable nature of combinatorial games, even when they have quite simple rules.

12. Extend the analysis of **All Square** (WW, 385).

[This game involves **complimenting moves** after which the same player has an extra **bonus move**. Note that this happens in Dots-and-Boxes when a box is completed. All Square is played with heaps of beans. A move splits a heap into two smaller ones. If both heap sizes are perfect squares, the player must move again: if he can't he loses!]

13. Extend the misère analysis of various octal games, e.g., **Officers**, **Dawson's Chess**, ..., and of **Grundy's Game**.

[See Allemang (1984), WW, 411–421. William L. Sibert made a breakthrough by completing the analysis of misère Kayles; see the post-script to Sibert & Conway (1990). Plambeck (1992) has used their method to analyse a few other games, but there's a wide open field here.]

14. **Moebius**, when played on 18 coins has a remarkable pattern. Is there any trace of pattern for larger numbers of coins? Can any estimate be made for the rate of growth of the nim-values?

[See Coin-turning games in WW, 432–435; and Vera Pless's lecture in PSAM **43**. Played with a row of coins. A move turns 1, 2, 3, 4 or 5 coins, of which the rightmost must go from heads to tails (to make sure the game satisfies the Ending Condition). The winner is the player who makes all coins tails.]

15. **Mogul** has an even more striking pattern when played on 24 coins, which has some echoes when played on 40, 56 or 64 coins. Thereafter, is there complete chaos?

[See references for problem 14. A move turns 1, 2, ..., 7 coins.]

16. Find an analysis of **Antonim** with four or more coins (WW, 459–462).

[Played with coins on a strip of squares. A move moves a coin from one square to a smaller-numbered square. Only one coin to a square, except that square zero can have any number of coins. It is known that (a, b, c) is a $\mathcal{P}$-position in Antonim just if $(a+1, b+1, c+1)$ is a $\mathcal{P}$-position in Nim, but for more than 3 coins much remains to be discovered.]

17. Extend the analysis of **Kotzig's Nim** (WW, 481–483). Is the game eventually periodic in terms of the length of the circle for every finite move set? Analyze the misère version of Kotzig's Nim.

[Players alternately place coins on a circular strip, at most one coin on a square. Each coin must be placed m squares clockwise from the previously placed coin, provided m is in the given move set, and provided the square is not already occupied. The complete analysis is known only for a few small move sets.]

There is a paper by Aviezri S. Fraenkel, Alan Jaffray, Anton Kotzig & Gert Sabidussi (1995), but a great deal remains to be discovered.

18. Obtain asymptotic estimates for the proportions of $\mathcal{N}$-, $\mathcal{O}$- and $\mathcal{P}$-positions in Epstein's **Put-or-Take-a-Square** game (WW, 484–486).

[Played with one heap of beans. At each turn there are just two options, to take away or add the largest perfect square number of beans that there is in the heap. 5 is a $\mathcal{P}$-position, because 5 ± 4 are both squares; 2 and 3 are $\mathcal{O}$-positions, a win for neither player, since the best play is to go from one to the other, and not to 1 or 4 which are $\mathcal{N}$-positions.]

19. Simon Norton's game of **Tribulations** is similar to Epstein's game, but squares are replaced by triangular numbers. Norton conjectures that there are no $\mathcal{O}$-positions, and that the $\mathcal{N}$-positions outnumber the $\mathcal{P}$-positions in golden ratio. True up to 5000 beans.

20. Complete the analysis of **D.U.D.E.N.E.Y.**

[Played with a single heap of beans. Either player may take any number of beans from 1 to Y, except that the immediately previous move mustn't be repeated. When you can't move you lose. Analysis easy for Y even, and known (WW, 487–489) for 53/64 of the odd values of Y.]

Marc Wallace and Alan Jaffray have made a little progress here, but is the situation one in which there is always a small fraction of cases remaining, no matter how far the analysis is pursued?

21. Schuhstrings is the same as D.U.D.E.N.E.Y., except that a deduction of zero is also allowed, but cannot be immediately repeated (WW, 489–490).

22. Analyze Nim in which you are not allowed to repeat a move. There are at least five possible forms (assume that b beans have been taken from heap H):

medium local: b beans may not be taken from heap H until some other move is made in heap H.

short local: b beans may not be taken from heap H on the next move.

long local: b beans may never again be taken from heap H.

short global: b beans may not be taken from any heap on the next move.

long global: b beans may never again be taken from any heap.

23. Burning the candle at both ends. John Conway & Aviezri Fraenkel ask us to analyze Nim played with a row of heaps. A move may only be made in the leftmost or in the rightmost heap. When a heap becomes empty, then of course its neighbor becomes the end heap.

24. Continue the analysis of **The Princess and the Roses** (WW, 490–494).

[Played with heaps of beans. Take one bean, or two beans, one from each of two different heaps. The rules seem trivially simple, but the analysis takes on remarkable ramifications.]

25. Analyze the Conway-Paterson game of **Sprouts** with 7 or more spots, or the misère form with 5 or more spots (WW, 564–568).

[A move joins two spots, or a spot to itself by a curve which doesn't meet any other spot or previously drawn curve. When a curve is drawn, a new spot must be placed on it. The valence of any spot must not exceed three. Since this question was asked, Daniel Sleator of A.T.&T. Bell Laboratories has pushed the normal analysis to about 10 spots and the misère analysis to about 8.]

26. Extend the analysis of **Sylver Coinage** (WW, 575–597).

[Players alternately name different positive integers, but may not name a number which is the sum of previously named ones, with repetitions allowed. Whoever names 1 loses. See Section 3. of Richard Nowakowski's chapter of PSAM **43**.]

27. Extend the analysis of **Chomp** (WW, 598–599).

[Players alternately name divisors of N, which may not be multiples of previously named numbers. Whoever names 1 loses. David Gale offers a prize of US$100.00 for the first complete analysis of 3D-Chomp, i.e., where N has three distinct prime divisors, raised to arbitrarily high powers.]

28. Extend Uléhla's or Berlekamp's analysis of von Neumann's game from diforests to directed acyclic graphs (WW, 570–572; Uléhla [1980]). Since Chomp and the superset game [Gale & Neyman (1982)] can be described by directed acyclic graphs but not by diforests, the proposed extension could presumably throw some light on these two unsolved games.

29. Prove that Black doesn't have a forced win in Chess.

30. A King and Rook versus King problem. Played on a quarter-infinite board, with initial position WKa1, WRb2 and BKc3. Can White win? If so, in how few moves? It may be better to ask, "what is the smallest board (if any) that White can win on if Black is given a win if he walks off the North or East edges of the board?" Is the answer 9×11? In an earlier edition of this paper I attributed this problem to Simon Norton, but it was proposed as a kriegsspiel problem, with unspecified position of the WK, and with W to win with probability 1, by Lloyd Shapley around 1960.

31. David Gale's version of Lion and Man. L and M are confined to the non-negative quadrant of the plane. They move alternately a distance of at most one unit. For which initial positions can L catch M?

Conjecture: L wins if he is northeast of M. This condition is clearly necessary, but the answer is not known even if M is at the origin and L is on the diagonal.

Variation. Replace quadrant by wedge-shaped region.

32. Gale's Vingt-et-un. Cards numbered 1 through 10 are laid on the table. L chooses a card. Then R chooses cards until his total of chosen cards exceeds the card chosen by L. Then L chooses until her cumulative total exceeds that of R, etc. The first player to get 21 wins. Who is it?

[As posed here it is not clear if the object is to get 21 exactly or 21-or-more. Jeffery Magnoli, a student of Julian West, thought that the latter rule was the more interesting and found a first-player win in 6-card Onze and in 8-card Dix-sept.]

33. Subset Take-away. Given a finite set, players alternately choose proper subsets subject to the rule that once a subset has been chosen, none of *its* subsets may be chosen subsequently by either player. Last player wins.

[David Gale conjectures that it's a second player win—this is true for sets of less than six elements.]

34. R. B. Eggleton & A. S. Fraenkel ask for a theory of **Cannibal Games** or an analysis of special families of positions. They are played on an arbitrary finite digraph. Place any numbers of "cannibals" on any vertices. A move is to select a cannibal and move it along a directed edge to a neighboring vertex. If this is occupied, the incoming cannibal eats the whole population (**Greedy Cannibals**) or just one cannibal (**Polite Cannibals**). A player unable to move loses. Draws are possible. A partizan version can be played with cannibals of two colors, each eating only the opposite color.

35. Welter's Game on an arbitrary digraph. Place a number of monochromatic tokens on distinct vertices of a directed acyclic graph. A token may be moved to any *unoccupied* immediate follower. Last player wins. Make a dictionary of $\mathcal{P}$-positions and formulate a winning strategy for other positions.

36. Restricted Positrons & Electrons. Place a number of Positrons (Pink tokens) and Electrons (Ebony tokens) on distinct vertices of a Welter strip. Any particle can be moved by either player leftward to any square u provided that u is either unoccupied or occupied by a particle of the opposite type. In the latter case, naturally, both particles become annihilated (i.e., they are removed from the strip), as physicists tell us positrons and electrons do. Play ends when the excess particles of one type over the other are jammed in the lowest positions of the strip. Last player wins. Formulate a winning strategy for those positions where one exists. Note that if the particles are of one type only, this is Welter's Game. As a strategy is known for Misère Welter [WW, 480–481] it may not be unreasonable to ask for a misère analysis as well.

37. General Positrons & Electrons. As 36. but played on an arbitrary digraph. Last player wins.

38. Fulves's Merger. Karl Fulves, Box 433, Teaneck NJ 07666, sent the following game. Start with heaps of 1, 2, 3, 4, 5, 6 and 7 beans. Two players alternately transfer any number of beans from one heap to another, except that beans may not be transferred from a larger to a smaller heap. The player who makes all the heaps *even* in size is the winner.

The total number of beans remains constant, and is even (28 in this case, though one is interested in even numbers in general: a similar game can be played in which the total number is odd and the object is to make all the heaps odd in size). As the moves are nondisjunctive the Sprague-Grundy theory is not likely to be of help. Fulves's particular game can be solved using Steinhaus's remoteness function.

The number of odd heaps is even. If the number of odd heaps is two, there is an immediate win: such a position is an $\mathcal{N}$-position of remoteness 1. [In normal play, $\mathcal{P}$-positions have even remoteness, and $\mathcal{N}$-positions have odd remoteness. To calculate the remoteness of a position:

Add one to the least even remoteness of any option, if there is an option of even remoteness, else

Add one to the greatest odd remoteness among the options, else, if there are no options,

The remoteness is 0. WIN QUICKLY – LOSE SLOWLY.]

If the number of odd heaps is zero, the position is terminal. All other $\mathcal{P}$-positions must contain at least four odd heaps. Some examples, with their even remotenesses, r, are:

$\mathcal{P}$-position	r	$\mathcal{P}$-position	r	$\mathcal{P}$-position	r
1 1 1 25	2	1 1 1 1 1 23	4	1 1 1 1 2 4 18	10
1 3 3 21	4	1 1 1 2 2 21	6	1 1 1 1 3 5 16	12
1 5 5 17	6	1 1 1 3 3 19	8	1 1 1 1 6 8 10	18
1 7 7 13	8	1 1 1 4 4 17	10	1 1 1 2 3 8 12	14
1 9 9 9	10	1 1 1 5 5 15	12	1 1 1 2 5 9 9	20
3 5 9 11	10	1 1 1 6 6 13	14	1 1 1 3 6 7 9	22
		1 1 1 7 7 11	16	1 1 2 2 2 3 17	10
1 1 2 5 19	6	1 1 1 8 8 9	18	1 1 2 2 4 5 13	14
1 1 3 6 17	8	1 1 2 3 7 14	12	1 1 2 3 5 5 11	18
1 1 4 7 15	10	1 1 2 4 9 11	14	1 1 2 4 6 7 7	20
1 2 3 9 13	12	1 1 3 5 8 10	18	1 1 3 3 3 4 13	16
1 3 5 5 14	12	1 2 2 3 5 15	12	1 1 3 4 4 7 8	24
1 5 6 7 9	14	1 2 2 7 7 9	14	1 1 4 4 4 5 9	24
2 3 3 3 17	8	1 2 3 3 8 11	14	1 2 2 3 3 3 14	16
2 5 5 5 11	14	1 2 5 5 7 8	16	1 2 3 3 5 7 7	22
2 5 7 7 7	14	1 3 3 3 3 15	14		
3 3 4 5 13	14	1 3 3 4 7 10	16		
3 3 7 7 8	16	1 3 3 6 6 9	18		
		1 3 4 4 5 11	18		
		2 2 3 3 9 9	16		
		2 3 3 5 5 10	18		
		3 3 3 5 6 8	20		

Here is a winning strategy from Fulves's starting position:

Rem	Position	
21	1 2 3 4 5 6 7	
20	1 1 2 4 6 7 7	The only other winning move is to
	/ \	1 2 3 3 5 7 7, and takes 2 moves longe
19	1 1 1 4 6 7 8 1 1 1 2 6 7 10 ··· *	

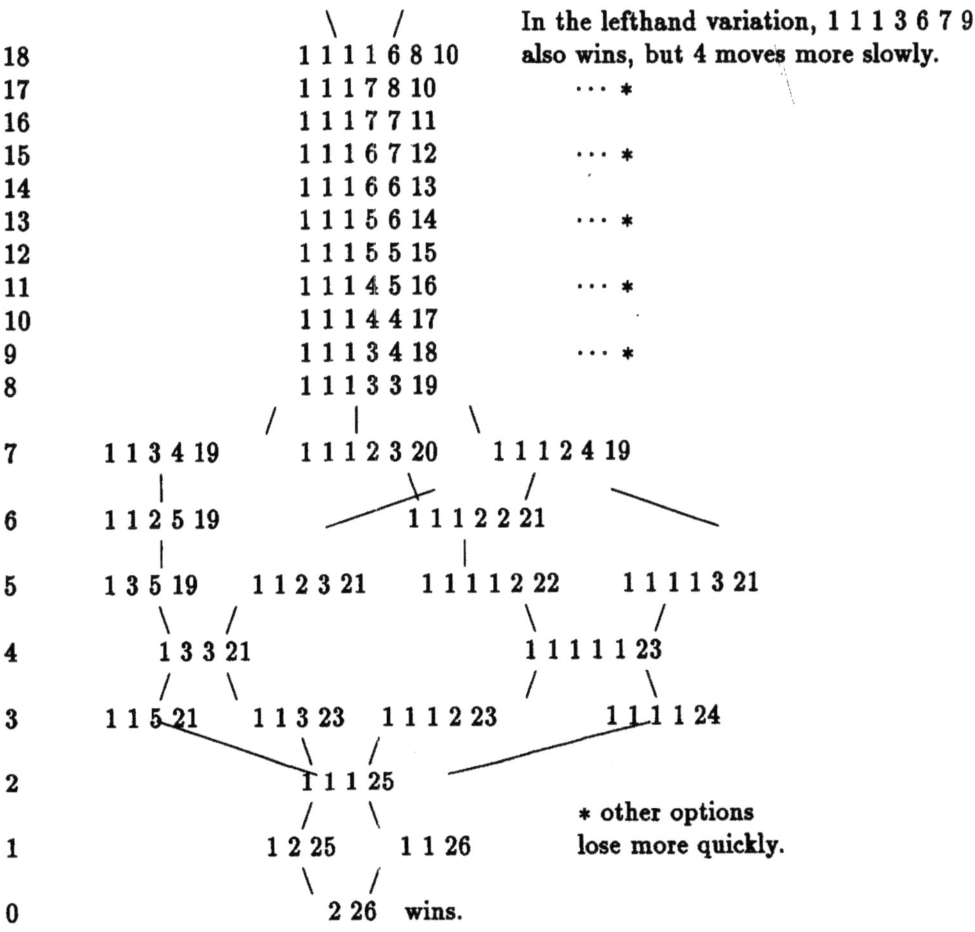

Richard Nowakowski has investigated some variants of Fulves's game:

1. Take a row of consecutive coins. A move is to take one or more coins from a heap and put them on an adjacent heap of coins provided that the second heap is at least as large as the first. Assume that we start with a row of adjacent coins — each heap consisting of one coin. The nim values are periodic with period 7 and the values are 0 1 2 0 3 1 0 — no exceptional values. To prove this one needs to know the nim values of the game with the first heap having two coins and also the game with both end heaps having two coins. The latter

is periodic, period 7 — 2 1 0 0 1 3 0 — no exceptions. The former also has period 7 — 5 5 4 2 5 6 4 — but this does not start until the game 2.1^{15}

2. Take a row of three consecutive heaps — sizes i, j, k. It is not too hard to see that

a. if $j > i$ and $j > k$ then the nim-value of the position $i.j.k$ is $i \overset{*}{+} k$.

b. if $i = j$ and $j \geq k$, then the nim-value is $i + k$.

c. if either $i > j$ or $k > j$, then the nim-value is j.

The proof follows easily by considering the tables of nim-values for

$$j = 0, 1, 2, 3, \ldots$$

The misère version of 1. might be interesting.

39. Sowing or Mancala Games

Kraitchik describes Ruma, which he attributes to Mr. Punga. Bell & Cornelius list Congklak, from Indonesia; Mankal'ah L'ib Al-Ghashim (the game of the unlearned); Leab El-Akil (the game of the wise or intelligent); Kiarabu, from Africa; as well as some solitaire games based on a similar idea. Botermans et al. describe Mefuhva from Malawi and Kiuthi from Kenya. Many of these games go back for thousands of years, but several should be susceptible to present day methods of analysis.

Conway starts with a line of heaps of beans. A typical move is to take (some of) a heap of size N and do something with it that depends on the game and on N. He regards the nicest move as what he calls the **African move** in which the beans are 'sowed' onto successive heaps, but subject to the condition that the last bean must land in a nonempty place. Beans are sowed to the right if you are Left, to the left if you are Right, or either way if you're playing impartially.

In the partizan version, the position 1 has value 0, of course; the position $1.1 = \{0.2 \mid 2.0\}$ also has value $\{\ \mid\ \} = 0$; and so does $1.1.1 = \{1.0.2, 2.1 \mid 1.2, 2.0.1\}$, since $2.1 = \{\ \mid 3.0\}$, $3.0 = \{\ \mid\ \}$ and $1.0.2 = \{\ \mid 2.1\}$, so that '3' has value 0, 2.1 has value -1, 1.0.2 value -2, 1.1.1 value $\{-2, -1 \mid 1, 2\} = 0$.

40. Chess. Noam Elkies has found endgames with values 0, 1, $\frac{1}{2}$, $*$, $\uparrow$, $\uparrow *$, $\Uparrow *$, etc. See also problems **29** and **30** above.

41. Sequential compounds of games have been studied by Stromquist & Ullman. They mention a more general compound. Let $(P, <)$ be a finite poset and for each $x \in P$ let G_x be a game. Consider a game $G(P)$ played as follows. Moves are allowed in any single component G_x provided that no legal moves remain in any component G_y with $y > x$. A player unable to move loses. The sequential compound is the special case when $(P, <)$ is a chain (or linear order). The sum or disjunctive compound is the case where $(P, <)$ is an antichain. They have no coherent theory of games $G(P)$ for arbitrary posets. They list some more specific problems which may be more tractable.

42. Beanstalk and **Beans-Don't-Talk** are games invented by John Isbell and by John Conway. Beanstalk is played between Jack and the Giant. The Giant chooses a positive integer, n_0. Then J. and G. play alternately n_1, n_2, n_3, ... according to the rule $n_{i+1} = n_i/2$ if n_i is even, $= 3n_i \pm 1$ if n_i is odd; i.e. if n_i is even, there's only one option, while if n_i is odd there are just two. The winner is the person moving to 1. If the Giant chooses an odd number > 1, can Jack always win? Not by using the Greedy Strategy (always descend when it's safe to do so, e.g., play the moves that are underlined in the game below), as may be seen from $n_0 = 7$ (J.'s moves in parentheses).

7 (22) 11 (34) 17 (50) 25 (74) 37 (110) 55 (166) 83 (248) 124 (62) 31 (94) 47 (142 71 (214) 107 (322) 161 (482) 241 (722) 361 (1082) 541 (1624) 812 (406) 203 (608) 304 (152) 76 (38) 19 (56) 28 (14) 7 ...

[to win, Jack should answer 83 with (250!) 125 (374!) but it still needs care.]

In Beans-Don't-Talk, the move is from n to $(3n \pm 1)/2^*$ where 2^* is the highest power of two dividing the numerator; the winner is still the person moving to 1. Are there any drawn positions? There are certainly drawn **plays**, e.g., 7 (5) 7 (5) ..., but 5 is an $\mathcal{N}$-position because there is the immediate winning option $(5\times3+1)/2^4 = 1$, and 7 is a $\mathcal{P}$-position since the other option $(7\times3+1)/2 = 11$ is met by $(11\times3-1)/2^5 = 1$. What we want to know is: *are there any* $\mathcal{O}$-**positions** (positions of infinite remoteness)?

[For a definition of **remoteness** see **38.** above. There are several unanswered questions about the remotenesses of positions in these two games. Remoteness may also be the best tool we have for problems **18** and **19** above.]

REFERENCES

[1] Dean Thomas Allemang, Machine Computation with Finite Games, M.Sc. thesis, Cambridge University, 1984.

[2] L. V. Allis, H. J. van der Herik & M. P. H. Huntjens, Go-Moku and Threat-Space Search, CS-report, Univ. of Limburg, 1992

[3] Richard Bruce Austin, Impartial and Partisan Games, M.Sc. thesis, University of Calgary, 1976.

[4] Robbie Bell & Michael Cornelius, *Board Games Round the World*, Cambridge Univ. Press, 1988.

[5] Elwyn Ralph Berlekamp, Blockbusting and Domineering, *J. Combin. Theory Ser. A*, **49**(1988) 67–116.

[6] Elwyn Ralph Berlekamp, John Horton Conway & Richard Kenneth Guy, *Winning Ways for your Mathematical Plays*, Academic Press, 1982.

[7] Elwyn Ralph Berlekamp & David Wolfe, *Mathematical Go*, A. K. Peters, 1994.

[8] Jack Botermans, Tony Burrett, Pieter van Delft & Carla van Splunteren, *The World of Games*, Facts on File, New York & Oxford, 1989.

[9] Charles L. Bouton, Nim, a game with a complete mathematical theory, *Ann. of Math., Princeton* (2), **3**(1901–02), 35–39.

[10] John Horton Conway, *On Numbers and Games*, London Math. Soc. Monograph **6**, Academic Press, 1976.

[11] John Horton Conway & Neil J. A. Sloane, Lexicographic codes: error-correcting codes from game theory, *IEEE Trans. Info. Theory*, **IT–32** No. 3 (May, 1986) 337–348.

[12] Robert T. Curtis, A new combinatorial approach to M_{24}, *Math. Proc. Cambridge Philos. Soc.*, **79**(1976) 25–42.

[13] Robert T. Curtis, The maximal subgroups of M_{24}, *Math. Proc. Cambridge Philos. Soc.*, **81**(1977) 185–192.

[14] Robert T. Curtis, The Steiner system $S(5,6,12)$, the Mathieu group M_{12} and the "kitten", in Computational Group Theory (Durham, 1982) 353–358, Academic Press, London & New York, 1984).

[15] Thomas Rayner Dawson, Problem 1603, *Fairy Chess Review*, (Dec. 1934) p. 94.

[16] Thomas Rayner Dawson, *Caissa's Wild Roses*, 1935, p. 13.

[17] Blanche Descartes, Why are series musical? *Eureka*, **16**(1953) 18–20; reprint *ibid.* **27**(1964) 29–31.

[18] Henry Ernest Dudeney, *The Canterbury Puzzles and other Curious Problems*, Nelson, London, 1907; reprinted Dover, New York, 1958, pp. 118, 220.

[19] Noam D. Elkies, On numbers and endgames – work in progress, July 1994.

[20] Thomas S. Ferguson, On sums of graph games with the last player losing, *Internat. J. Game Theory*, **3**(1974) 159–167; MR **52** #5046.

[21] Aviezri S. Fraenkel, Selected bibliography on combinatorial games and some related material, in Richard K. Guy (editor), *Combinatorial Games*, *Proc. Sympos. Appl. Math.*, **43**(1991) 191–226.

[22] Aviezri S. Fraenkel, Alan Jaffray, Anton Kotzig & Gert Sabidussi, Modular Nim, *Theoret. Comput. Sci.*, (1995) (to appear)

[23] Aviezri S. Fraenkel & Anton Kotzig, Partizan octal games: partizan subtraction games, *Internat. J. Game Theory*, **16**(1987) 145–154.

[24] David Gale & A. Neyman, Nim-type games, *Internat. J. Game Theory*, **11**(1982) 17–20.

[25] Anil Gangolli & Thane Plambeck, A note on periodicity in some octal games, *Internat. J. Game Theory* **18**(1989) 311–320.

[26] Patrick Michael Grundy, Mathematics and games, *Eureka*, **2**(1939) 6–8; reprint *ibid.* **27**(1964) 9–11.

[27] Patrick Michael Grundy & Cedric Austen Bardell Smith, Disjunctive games with the last player losing, *Proc. Cambridge Philos. Soc.*, **52**(1956), 527–533; MR **18**, 546.

[28] Richard Kenneth Guy, Twenty questions concerning Conway's Sylver Coinage, *Amer. Math. Monthly*, **83**(1976) 634–637.

[29] Richard Kenneth Guy, John Isbell's Game of Beanstalk and John Conway's Game of Beans-Don't-Talk, *Math. Mag.*, **59**(1986) 259–269 [William Jockusch, Carleton College MN has corrected the tentative tables on p. 262].

[30] Richard Kenneth Guy (editor), *Combinatorial Games, Proc. Sympos. Appl. Math.*, **43**, Amer. Math. Soc., Providence RI, 1991.

[31] Richard Kenneth Guy & Cedric Austen Bardell Smith, The G-values for various games, *Proc. Cambridge Philos. Soc.*, **52**(1956) 514–526; MR **18**, 546.

[32] John Charles Kenyon, Nim-like games and the Sprague-Grundy Theory, M.Sc. thesis, Univ. of Calgary, 1967.

[33] Donald Ervin Knuth, *Surreal Numbers*, Addison-Wesley, Reading MA, 1974.

[34] Maurice Kraitchik, *Mathematical Recreations*, Geo. Allen & Unwin, 1941; 2nd ed. Dover, 1953.

[35] Hendrik Willem Lenstra, Nim multiplication, Séminaire de Théorie des Nombres, exposé No. 11, Université de Bordeaux, 1977.

[36] Sam Loyd, *Cyclopedia of Tricks and Puzzles*, New York, 1914, p. 232.

[37] David Moews, Sums of games born on days 2 and 3, *Theor. Comput. Sci.*, **91**(1991) 119–128.

[38] Richard Joseph Nowakowski, Welter's Game, Sylver Coinage, Dots-and-Boxes, in Richard K. Guy (editor), *Combinatorial Games, Proc. Sympos. Appl. Math.*, **43**(1991) 191–226.

[39] Thane E. Plambeck, Daisies, Kayles, and the Sibert-Conway decomposition in misère octal games, *Theoret. Comput. Sci.*, **96**(1992) 361–388.

[40] Vera Pless, Games and codes, in Richard K. Guy (editor), *Combinatorial Games, Proc. Sympos. Appl. Math.*, **43**(1991) 191–226.

[41] Lloyd S. Shapley, The invisible chessboard (preprint).

[42] William L. Sibert & John H. Conway, Mathematical Kayles (with a postscript by Richard K. Guy), *Internat. J. Game Theory*, **20**(1992) 237–246.

[43] Roland Percival Sprague, Bemerkungen über ein spezielle Abelsche Gruppe, *Math. Z.* **51**(1947) 82–84; MR **9**, 330–331.

[44] Roland Percival Sprague, Über mathematische Kampfspiele, *Tôhoku Math. J.*, **41**(1935–36), 438–444; *Zbl.*, **13**, 290.

[45] Hugo Steinhaus, Definicje potrzebne do teorji gry i pościgu, *Myśl. Akad. Lwów*, **1** (1925) #1, 13–14; reprinted, Definitions for a theory of games and pursuit, *Naval Res. Logist. Quart.*, **7**(1960) 105–108.

[46] Walter Stromquist & Daniel Ullman, Sequential compounds of combinatorial games, *Theoret. Comput. Sci.*, **119**(1993) 311–321.

[47] J. Uléhla, A complete analysis of von Neumann's Hackendot, *Internat. J. Game Theory*, **8**(1980) 107–113.

[48] David Wolfe, Snakes in Domineering games, *Theoret. Comput. Sci.*, **119**(1993) 311–321.

[49] Y. Yamasaki, On misère Nim-type games, *J. Math. Soc. Japan*, **32**(1980) 461–475.

$(F,2)$–ROTATIONAL STEINER TRIPLE SYSTEMS

Zhike Jiang

Combinatorics and Optimization
University of Waterloo
Waterloo, Ontario, CANADA N2L 3G1

ABSTRACT

A Steiner triple system of order v is called (f,k)-rotational if it admits an automorphism consisting of f fixed points and k cycles of length $(v-f)/k$. In this paper we deal with the existence of such Steiner triple systems when $k=2, f>1$ and show that an $(f,2)$-rotational Steiner triple system of order v with $f>1$ exists if and only if (i) $v \equiv 1,3 \pmod 6$, (ii) $f \equiv 1,3 \pmod 6$, (iii) if $v-f \equiv 2 \pmod 4$ then $v \geq 3f$; if $v-f \equiv 0 \pmod 4$ then $v=f$ or $v \geq 3f-2$, with the exceptions of $v=3f-2$ when $f \equiv 1 \pmod{12}$, $v=21$ when $f=3$, $v=3f-2$ or $v=3f+10$ when $f \equiv 15 \pmod{24}$, $v=3f+4$ when $f \equiv 3 \pmod{12}$, $v=25$ when $f=7$, $v=3f-2$ when $f \equiv 7 \pmod{12}$, $v=27$ when $f=9$, $v=3f-2$ when $f \equiv 21 \pmod{24}$, $v=3f+10$ when $f \equiv 9 \pmod{24}$ and $v=3f+4$ when $f \equiv 9 \pmod{12}$.

1 INTRODUCTION

A *Steiner triple system* of order v, denoted by STS(v), is an ordered pair $(V,\mathcal{T})$, where V is a v-set and $\mathcal{T}$ a collection of 3-subsets of V, called *triples*, such that each pair of elements in V is contained in exactly one triple of $\mathcal{T}$. It is well known that an STS(v) exists if and only if $v \equiv 1,3 \pmod 6$. A *subsystem* of order u of an STS(v) $(V,\mathcal{T})$ is an STS(u) $(U,\mathcal{S})$ such that $U \subseteq V$, $\mathcal{S} \subseteq \mathcal{T}$. Doyen and Wilson [4] proved that an STS(v) with a subsystem STS(u) exists if and only if $v \equiv 1,3 \pmod 6$, $u \equiv 1,3 \pmod 6$ and $v=u$ or $v \geq 2u+1$.

An *automorphism* of an STS(v) on V is a permutation of V which preserves the triples. $\mathrm{STS}_\pi(v)$ denotes an STS(v) which admits a permutation π on the

C. J. Colbourn and E. S. Mahmoodian (eds.), Combinatorics Advances, 181–206.

ground set as one of its automorphisms. The existence of Steiner triple systems with a given permutation as its automorphism has been the interest of many authors. In this paper, we restrict π to be such a permutation on a v-set V that, in its cycle representation, consists of f fixed points and k cycles of length $(v-f)/k$ and call an $STS_\pi(v)$ on V (f,k)-*rotational*. Solutions to the existence problem of an (f,k)–rotational STS(v) were first given when $f=0, k=1$, as the existence of cyclic Steiner triple systems [8] (see also [7], [10], [12]), when $f=1, k=1,2,3,4,6$, as that of 1–, 2–, 3–, 4– and 6–rotational Steiner triple systems [1], [9], and when $f=1, k=(v-1)/2$, as that of reverse Steiner triple systems [3], [11], [14]. Then further work [6, 5] dealt with the solutions for $f>1$ and $k=(v-f)/2$, $f=3$ and $k=2n, 3n$ (as the existence of $2n$–, $3n$–near-rotational Steiner triple systems). The reader is referred to [2] for a survey of this problem. In particular, we mention the following results which are used later in this paper:

A (0, 1)-rotational (namely, cyclic) STS(v) exists if and only if $v \equiv 1,3 \pmod 6$, $v \neq 9$ [8].

A (1, 1)-rotational (namely, 1-rotational) STS(v) exists if and only if $v \equiv 3,9 \pmod{24}$ [9].

A (1, 2)-rotational (namely, 2-rotational) STS(v) exists if and only if $v \equiv 1,3,7,9,15,19 \pmod{24}$ [9].

A (3, 2)-rotational (namely, 2-near-rotational) STS(v) exists if and only if $v \equiv 1,3 \pmod 6$, $v \neq 13,21$ [5].

2 SKOLEM SEQUENCES

Classical Skolem k–sequences, also known as (A,k)–, (B,k)–, (C,k)– and (D,k)–systems [10], were invented for constructing (cyclic) Steiner triple systems [12], [7], [10]. The terminology "Skolem k–sequence" first appeared in [9] and it was named after Th. Skolem, who was the first one to relate these sequences to Steiner triple systems. A classical Skolem k–sequence involves two sets $D(k)$ $(=\{1,2,\ldots,k\})$, $S(k)$ of distinct integers with $|S(k)|=2|D(k)|$ and is a partition of $S(k)$ into $|D(k)|$ ordered pairs (a_r,b_r) $(r\in D(k))$ such that $b_r-a_r=r$ for all r. The essential idea for using such a Skolem k–sequence (a_r,b_r) $(r\in D(k))$ in constructing a Steiner triple system is that this sequence can be converted to a set of triples (called *difference triples*) $\{r, k+a_r, k+b_r\}$

$(r \in D(k))$ which have the property that these triples are pairwise disjoint and form a partition of $D(k) \bigcup (k + S(k))$. In some cases (as in the constructions of the next section), a sequence like this is converted to a set of triples of other forms and, for this purpose, we sometimes allow repetitions of an integer in $D(k)$. Such a sequence is generally referred to a *generalized Skolem sequence.* To be precise, we give the following definition.

Definition 2.1 *A generalized Skolem sequence $(D(k), S(k))$, where $D(k)$ is a sequence of positive integers (repetitions allowed) and $S(k)$ a set of $2|D(k)|$ positive integers, is a sequence of $|D(k)|$ ordered pairs (a_r, b_r) $(r \in D(k))$ such that (i) $\bigcup_{r \in D(k)} \{a_r, b_r\} = S(k)$ and (ii) $b_r - a_r = r$ for all r.*

In this section, we establish a few more generalized Skolem sequences which, with those classical Skolem sequences as mentioned above, are used repeatedly in the constructions of next section. For the sake of simplicity, we first summarize the relevant generalized Skolem sequences and their existence conditions in Table 1 and we assign each of these sequences a letter in the first column of the table in accordance with tradition (An (E,k)–system and an (F,k)–system have appeared, separately, in [1] and [9]). Then we settle the existence of the newly found sequences in Theorems 2.2–2.5. We can always obtain the necessary conditions, by observing that $\sum_{r \in D(k)} a_r = \frac{1}{2}(\sum_{r \in D(k)} (a_r + b_r) - \sum_{r \in D(k)} r) = \frac{1}{2}(\sum_{r \in S(k)} r - \sum_{r \in D(k)} r)$ is an integer.

Theorem 2.2 *A (G,k)–system exists if and only if $k \equiv 1, 2 \pmod 4$.*

Proof. For the sufficiency we take the following pairs.

Suppose $k \equiv 1 \pmod 4$.

$k = 1$: (1, 2);

$k \geq 5$ and let $k = 4s + 1$ where $s \geq 1$:

$$\begin{array}{ll} (r, 4s + 2 - r) & r = 1, 2, \ldots, 2s, \\ (4s + 3 + r, 8s + 4 - r) & r = 1, 2, \ldots, s - 1 \ (s \geq 2), \\ (5s + 2 + r, 7s + 3 - r) & r = 1, 2, \ldots, s - 1 \ (s \geq 2), \end{array}$$

$(7s + 3, 7s + 4)$, $(4s + 2, 6s + 3)$, $(2s + 1, 6s + 2)$.

Table 1 Generalized Skolem Sequences and their Existence Conditions

Name Used	$D(k)$, $S(k)$	Existence (mod 4)	Reference
(A,k)–system	$(1,2,\ldots,k)$, $\{1,2,\ldots,2k\}$	$k \equiv 0,1$	[12]
(B,k)–system	$(1,2,\ldots,k)$, $\{1,2,\ldots,2k-1,2k+1\}$	$k \equiv 2,3$	[7]
(C,k)–system	$(1,2,\ldots,k)$, $\{1,2,\ldots,2k+1\} \setminus \{k+1\}$	$k \equiv 0,3$	[10]
(D,k)–system	$(1,2,\ldots,k)$, $\{1,2,\ldots,2k,2k+2\} \setminus \{k+1\}$	$k \equiv 1,2$ $k \neq 1$	[10]
(G,k)–system	$(1,2,\ldots,k)$, $\{1,2,\ldots,2k+1\} \setminus \{k+2\}$	$k \equiv 1,2$	
(H,k)–system	$(1,2,\ldots,k)$, $\{1,2,\ldots,2k,2k+2\} \setminus \{k+2\}$	$k \equiv 0,3$	
(I,k)–system	$(1,3,4,\ldots,k-2,k-1,k-1,k)$, $\{1,2,\ldots,2k\}$	$k \equiv 1,2$ $k \neq 1,2$	
(J,k)–system	$(1,3,4,\ldots,k-2,k-1,k,k)$, $\{1,2,\ldots,2k-1,2k+1\}$	$k \equiv 1,2$ $k \neq 1,2$	

Suppose $k \equiv 2 \pmod 4$.

$k = 2$: (1, 2), (3, 5);

$k \geq 6$ and let $k = 4s + 2$ where $s \geq 1$:

$$\begin{array}{ll}
(r, 4s+4-r) & r = 1, 2, \ldots, 2s+1, \\
(4s+4+r, 8s+5-r) & r = 1, 2, \ldots, s-1 \ (s \geq 2), \\
(5s+3+r, 7s+4-r) & r = 1, 2, \ldots, s-1 \ (s \geq 2),
\end{array}$$

$(7s+4, 7s+5)$, $(6s+4, 8s+5)$, $(2s+2, 6s+3)$. □

Theorem 2.3 *An (H,k)–system exists if and only if* $k \equiv 0, 3 \pmod 4$.

Proof. For the sufficiency we take the following pairs.

Suppose $k \equiv 0 \pmod 4$.

$k = 4$: (2, 3), (8, 10), (4, 7), (1, 5);

$k \geq 8$ and let $k = 4s$ where $s \geq 2$:

$$\begin{array}{ll}
(r, 4s+1-r) & r = 1, 2, \ldots, s-1, \\
(s+1+r, 3s+2-r) & r = 1, 2, \ldots, s-1, \\
(4s+2+r, 8s-r) & r = 1, 2, \ldots, 2s-3,
\end{array}$$

$(s, s+1)$, $(8s, 8s+2)$, $(4s+1, 6s+2)$, $(2s+2, 6s)$, $(2s+1, 6s+1)$.

Suppose $k \equiv 3 \pmod 4$.

$k = 3$: (2, 3), (6, 8), (1, 4);

$k \geq 7$ and let $k = 4s - 1$ where $s \geq 2$:

$$\begin{array}{ll}
(r, 4s-r) & r = 1, 2, \ldots, s-1, \\
(s+1+r, 3s+1-r) & r = 1, 2, \ldots, s-2 \ (s \geq 3), \\
(4s+1+r, 8s-2-r) & r = 1, 2, \ldots, 2s-3,
\end{array}$$

$(s, s+1)$, $(8s-2, 8s)$, $(2s, 4s)$, $(2s+2, 6s-1)$, $(2s+1, 6s)$. □

Theorem 2.4 *An (I,k)–system exists if and only if $k \equiv 1,2 \pmod 4$, $k \neq 1,2$.*

Proof. For the sufficiency we take the following pairs.

Suppose $k \equiv 1 \pmod 4$.

$k = 5$: (2, 3), (7, 10), (4, 8), (5, 9), (1, 6);

$k \geq 9$ and let $k = 4s+1$ where $s \geq 2$:

$$\begin{array}{ll} (r, 4s+3-r) & r = 1,2,\ldots,2s, \\ (4s+2+r, 8s+2-r) & r = 1,2,\ldots,s-2 \ (s \geq 3), \\ (5s+r, 7s+2-r) & r = 1,2,\ldots,s-1, \end{array}$$

$(7s+2, 7s+3)$, $(6s, 8s+2)$, $(2s+1, 6s+1)$, $(2s+2, 6s+2)$.

Suppose $k \equiv 2 \pmod 4$.

$k \geq 6$ and let $k = 4s+2$ where $s \geq 1$:

$$\begin{array}{ll} (r, 4s+3-r) & r = 1,2,\ldots,s, \\ (s+2+r, 3s+3-r) & r = 1,2,\ldots,s-1 \ (s \geq 2), \\ (4s+3+r, 8s+5-r) & r = 1,2,\ldots,2s-1, \end{array}$$

$(s+1, s+2)$, $(4s+3, 6s+4)$, $(2s+2, 6s+3)$, $(2s+3, 6s+5)$. □

Theorem 2.5 *A (J,k)–system exists if and only if $k \equiv 1,2 \pmod 4$, $k \neq 1,2$.*

Proof. For the sufficiency we take the following pairs.

Suppose $k \equiv 1 \pmod 4$.

$k \geq 5$ and let $k = 4s+1$ where $s \geq 1$:

$$\begin{array}{ll} (r, 4s+1-r) & r = 1,2,\ldots,s-1 \ (s \geq 2), \\ (s-1+r, 3s-r) & r = 1,2,\ldots,s-1 \ (s \geq 2), \\ (4s+r, 8s+2-r) & r = 1,2,\ldots,2s-1, \end{array}$$

$(3s, 3s+1)$, $(6s+2, 8s+3)$, $(2s-1, 6s)$, $(2s, 6s+1)$.

Suppose $k \equiv 2 \pmod 4$.

$k \geq 6$ and let $k = 4s+2$ where $s \geq 1$:

$(r, 4s+3-r)$	$r = 1, 2, \ldots, s,$
$(s+r, 3s+1-r)$	$r = 1, 2, \ldots, s-1$ $(s \geq 2),$
$(4s+2+r, 8s+4-r)$	$r = 1, 2, \ldots, 2s-1,$

$(3s+1, 3s+2)$, $(6s+4, 8s+5)$, $(2s, 6s+2)$, $(2s+1, 6s+3)$. □

The sequences in the next two theorems form a further generalization of generalized Skolem sequences.

Theorem 2.6 *Let*

$$D(n) = (1, 2, \ldots, \frac{n}{3}-1, \frac{n}{3}+1, \ldots, \frac{n}{2}-3, \frac{n}{2}-2, \frac{n}{2}-2, \frac{n}{2}-1),$$

and

$$S(n) = \{1, 2, \ldots, n\}.$$

If $n \equiv 0, 6 \pmod{24}$, $n > 6$, *then there exist* $p, q \in S(n)$ *such that* $S(n) \backslash \{p, q\}$ *has a partition into* $\frac{n}{2}-1$ *ordered pairs* (a_{d_r}, b_{d_r}) $(r = 1, 2, \ldots, \frac{n}{2}-1)$ *with the property that* $b_{d_r} - a_{d_r} = d_r$ *for any* $1 \leq r \leq \frac{n}{2}-1$, *where* d_r *is the* r*th term of* $D(n)$.

Proof.

Suppose $n \equiv 0 \pmod{24}$.

$n \geq 24$ and let $n = 24s$ where $s \geq 1$:

$(r, 12s+1-r)$	$r = 1, 2, \ldots, 6s-1,$
$(12s+r, 24s-r)$	$r = 1, 2, \ldots, 2s-1,$
$(14s-1+r, 22s-1-r)$	$r = 1, 2, \ldots, 4s-1,$

$(22s-1, 22s)$, $(6s+1, 18s-1)$, and $\{p, q\} = \{6s, 24s\}$.

Suppose $n \equiv 6 \pmod{24}$.

$n \geq 30$ and let $n = 24s + 6$ where $s \geq 1$:

$$\begin{array}{ll}
(r, 12s+3-r) & r = 1, 2, \ldots, 5s, \\
(5s+2+r, 7s+3-r) & r = 1, 2, \ldots, s-1 \ (s \geq 2), \\
(12s+2+r, 24s+6-r) & r = 1, 2, \ldots, 2s, \\
(14s+2+r, 22s+4-r) & r = 1, 2, \ldots, 4s,
\end{array}$$

$(5s+1, 5s+2)$, $(22s+5, 24s+6)$, $(6s+2, 18s+3)$, and $\{p, q\} = \{6s+3, 22s+4\}$.
□

Theorem 2.7 *Let*

$$D(n) = (1, 2, \ldots, \frac{n}{3} - 1, \frac{n}{3} + 1, \ldots, \frac{n}{2} - 2, \frac{n}{2} - 1, \frac{n}{2} - 1),$$

and

$$S(n) = \{1, 2, \ldots, n\}.$$

If $n \equiv 12, 18 \pmod{24}$, then there exist $p, q \in S(n)$ such that $S(n) \setminus \{p, q\}$ has a partition into $\frac{n}{2} - 1$ ordered pairs $(a_{d_r}, b_{d_r})(r = 1, 2, \ldots, \frac{n}{2} - 1)$ with the property that $b_{d_r} - a_{d_r} = d_r$ for any $1 \leq r \leq \frac{n}{2} - 1$, where d_r is the rth term of $D(n)$.

Proof.

Suppose $n \equiv 12 \pmod{24}$.

$n = 12$: (1, 2), (5, 7), (9, 12), (3, 8), (6, 11), and $\{p, q\} = \{4, 10\}$;

$n \geq 36$ and let $n = 24s + 12$ where $s \geq 1$:

$$\begin{array}{ll}
(r, 12s+7-r) & r = 1, 2, \ldots, 5s+2, \\
(5s+4+r, 7s+5-r) & r = 1, 2, \ldots, s-1 \ (s \geq 2), \\
(12s+6+r, 24s+12-r) & r = 1, 2, \ldots, 2s, \\
(14s+7+r, 22s+11-r) & r = 1, 2, \ldots, 4s+1,
\end{array}$$

$(5s+3, 5s+4)$, $(22s+11, 24s+12)$, $(6s+4, 18s+9)$, and $\{p, q\} = \{6s+5, 14s+7\}$.

Suppose $n \equiv 18 \pmod{24}$.

$n \geq 18$ and let $n = 24s + 18$ where $s \geq 0$:

$$\begin{array}{ll}(r, 12s+9-r) & r = 1, 2, \ldots, 6s+3, \\ (12s+8+r, 24s+18-r) & r = 1, 2, \ldots, 2s+1, \\ (14s+9+r, 22s+15-r) & r = 1, 2, \ldots, 4s+2,\end{array}$$

$(22s+15, 22s+16)$, $(6s+4, 18s+12)$, and $\{p, q\} = \{6s+5, 24s+18\}$. □

3 CONSTRUCTIONS

In the remainder of the paper, we take $V = (Z_n \times \{1,2\}) \bigcup \{\infty_1, \infty_2, \ldots, \infty_f\}$, where $n = (v-f)/2$, to be the ground set of the Steiner triple system STS(v). For brevity we denote an element $(i,j) \in Z_n \times \{1,2\}$ by i_j. We then take $\pi = (\infty_1)(\infty_2)\ldots(\infty_f)(0_1 1_1 \ldots (n-1)_1)(0_2 1_2 \ldots (n-1)_2)$ to be the permutation on V under consideration.

By a *base triple* of a Steiner triple system $(V, \mathcal{T})$ under an automorphism we mean a triple which is a representative of an orbit of the automorphism on $\mathcal{T}$. Assuming one of the nonfixed points to be 0_i, a base triple of an STS$_\pi(v)$ on V is of one of the following types:

Type 1. $\{0_i, x_i, y_i\}$ $(x, y, y-x \neq \pm\frac{n}{2}$ if $2|n)$;

Type 2. $\{0_i, (\frac{n}{3})_i, (\frac{2n}{3})_i\}$ (if $3|n$);

Type 3. $\{0_i, x_{i+1}, y_{i+1}\}$ or $\{0_i, x_i, y_{i+1}\}$;

Type 4. $\{\infty_j, 0_i, (\frac{n}{2})_i\}$ (if $2|n$);

Type 5. $\{\infty_j, 0_i, x_{i+1}\}$;

Type 6. $\{\infty_j, \infty_k, \infty_\ell\}$, where $i \in \{1,2\}, j, k, \ell \in \{1, 2, \ldots, f\}$.

Then constructing an STS$_\pi(v)$ involves constructing base triples of these types which cover each pure (nonzero) and mixed difference of $Z_n \times \{1,2\}$ exactly once. A base triple of Type 1, 3 or 5 induces an orbit of length n, of Type 2 an orbit of length $\frac{n}{3}$, of Type 4 an orbit of length $\frac{n}{2}$ and of Type 6 an orbit of length one, consisting of the single triple itself.

We partition our constructions in the following sixteen lemmas according to $v \equiv 1,3,7,9 \pmod{12}$ and $f \equiv 1,3,7,9 \pmod{12}$. In each nontrivial case, we start with building up an $(f',2)$–rotational STS$(v-(f-f'))$, for $f'=1,3$ or 7 whichever is proper, by giving the set of base triples, on a subset V' of V,

$$V' = (Z_n \times \{1,2\}) \bigcup \{\infty_1, \infty_2, \ldots, \infty_{f'}\},$$

with the automorphism to be the permutation obtained by restricting π to V'. Then we modify the base triples of this $(f',2)$–rotational STS$(v-(f-f'))$ to yield those for a required STS$_\pi(v)$ on V. An $(f',2)$–rotational STS$(v-(f-f'))$ is *extendible* (to an $(f,2)$–rotational STS(v)) if its base triples can be modified for this purpose. The base triples of the $(f',2)$–rotational STS$(v-(f-f'))$ constructed in the following lemmas can be roughly classified as

C1 triples involved in only fixed points (these triples can be taken to be those of an STS(f') on $\{\infty_1, \infty_2, \ldots, \infty_{f'}\}$),

C2 triples of Types 1, 2 and 4 with $i=1$ (triples of this kind are designed to cover most of the pure (1, 1)-differences and, in particular, they form the base triples of a cyclic STS(n) on $Z_n \times \{1\}$ when $n \equiv 1,3 \pmod 6$, $n \neq 9$ or those of a 1-rotational STS$(n+1)$ on $(Z_n \times \{1\}) \bigcup \{\infty_1\}$ when $n \equiv 2,8 \pmod{24}$),

C3 triples of the form $\{0_1, x_2, y_2\}$ (Type 3) (these triples should cover most of the pure (2, 2)- and mixed (1, 2)-differences) and

C4 triples of Type 5 and of some other types (which deal with the remaining fixed points and differences).

To add the new fixed points $\infty_{f'+1}, \infty_{f'+2}, \ldots, \infty_f$ into an $(f',2)$–rotational STS$(v-(f-f'))$ constructed in this way, consider any three triples $T_r = \{0_1, (x_r)_2, (y_r)_2\}$ $(r=1,2,3)$, where $x_r, y_r \in Z_n$, of C3. They have the *projection property* if two of $y_r - x_r$ $(r=1,2,3)$ sum to the third, say $(y_1-x_1)+(y_2-x_2)=(y_3-x_3)$, or all of the three sum to n. These three triples, if they have the projection property, can be *destroyed* and *replaced* by new triples $\{0_2, (y_1-x_1)_2, (y_3-x_3)_2\}$ or $\{0_2, (y_1-x_1)_2, (-(y_3-x_3))_2\}$ and $\{\infty_{j_r}, 0_1, (x_r)_2\}$ $(r=1,2,3)$, $\{\infty_{j_s}, 0_1, (y_s)_2\}$ $(s=1,2,3)$, where $\infty_{j_r}, \infty_{j_s} \in \{\infty_{f'+1}, \infty_{f'+2}, \ldots, \infty_f\}$.

The consequence of this destroying and replacing procedure is that all the pure (2, 2)- and mixed (1, 2)-differences in the original three triples remain covered while six more fixed points are introduced. Therefore, in order to extend the $(f',2)$–rotational STS$(v-(f-f'))$, we may just keep the triples of C2 and C4, extend C1 to an STS(f) on the f fixed points, which can be done by the Doyen-Wilson theorem, and apply the destroying and replacing procedure on $(f-f')/6$ sets $\{T_1, T_2, T_3\}$ of triples of C3 with the projection property. Consequently, to guarantee the extendibility of an established $(f',2)$–rotational STS$(v-(f-f'))$, we only need to guarantee that there do exist $(f-f')/6$ sets $\{T_1, T_2, T_3\}$ of triples of C3 with the projection property.

Lemma 3.1 *Let $v \equiv 9$ (mod 12), $f \equiv 3$ (mod 12). If $v \geq 3f$ with $v \neq 21$ when $f = 3$, then there exists an $STS_\pi(v)$.*

Proof. Now, $n \equiv 3 \pmod 6$ and $f - 3 \equiv 0 \pmod 6$. Moreover, $n \geq f \geq 3$, $n \neq 9$ and therefore $\frac{n-3}{6} \geq \frac{f-3}{6}$, $\frac{n-3}{6} \neq 1$. Take (a_r, b_r) $(r = 1, 2, \ldots, \frac{n-3}{6})$ to be a $(C, \frac{n-3}{6})$-system when $n \equiv 3, 21 \pmod{24}$ or a $(D, \frac{n-3}{6})$-system when $n \equiv 9, 15 \pmod{24}$, and (c_r, d_r) $(r = 1, 2, \ldots, \frac{n-1}{2})$ to be an $(A, \frac{n-1}{2})$-system when $n \equiv 3, 9 \pmod{24}$ or a $(B, \frac{n-1}{2})$-system when $n \equiv 15, 21 \pmod{24}$. Then the set of base triples of a (3, 2)-rotational STS$(v-(f-3))$ consists of

(i) $\{\infty_1, \infty_2, \infty_3\}$,

(ii) the base triples of a cyclic STS(n) on $Z_n \times \{1\}$:

$\{0_1, r_1, (\frac{n-3}{6} + b_r)_1\}$ $(r = 1, 2, \ldots, \frac{n-3}{6})$ $(n > 3)$, $\{0_1, (\frac{n}{3})_1, (\frac{2n}{3})_1\}$,

(iii) $\{\infty_1, 0_1, x_2\}$ ($x = 0$ if $n \equiv 3, 9 \pmod{24}$ and $x = n-1$ if $n \equiv 15, 21 \pmod{24}$),

$\{0_1, (c_r)_2, (d_r)_2\}$ $(r = 1, 2, \ldots, \frac{n-1}{2})$

with $\{0_1, (c_{\frac{n}{3}})_2, (d_{\frac{n}{3}})_2\}$ to be destroyed and replaced by

$\{0_2, (\frac{n}{3})_2, (\frac{2n}{3})_2\}$, $\{\infty_2, 0_1, (c_{\frac{n}{3}})_2\}$, $\{\infty_3, 0_1, (d_{\frac{n}{3}})_2\}$.

This system is extendible since the $(f-3)/6$ sets of triples satisfying the projection property can be taken as

$$\{\{0_1, (c_r)_2, (d_r)_2\}, \{0_1, (c_{\frac{n-3}{6}+a_r})_2, (d_{\frac{n-3}{6}+a_r})_2\}, \{0_1, (c_{\frac{n-3}{6}+b_r})_2, (d_{\frac{n-3}{6}+b_r})_2\}\}$$

$(r = 1, 2, \ldots, \frac{f-3}{6})$.

When $\frac{n-3}{6} + b_r = \frac{n+1}{2}$ for some r, when $n \equiv 9, 15 \pmod{24}$, we use the destroying and replacing procedure on $\{0_1, (c_r)_2, (d_r)_2\}$, $\{0_1, (c_{\frac{n-3}{6}+a_r})_2, (d_{\frac{n-3}{6}+a_r})_2\}$,

$\{0_1, (c_{\frac{n-1}{2}})_2, (d_{\frac{n-1}{2}})_2\}$ (instead of $\{0_1, (c_{\frac{n+1}{2}})_2, (d_{\frac{n+1}{2}})_2\}$, which does not exist).
□

Lemma 3.2 *Let $v \equiv 3 \pmod{12}$, $f \equiv 9 \pmod{12}$. If $v \geq 3f$ with $v \neq 27$ when $f = 9$, then there exists an $STS_\pi(v)$.*

Proof. Now, $n \equiv 3 \pmod 6$ and $f - 3 \equiv 0 \pmod 6$. Moreover, $n \geq f \geq 9$, $n \neq 9$ and therefore $\frac{n-3}{6} \geq \frac{f-3}{6}$, $\frac{n-3}{6} \neq 1$. So, the construction in the proof of Lemma 3.1 can be used to give an $\text{STS}_\pi(v)$ as required. □

Lemma 3.3 *Let $v \equiv 1 \pmod{12}$, $f \equiv 7 \pmod{12}$. If $v \geq 3f + 4$ with $v \neq 25$ when $f = 7$, then there exists an $STS_\pi(v)$.*

Proof. Now, $n \equiv 3 \pmod 6$, $f - 1 \equiv 0 \pmod 6$. Moreover, $n \geq f + 2$, $n > 9$ and therefore $\frac{n-3}{6} \geq \frac{f-1}{6}$ since $v \neq 25$ when $f = 7$ and $v \geq 3f + 4$. Take (a_r, b_r) $(r = 1, 2, \ldots, \frac{n-3}{6})$ to be an $(A, \frac{n-3}{6})$-system when $n \equiv 3, 9 \pmod{24}$ or a $(B, \frac{n-3}{6})$-system when $n \equiv 15, 21 \pmod{24}$, and (c_r, d_r) $(r = 1, 2, \ldots, \frac{n-1}{2})$ to be an $(A, \frac{n-1}{2})$-system when $n \equiv 3, 9 \pmod{24}$ or a $(B, \frac{n-1}{2})$-system when $n \equiv 15, 21 \pmod{24}$. Then the set of base triples of a (1, 2)-rotational STS($v - (f-1)$) consists of

(i) the base triples of a cyclic STS(n) on $Z_n \times \{1\}$, which can be obtained in the same way as in the proof of Lemma 3.1 by using a $(C, \frac{n-3}{6})$-system or a $(D, \frac{n-3}{6})$-system whichever is proper,

(ii) $\{\infty_1, 0_1, x_2\}$ ($x = 0$ if $n \equiv 3, 9 \pmod{24}$ and $x = n - 1$ if $n \equiv 15, 21 \pmod{24}$),
$\{0_1, (c_r)_2, (d_r)_2\}$ $(r = 1, 2, \ldots, \frac{n-1}{2})$.

This system is extendible since the $(f-1)/6$ sets of triples satisfying the projection property can be taken as

$$\{\{0_1, (c_r)_2, (d_r)_2\}, \{0_1, (c_{\frac{n-3}{6}+a_r})_2, (d_{\frac{n-3}{6}+a_r})_2\}, \{0_1, (c_{\frac{n-3}{6}+b_r})_2, (d_{\frac{n-3}{6}+b_r})_2\}\}$$

$(r = 1, 2, \ldots, \frac{f-1}{6})$. □

Lemma 3.4 *Let $v \equiv 7 \pmod{12}$, $f \equiv 1 \pmod{12}$. If $v \geq 3f + 4$, then there exists an $STS_\pi(v)$.*

Proof. A (1, 2)-rotational (namely, 2-rotational) STS(19) does exist [9]. So we assume $v \neq 19$ when $f = 1$. Then we have $n \equiv 3 \pmod 6$, $f - 1 \equiv 0 \pmod 6$, $n \neq 9$ and $n \geq f + 2 \geq 3$. Consequently, $\frac{n-3}{6} \geq \frac{f-1}{6}$. Then the construction in the proof of Lemma 3.3 can be adapted to this case to yield an $\mathrm{STS}_\pi(v)$ as required. (We do not require $n > 9$ in the proof of Lemma 3.3, although this is true there.) □

Lemma 3.5 *Let* $v \equiv 9 \pmod{12}$, $f \equiv 7 \pmod{12}$. *If* $v \geq 3f$, *then there exists an* $STS_\pi(v)$.

Proof. Now, $n \equiv 1 \pmod 6$, $f - 1 \equiv 0 \pmod 6$. Moreover, $n \geq f \geq 7$ and therefore $\frac{n-1}{6} \geq \frac{f-1}{6}$. Take (a_r, b_r) $(r = 1, 2, \ldots, \frac{n-1}{6})$ to be an $(A, \frac{n-1}{6})$-system when $n \equiv 1, 7 \pmod{24}$ or a $(B, \frac{n-1}{6})$-system when $n \equiv 13, 19 \pmod{24}$, and (c_r, d_r) $(r = 1, 2, \ldots, \frac{n-1}{2})$ to be an $(A, \frac{n-1}{2})$-system when $n \equiv 1, 19 \pmod{24}$ or a $(B, \frac{n-1}{2})$-system when $n \equiv 7, 13 \pmod{24}$. Then the set of base triples of a (1, 2)-rotational STS$(v - (f - 1))$ consists of

(i) the base triples of a cyclic STS(n) on $Z_n \times \{1\}$:
$\{0_1, r_1, (\frac{n-1}{6} + b_r)_1\}$ $(r = 1, 2, \ldots, \frac{n-1}{6})$,

(ii) $\{\infty_1, 0_1, x_2\}$ ($x = 0$ if $n \equiv 1, 19 \pmod{24}$ and $x = n - 1$ if $n \equiv 7, 13 \pmod{24}$),
$\{0_1, (c_r)_2, (d_r)_2\}$ $(r = 1, 2, \ldots, \frac{n-1}{2})$.

This system is extendible since the $(f - 1)/6$ sets of triples satisfying the projection property can be taken as

$$\{\{0_1, (c_r)_2, (d_r)_2\}, \{0_1, (c_{\frac{n-1}{6}+a_r})_2, (d_{\frac{n-1}{6}+a_r})_2\}, \{0_1, (c_{\frac{n-1}{6}+b_r})_2, (d_{\frac{n-1}{6}+b_r})_2\}\}$$

$(r = 1, 2, \ldots, \frac{f-1}{6})$. When $n \equiv 13, 19 \pmod{24}$, we may have $\frac{n-1}{6} + b_r = \frac{n+1}{2}$ for some r. If this happens, we apply the destroying and replacing procedure on $\{0_1, (c_r)_2, (d_r)_2\}$, $\{0_1, (c_{\frac{n-1}{6}+a_r})_2, (d_{\frac{n-1}{6}+a_r})_2\}$, $\{0_1, (c_{\frac{n-1}{2}})_2, (d_{\frac{n-1}{2}})_2\}$. □

Lemma 3.6 *Let* $v \equiv 3 \pmod{12}$, $f \equiv 1 \pmod{12}$. *If* $v \geq 3f$, *then there exists an* $STS_\pi(v)$.

Proof. Now, $n \equiv 1 \pmod 6$, $f - 1 \equiv 0 \pmod 6$. Moreover, $n \geq f \geq 1$ and therefore $\frac{n-1}{6} \geq \frac{f-1}{6}$. Then the construction in the proof of Lemma 3.5 can be adopted here (We do not require $n \geq 7$ and therefore a (1, 2)-rotational STS(3) also follows from the construction). $\square$

Lemma 3.7 *Let $v \equiv 7 \pmod{12}$, $f \equiv 9 \pmod{12}$. If $v \geq 3f + 16$, then there exists an $STS_\pi(v)$.*

Proof. Now, $n \equiv 5 \pmod 6$, $f - 3 \equiv 0 \pmod 6$. Moreover, $n \geq f + 8 \geq 17$ and therefore $\frac{n-5}{6} - 1 \geq \frac{f-3}{6}$. Take (a_r, b_r) $(r = 1, 2, \ldots, \frac{n-5}{6})$ to be an $(A, \frac{n-5}{6})$-system when $n \equiv 5, 11 \pmod{24}$ or a $(B, \frac{n-5}{6})$-system when $n \equiv 17, 23 \pmod{24}$, and (c_r, d_r) $(r = 1, 2, \ldots, \frac{n-1}{2})$ to be an $(A, \frac{n-1}{2})$-system when $n \equiv 11, 17 \pmod{24}$ or a $(B, \frac{n-1}{2})$-system when $n \equiv 5, 23 \pmod{24}$. Then the set of base triples of a (3, 2)-rotational STS($v - (f-3)$) consists of

(i) $\{\infty_1, \infty_2, \infty_3\}$,

(ii) $\{0_1, r_1, (\frac{n-5}{6} + b_r)_1\}$ $(r = 1, 2, \ldots, \frac{n-5}{6})$,

(iii) $\{\infty_1, 0_1, x_2\}$ ($x = 0$ if $n \equiv 11, 17 \pmod{24}$ and $x = n-1$ if $n \equiv 5, 23 \pmod{24}$),

(iv) $\{0_1, (c_r)_2, (d_r)_2\}$ $(r = 1, 2, \ldots, \frac{n-1}{2})$ with

when $n \equiv 5, 11 \pmod{24}$, triples
$\{0_1, (c_{\frac{n-3}{2}})_2, (d_{\frac{n-3}{2}})_2\}$, $\{0_1, (c_{\frac{n-1}{2}})_2, (d_{\frac{n-1}{2}})_2\}$, $\{0_1, (c_1)_2, (d_1)_2\}$
to be destroyed and replaced by
$\{0_1, (\frac{n-3}{2})_1, (d_{\frac{n-3}{2}})_2\}$, $\{0_1, (\frac{n-1}{2})_1, (d_{\frac{n-1}{2}})_2\}$, $\{0_2, 1_2, (\frac{n-1}{2})_2\}$,
$\{\infty_2, 0_1, (c_1)_2\}$, $\{\infty_3, 0_1, (d_1)_2\}$;
when $n \equiv 17, 23 \pmod{24}$, triples
$\{0_1, (c_{\frac{n-5}{2}})_2, (d_{\frac{n-5}{2}})_2\}$, $\{0_1, (c_{\frac{n-1}{2}})_2, (d_{\frac{n-1}{2}})_2\}$, $\{0_1, (c_2)_2, (d_2)_2\}$
to be destroyed and replaced by
$\{0_1, (\frac{n-5}{2})_1, (d_{\frac{n-5}{2}})_2\}$, $\{0_1, (\frac{n-1}{2})_1, (d_{\frac{n-1}{2}})_2\}$, $\{0_2, 2_2, (\frac{n-1}{2})_2\}$,
$\{\infty_2, 0_1, (c_2)_2\}$, $\{\infty_3, 0_1, (d_2)_2\}$.

This system is extendible since the $(f-3)/6$ sets of triples satisfying the projection property can be taken as

$$\{\{0_1, (c_r)_2, (d_r)_2\}, \{0_1, (c_{\frac{n-5}{6}+a_r})_2, (d_{\frac{n-5}{6}+a_r})_2\}, \{0_1, (c_{\frac{n-5}{6}+b_r})_2, (d_{\frac{n-5}{6}+b_r})_2\}\},$$

where $r = 2, 3, \ldots, \frac{f-3}{6}+1$ when $n \equiv 5, 11 \pmod{24}$ and $r = 1, 3, 4, \ldots, \frac{f-3}{6}+1$ when $n \equiv 17, 23 \pmod{24}$. □

Lemma 3.8 *Let $v \equiv 1 \pmod{12}$, $f \equiv 3 \pmod{12}$. If $v \geq 3f + 16$, then there exists an $STS_\pi(v)$.*

Proof. Now, $n \equiv 5 \pmod 6$, $f - 3 \equiv 0 \pmod 6$. Moreover, $n \geq f + 8 \geq 11$ and therefore $\frac{n-5}{6} - 1 \geq \frac{f-3}{6}$. So the construction in the proof of Lemma 3.7 can be adapted to this case. (Although $n \geq 17$ in Lemma 3.7, the construction also works for $n = 11$ in this case.) □

Lemma 3.9 *Let $v \equiv 3 \pmod{12}$, $f \equiv 7 \pmod{12}$. If $v \geq 3f + 6$, then there exists an $STS_\pi(v)$.*

Proof. Now, $n \equiv 4 \pmod 6$, $f - 7 \equiv 0 \pmod 6$. Moreover, $n \geq f + 3 \geq 10$ and therefore $\frac{n-4}{6} - 1 \geq \frac{f-7}{6}$. Take (a_r, b_r) $(r = 1, 2, \ldots, \frac{n-4}{6})$ to be an $(A, \frac{n-4}{6})$-system when $n \equiv 4, 10 \pmod{24}$ or a $(B, \frac{n-4}{6})$-system when $n \equiv 16, 22 \pmod{24}$, and (c_r, d_r) $(r = 1, 2, \ldots, \frac{n}{2} - 1)$ to be a $(C, \frac{n}{2} - 1)$-system when $n \equiv 10, 16 \pmod{24}$ or a $(D, \frac{n}{2} - 1)$-system when $n \equiv 4, 22 \pmod{24}$ (note that $\frac{n}{2} - 1 > 1$). Then the set of base triples of a (7, 2)-rotational STS$(v - (f - 7))$ consists of

(i) the triples of an STS(7) on $\{\infty_1, \infty_2, \ldots, \infty_7\}$,

(ii) $\{\infty_1, 0_1, (\frac{n}{2})_1\}$, $\{0_1, r_1, (\frac{n-4}{6} + b_r)_1\}$ $(r = 1, 2, \ldots, \frac{n-4}{6})$,

(iii) $\{\infty_1, 0_2, (\frac{n}{2})_2\}$, $\{0_1, (c_r)_2, (d_r)_2\}$ $(r = 1, 2, \ldots, \frac{n}{2} - 1)$, and

$\{\infty_2, 0_1, x_2\}$ ($x = 0$ if $n \equiv 10, 16 \pmod{24}$ and $x = n - 1$ if $n \equiv 4, 22 \pmod{24}$)

with triples

$\{0_1, (c_\ell)_2, (d_\ell)_2\}$, $\{0_1, (c_{\frac{n-4}{6}+a_\ell})_2, (d_{\frac{n-4}{6}+a_\ell})_2\}$,

$\{0_1, (c_{\frac{n-4}{6}+b_\ell})_2, (d_{\frac{n-4}{6}+b_\ell})_2\}$

to be destroyed and replaced by

$\{\infty_j, 0_1, (y_{j-2})_2\}$ $(j = 3, 4, \ldots, 7)$, $\{0_2, \ell_2, (\frac{n-4}{6} + b_\ell)_2\}$

and by

$\{0_1, (\frac{n}{2} - 1)_1, (\frac{n}{2})_2\}$ (when $n \equiv 4, 10 \pmod{24}$) or

$\{0_1, (\frac{n}{2}-2)_1, (\frac{n}{2})_2\}$ (when $n \equiv 16, 22 \pmod{24}$),
where $1 \leq \ell \leq \frac{n-4}{6}$ such that some element in

$$\{c_\ell, d_\ell, c_{\frac{n-4}{6}+a_\ell}, d_{\frac{n-4}{6}+a_\ell}, c_{\frac{n-4}{6}+b_\ell}, d_{\frac{n-4}{6}+b_\ell}\}$$

is 1 when $n \equiv 4, 10 \pmod{24}$ or 2 when $n \equiv 16, 22 \pmod{24}$, and $y_1, y_2, \ldots, y_5$ are defined to be the other elements in the set (the existence of such an ℓ is guaranteed by the range of $\{r, \frac{n-4}{6}+a_r, \frac{n-4}{6}+b_r\}$ $(r = 1, 2, \ldots, \frac{n-4}{6})$.

This system is extendible since the $(f-7)/6$ sets of triples satisfying the projection property can be taken to be

$$\{\{0_1, (c_r)_2, (d_r)_2\}, \{0_1, (c_{\frac{n-4}{6}+a_r})_2, (d_{\frac{n-4}{6}+a_r})_2\}, \{0_1, (c_{\frac{n-4}{6}+b_r})_2, (d_{\frac{n-4}{6}+b_r})_2\}\}$$

$(r = 1, 2, \ldots, \ell-1, \ell+1, \ldots, \frac{f-7}{6}+1)$. □

Lemma 3.10 *Let $v \equiv 9 \pmod{12}$, $f \equiv 1 \pmod{12}$. If $f \geq 13$ and $v \geq 3f+6$, then there exists an $STS_\pi(v)$.*

Proof. Now, $n \equiv 4 \pmod 6$, $f - 7 \equiv 0 \pmod 6$. Moreover, $n \geq f+3 \geq 16$ and therefore $\frac{n-4}{6} - 1 \geq \frac{f-7}{6}$. So the construction in the proof of Lemma 3.9 can be adapted to this case. □

Lemma 3.11 *Let $v \equiv 1 \pmod{12}$, $f \equiv 9 \pmod{12}$. If $v \geq 3f-2$ with $v \neq 3f-2$ when $f \equiv 21 \pmod{24}$ and $v \neq 3f+10$ when $f \equiv 9 \pmod{24}$, then there exists an $STS_\pi(v)$.*

Proof. Now, $n \equiv 2 \pmod 6$, $f - 3 \equiv 0 \pmod 6$. Furthermore, since $n = f-1$ when $v = 3f-2$ and $n = f+5$ when $v = 3f+10$, the assumption implies $v \geq 3f-2$ when $n \equiv 2, 8 \pmod{24}$, $v \geq 3f+22$ when $n \equiv 14, 20 \pmod{24}$ and therefore $n \geq f-1 \geq 8$ when $n \equiv 2, 8 \pmod{24}$, $n \geq f+11 \geq 20$ when $n \equiv 14, 20 \pmod{24}$. Consequently, $\frac{n-2}{6} \geq \frac{f-3}{6}$ when $n \equiv 2, 8 \pmod{24}$, $\frac{n-8}{6} - 1 \geq \frac{f-3}{6}$ when $n \equiv 14, 20 \pmod{24}$.

First suppose $n \equiv 2, 8 \pmod{24}$. Take (a_r, b_r) $(r = 1, 2, \ldots, \frac{n-2}{6})$, (c_r, d_r) $(r = 1, 2, \ldots, \frac{n}{2})$ to be an $(A, \frac{n-2}{6})$-system and an $(A, \frac{n}{2})$-system, respectively. Then the set of base triples of a (3, 2)-rotational STS$(v-(f-3))$ consists of

(i) $\{\infty_1, \infty_2, \infty_3\}$,

(ii) the base triples of a 1-rotational STS$(n+1)$ on $(Z_n \times \{1\}) \bigcup \{\infty_1\}$:
$\{\infty_1, 0_1, (\frac{n}{2})_1\}$, $\{0_1, r_1, (\frac{n-2}{6} + b_r)_1\}$ $(r = 1, 2, \ldots, \frac{n-2}{6})$,

(iii) $\{0_1, (c_r)_2, (d_r)_2\}$ $(r = 1, 2, \ldots, \frac{n}{2})$
with triple $\{0_1, (c_{\frac{n}{2}})_2, (d_{\frac{n}{2}})_2\}$ to be destroyed and replaced by
$\{\infty_1, 0_2, (\frac{n}{2})_2\}$, $\{\infty_2, 0_1, (c_{\frac{n}{2}})_2\}$, $\{\infty_3, 0_1, (d_{\frac{n}{2}})_2\}$.

This system is extendible since the $(f-3)/6$ sets of triples satisfying the projection property can be taken as

$$\{\{0_1, (c_r)_2, (d_r)_2\}, \{0_1, (c_{\frac{n-2}{6}+a_r})_2, (d_{\frac{n-2}{6}+a_r})_2\}, \{0_1, (c_{\frac{n-2}{6}+b_r})_2, (d_{\frac{n-2}{6}+b_r})_2\}\}$$

$(r = 1, 2, \ldots, \ldots, \frac{f-3}{6})$.

Now suppose that $n \equiv 14, 20 \pmod{24}$. Take (a_r, b_r) $(r = 1, 2, \ldots, \frac{n-8}{6})$ to be an $(A, \frac{n-8}{6})$-system when $n \equiv 14 \pmod{24}$ or a $(B, \frac{n-8}{6})$-system when $n \equiv 20 \pmod{24}$, and (c_{d_r}, e_{d_r}) $(r = 1, 2, \ldots, \frac{n}{2} - 1)$ to be an $(I, \frac{n}{2} - 1)$-system when $n \equiv 14 \pmod{24}$ or a $(J, \frac{n}{2} - 1)$-system when $n \equiv 20 \pmod{24}$, where d_r is the *rth term* in $D(\frac{n}{2} - 1)$ in each case. Since $\frac{n}{2} - 1 \geq 9$ (since $n \geq 20$) in both cases, there do exist $(I, \frac{n}{2} - 1)$- and $(J, \frac{n}{2} - 1)$-systems. The set of base triples of a (3, 2)-rotational STS$(v - (f-3))$ consists of the following triples (a few triples in (iv) are obtained also by applying the destroying and replacing procedure but for simplicity we state them directly).

(i) $\{\infty_1, \infty_2, \infty_3\}$,

(ii) $\{\infty_1, 0_1, (\frac{n}{2})_1\}$, $\{0_1, r_1, (\frac{n-8}{6} + b_r)_1\}$ $(r = 1, 2, \ldots, \frac{n-8}{6})$,

(iii) $\{\infty_1, 0_2, (\frac{n}{2})_2\}$, $\{\infty_2, 0_1, 0_2\}$,
$\{\infty_3, 0_1, x_2\}$ $(x = n - 1 \text{ if } n \equiv 14 \pmod{24}$,
$x = n - 2 \text{ if } n \equiv 20 \pmod{24})$,

$\{0_2, 2_2, y_2\}$ $(y = \frac{n}{2} - 1 \text{ if } n \equiv 14 \pmod{24}$,
$y = \frac{n}{2} - 2 \text{ if } n \equiv 20 \pmod{24})$,

(iv) $\{0_1, (c_{d_r})_2, (e_{d_r})_2\}$,
where $r = 1, 2, \ldots, \frac{n}{2} - 5, \frac{n}{2} - 3$ when $n \equiv 14 \pmod{24}$, $r = 1, 2, \ldots, \frac{n}{2} - 6, \frac{n}{2} - 4, \frac{n}{2} - 2$ when $n \equiv 20 \pmod{24}$,

$\{0_1, (d_r)_1, (e_{d_r})_2\}$,
where $r = \frac{n}{2}-4, \frac{n}{2}-2, \frac{n}{2}-1$ when $n \equiv 14 \pmod{24}$, $r = \frac{n}{2}-5, \frac{n}{2}-3, \frac{n}{2}-1$ when $n \equiv 20 \pmod{24}$.

This system is extendible since the $(f-3)/6$ sets of triples satisfying the projection property can be taken to be

$$\{\{0_1, (c_\ell)_2, (e_\ell)_2\}, \{0_1, (c_{\frac{n-8}{6}+a_\ell})_2, (e_{\frac{n-8}{6}+a_\ell})_2\}, \{0_1, (c_{\frac{n-8}{6}+b_\ell})_2, (e_{\frac{n-8}{6}+b_\ell})_2\}\},$$

where $\ell = 1, 3, 4, \ldots, \frac{f-3}{6}+1$. Note that $\{\ell, \frac{n-8}{6}+a_\ell, \frac{n-8}{6}+b_\ell\} \subseteq \{1, 2, \ldots, \frac{n}{2}-4\} \setminus \{2\} \subseteq \{1, 3, 4, \ldots, \frac{n}{2}-4, \frac{n}{2}-2\}$ when $n \equiv 14 \pmod{24}$ and similarly $\{\ell, \frac{n-8}{6}+a_\ell, \frac{n-8}{6}+b_\ell\} \subseteq \{1, 3, 4, \ldots, \frac{n}{2}-5, \frac{n}{2}-3, \frac{n}{2}-1\}$ when $n \equiv 20 \pmod{24}$, for any $\ell = 1, 3, 4, \ldots, \frac{f-3}{6}+1$. Therefore these triples do exist in (iv). □

Lemma 3.12 *Let $v \equiv 7 \pmod{12}$, $f \equiv 3 \pmod{12}$. If $v \geq 3f-2$ with $v \neq 3f-2, 3f+10$ when $f \equiv 15 \pmod{24}$, then there exists an $STS_\pi(v)$.*

Proof. Now, $n \equiv 2 \pmod 6$, $f-3 \equiv 0 \pmod 6$. Here, an argument similar to that in the proof of Lemma 3.11 establishes $n \geq f-1 \geq 2$ when $n \equiv 2, 8 \pmod{24}$ and $n \geq f+11 \geq 14$ when $n \equiv 14, 20 \pmod{24}$. Although it is specified in the proof that the bounds of n are $n \geq 8$ when $n \equiv 2, 8 \pmod{24}$ and $n \geq 20$ when $n \equiv 14, 20 \pmod{24}$, the constructions there do yield a (3, 2)-rotational STS(7) (corresponding to $n = 2$) and (3, 2)-rotational STS(31) (corresponding to $n = 14$) for this lemma. For other cases in this lemma, the bounds on n as for Lemma 3.11 are satisfied and therefore the constructions can be adopted. □

Lemma 3.13 *Let $v \equiv 3 \pmod{12}$, $f \equiv 3 \pmod{12}$. If $v \geq 3f+6$, then there exists an $STS_\pi(v)$.*

Proof. Now, $n \equiv 0 \pmod 6$, $f-3 \equiv 0 \pmod 6$. A (3, 2)-rotational STS(15) on $(Z_6 \times \{1,2\}) \bigcup \{\infty_1, \infty_2, \infty_3\}$ can be given by base triples $\{\infty_1, \infty_2, \infty_3\}$, $\{\infty_1, 0_1, 3_1\}$, $\{\infty_1, 0_2, 3_2\}$, $\{\infty_2, 0_1, 0_2\}$, $\{\infty_3, 0_1, 5_2\}$, $\{0_1, 2_1, 4_1\}$, $\{0_2, 2_2, 4_2\}$, $\{0_1, 1_1, 2_2\}$, $\{0_1, 3_2, 4_2\}$. So we assume $v \geq 27$ when $f = 3$ in addition. Then $n \geq 12$, $n \geq f+3$ and therefore $\frac{n-6}{6} \geq \frac{f-3}{6}$. Take (a_r, b_r) $(r = 1, 2, \ldots, \frac{n-6}{6})$ to be a $(G, \frac{n-6}{6})$-system when $n \equiv 12, 18 \pmod{24}$ or an $(H, \frac{n-6}{6})$-system when $n \equiv 0, 6 \pmod{24}$. Consider

$$S(n) = \{1, 2, \ldots, n\}$$

and

$$D(n)=(1,2,\ldots,\frac{n}{3}-1,\frac{n}{3}+1,\ldots,\frac{n}{2}-2,\frac{n}{2}-1,\frac{n}{2}-1)$$

when $n\equiv 12,18 \pmod{24}$, or

$$D(n)=(1,2,\ldots,\frac{n}{3}-1,\frac{n}{3}+1,\ldots,\frac{n}{2}-3,\frac{n}{2}-2,\frac{n}{2}-2,\frac{n}{2}-1)$$

when $n\equiv 0,6 \pmod{24}$. Let (s_{d_r},t_{d_r}) $(r=1,2,\ldots,\frac{n}{2}-1)$ and p,q be such that $p,q\in S(n)$, and $\{s_{d_r},t_{d_r}\}$ $(r=1,2,\ldots,\frac{n}{2}-1)$ partitions $S(n)\setminus\{p,q\}$ with the further property that $t_{d_r}-s_{d_r}=d_r$ for any r, where d_r is the rth term of $D(n)$ in each case. The existence of these (s_{d_r},t_{d_r}) and p,q in each case is guaranteed by Theorem 2.7 or Theorem 2.6. Then the set of base triples of a (3, 2)-rotational STS$(v-(f-3))$ consists of

(i) $\{\infty_1,\infty_2,\infty_3\}$,

(ii) $\{0_1,(\frac{n}{3})_1,(\frac{2n}{3})_1\}$, $\{\infty_1,0_1,(\frac{n}{2})_1\}$, $\{0_1,r_1,(\frac{n-6}{6}+b_r)_1\}$ $(r=1,2,\ldots,\frac{n-6}{6})$,

(iii) $\{0_2,(\frac{n}{3})_2,(\frac{2n}{3})_2\}$, $\{\infty_1,0_2,(\frac{n}{2})_2\}$, $\{\infty_2,0_1,p_2\}$, $\{\infty_3,0_1,q_2\}$,
$\{0_1,(s_{d_r})_2,(t_{d_r})_2\}$,
where $r=1,2,\ldots,\frac{n}{2}-2$ when $n\equiv 12,18 \pmod{24}$ and $r=1,2,\ldots,\frac{n}{2}-3,\frac{n}{2}-1$ when $n\equiv 0,6 \pmod{24}$, and a triple
$\{0_1,(\frac{n}{2}-1)_1,(t_{d_{\frac{n}{2}-1}})_2\}$ $(n\equiv 12,18 \pmod{24})$,
or
$\{0_1,(\frac{n}{2}-2)_1,(t_{d_{\frac{n}{2}-2}})_2\}$ $(n\equiv 0,6 \pmod{24})$.

This system is extendible since the $(f-3)/6$ sets of triples satisfying the projection property can be taken as

$$\{\{0_1,(s_\ell)_2,(t_\ell)_2\},\{0_1,(s_{\frac{n-6}{6}+a_\ell})_2,(t_{\frac{n-6}{6}+a_\ell})_2\},\{0_1,(s_{\frac{n-6}{6}+b_\ell})_2,(t_{\frac{n-6}{6}+b_\ell})_2\}\},$$

where $\ell=1,2,\ldots,\frac{f-3}{6}$. Note that $\{\ell,\frac{n-6}{6}+a_\ell,\frac{n-6}{6}+b_\ell\}\subseteq\{1,2,\ldots,\frac{n}{2}-2\}\setminus\{\frac{n}{3}\}\subseteq\{1,2,\ldots,\frac{n}{2}-1\}\setminus\{\frac{n}{3}\}$ when $n\equiv 12,18 \pmod{24}$ and similarly $\{\ell,\frac{n-6}{6}+a_\ell,\frac{n-6}{6}+b_\ell\}\subseteq\{1,2,\ldots,\frac{n}{2}-1\}\setminus\{\frac{n}{3}\}$ when $n\equiv 0,6 \pmod{24}$, for any ℓ, and therefore these triples exist in (iii). □

Lemma 3.14 *Let $v\equiv 9$ (mod 12), $f\equiv 9$ (mod 12). If $v\geq 3f+6$, then there exists an $STS_\pi(v)$.*

Proof. Now, $n \equiv 0 \pmod 6$, $f-3 \equiv 0 \pmod 6$. Moreover, $n \geq f+3 \geq 12$ and therefore $\frac{n-6}{6} \geq \frac{f-3}{6}$. So the construction for the nontrivial cases in Lemma 3.13 can be adapted to this case. □

Lemma 3.15 *Let $v \equiv 7 \pmod{12}$, $f \equiv 7 \pmod{12}$. If $v \geq 3f+10$, then there exists an $STS_\pi(v)$.*

Proof. Now, $n \equiv 0 \pmod 6$, $f-7 \equiv 0 \pmod 6$. Moreover, $n \geq f+5 \geq 12$ and therefore $\frac{n-6}{6} - 1 \geq \frac{f-7}{6}$. Take (a_r, b_r) $(r = 1, 2, \ldots, \frac{n-6}{6})$ to be a $(G, \frac{n-6}{6})$-system when $n \equiv 12, 18 \pmod{24}$ or an $(H, \frac{n-6}{6})$-system when $n \equiv 0, 6 \pmod{24}$, and (c_r, d_r) $(r = 1, 2, \ldots, \frac{n}{2} - 1)$ to be an $(A, \frac{n}{2} - 1)$-system when $n \equiv 12, 18 \pmod{24}$ or a $(B, \frac{n}{2} - 1)$-system when $n \equiv 0, 6 \pmod{24}$. We give the constructions as in the following two cases.

Case 1. $n \equiv 12, 18 \pmod{24}$. The set of base triples of a (7, 2)-rotational STS($v - (f-7)$) consists of

(i) the triples of an STS(7) on $\{\infty_1, \infty_2, \ldots, \infty_7\}$,

(ii) $\{0_1, (\frac{n}{3})_1, (\frac{2n}{3})_1\}$, $\{\infty_1, 0_1, (\frac{n}{2})_1\}$, $\{0_1, r_1, (\frac{n-6}{6} + b_r)_1\}$ $(r = 1, 2, \ldots, \frac{n-6}{6})$,

(iii) $\{\infty_1, 0_2, (\frac{n}{2})_2\}$, $\{\infty_2, 0_1, 0_2\}$, $\{\infty_3, 0_1, (n-1)_2\}$,

(iv) $\{0_1, (c_r)_2, (d_r)_2\}$ $(r = 1, 2, \ldots, \frac{n}{2} - 1)$

with, when $n \equiv 12 \pmod{24}$, triples

$\{0_1, (c_{\frac{n}{2}-1})_2, (d_{\frac{n}{2}-1})_2\}$, $\{0_1, (c_1)_2, (d_1)_2\}$, $\{0_1, (c_{\frac{n}{2}-2})_2, (d_{\frac{n}{2}-2})_2\}$

to be destroyed and replaced by

$\{0_1, (\frac{n}{2} - 1)_1, (d_{\frac{n}{2}-1})_2\}$, $\{0_2, 1_2, (\frac{n}{2} - 1)_2\}$, $\{\infty_4, 0_1, (c_1)_2\}$,

$\{\infty_5, 0_1, (d_1)_2\}$, $\{\infty_6, 0_1, (c_{\frac{n}{2}-2})_2\}$, $\{\infty_7, 0_1, (d_{\frac{n}{2}-2})_2\}$;

or with , when $n \equiv 18 \pmod{24}$, triples

$\{0_1, (c_{\frac{n}{2}-1})_2, (d_{\frac{n}{2}-1})_2\}$, $\{0_1, (c_2)_2, (d_2)_2\}$, $\{0_1, (c_{\frac{n}{2}-3})_2, (d_{\frac{n}{2}-3})_2\}$

to be destroyed and replaced by

$\{0_1, (\frac{n}{2} - 1)_1, (d_{\frac{n}{2}-1})_2\}$, $\{0_2, 2_2, (\frac{n}{2} - 1)_2\}$, $\{\infty_4, 0_1, (c_2)_2\}$

$\{\infty_5, 0_1, (d_2)_2\}$, $\{\infty_6, 0_1, (c_{\frac{n}{2}-3})_2\}$, $\{\infty_7, 0_1, (d_{\frac{n}{2}-3})_2\}$.

This system is extendible. To prove this, we take (s_r, t_r) $(r = 1, 2, \ldots, \frac{n-6}{6})$ to be an $(A, \frac{n-6}{6})$-system when $n \equiv 12 \pmod{24}$ or a $(B, \frac{n-6}{6})$-system when

$n \equiv 18 \pmod{24}$. It can be seen that $\{r, s_r, t_r\} \subseteq \{2, 3, \ldots, \frac{n}{2} - 3\}$, for each $r = 2, 3, \ldots, \frac{f-7}{6} + 1$, when $n \equiv 12 \pmod{24}$, and $\{r, s_r, t_r\} \subseteq \{1, 3, 4, \ldots, \frac{n}{2} - 4, \frac{n}{2} - 2\}$, for each $r = 1, 3, 4, \ldots, \frac{f-7}{6} + 1$, when $n \equiv 18 \pmod{24}$. Therefore, for each of these rs, the three triples $\{0_1, (c_r)_2, (d_r)_2\}$, $\{0_1, (c_{\frac{n-6}{6}+s_r})_2, (d_{\frac{n-6}{6}+s_r})_2\}$, $\{0_1, (c_{\frac{n-6}{6}+t_r})_2, (d_{\frac{n-6}{6}+t_r})_2\}$ are among those constructed in (iv) and they certainly satisfy the projection property. There are $(f-7)/6$ choices for r in each case, thus establishing the claim.

Case 2. $n \equiv 0, 6 \pmod{24}$. We first deal with the trivial case where $v = 3f + 10$. It can be seen that $v = 3f + 10$ happens only when $n \equiv 0 \pmod{24}$. Therefore, $n = f + 5 (\geq 24)$. An $(f, 2)$-rotational STS$(3f + 10)$ is given by the following base triples:

(i) the triples of an STS(f) on $\{\infty_1, \infty_2, \ldots, \infty_f\}$,

(ii) $\{0_1, (\frac{n}{3})_1, (\frac{2n}{3})_1\}$, $\{\infty_1, 0_1, (\frac{n}{2})_1\}$, $\{0_1, r_1, (\frac{n-6}{6} + b_r)_1\}$ $(r = 1, 2, \ldots, \frac{n-6}{6})$,

(iii) $\{\infty_1, 0_2, (\frac{n}{2})_2\}$,

(iv) $\{0_2, r_2, (\frac{n-6}{6} + b_r)_2\}$ $(r = 1, 2, \ldots, \frac{n-6}{6})$,
$\{0_1, (\frac{n}{2} - 2)_1, 0_2\}$, $\{0_1, 1_2, (\frac{n}{3} + 1)_2\}$, $\{0_1, (\frac{n}{2} + 1)_2, (n-1)_2\}$,
$\{\infty_j, 0_1, (x_j)_2\}$ $(j = 2, 3, \ldots, f)$,
where $\{x_2, x_3, \ldots, x_f\} = \{0, 1, \ldots, n-1\} \setminus \{0, 1, \frac{n}{3} + 1, \frac{n}{2} + 1, \frac{n}{2} + 2, n - 1\}$. Since $n \geq 24$, we have $n = f + 5$ and $\frac{n}{2} + 2 < n - 1$; therefore these two sets are indeed of the same size.

So without loss of generality, we assume $v \geq 3f + 22$ when $n \equiv 0 \pmod{24}$, in addition. Then we have $n \geq f + 11$ $(n \geq 24)$ when $n \equiv 0 \pmod{24}$ and $n \geq f + 5$ $(n \geq 30)$ when $n \equiv 6 \pmod{24}$. Consequently, $\frac{n-12}{6} - 1 \geq \frac{f-7}{6}$ when $n \equiv 0 \pmod{24}$ and $\frac{n-6}{6} - 1 \geq \frac{f-7}{6}$ when $n \equiv 6 \pmod{24}$. We now construct a (7, 2)-rotational STS$(v - (f-7))$ in the following, with the set of its base triples consisting of the triples in the above (ii), (iii) and

(v) the triples of an STS(7) on $\{\infty_1, \infty_2, \ldots, \infty_7\}$,

(vi) $\{\infty_2, 0_1, 0_2\}$, $\{\infty_3, 0_1, (n-2)_2\}$,

(vii) $\{0_1, (c_r)_2, (d_r)_2\}$ $(r = 1, 2, \ldots, \frac{n}{2} - 1)$,
with triples

$\{0_1, (c_{\frac{n}{2}-2})_2, (d_{\frac{n}{2}-2})_2\}$, $\{0_1, (c_1)_2, (d_1)_2\}$, $\{0_1, (c_{\frac{n}{2}-1})_2, (d_{\frac{n}{2}-1})_2\}$
to be destroyed and replaced by
$\{0_1, (\frac{n}{2}-2)_1, (d_{\frac{n}{2}-2})_2\}$, $\{0_2, 1_2, (\frac{n}{2}-1)_2\}$, $\{\infty_4, 0_1, (c_1)_2\}$, $\{\infty_5, 0_1, (d_1)_2\}$, $\{\infty_6, 0_1, (c_{\frac{n}{2}-1})_2\}$, $\{\infty_7, 0_1, (d_{\frac{n}{2}-1})_2\}$.

The extendibility of this system is shown below. When $n \equiv 0 \pmod{24}$, take (s_r, t_r) ($r = 1, 2, \ldots, \frac{n-12}{6}$) to be a $(B, \frac{n-12}{6})$-system. Then, for each $r = 2, 3, \ldots, \frac{f-7}{6} + 1$, the three triples

$$\{0_1, (c_r)_2, (d_r)_2\}, \{0_1, (c_{\frac{n-12}{6}+s_r})_2, (d_{\frac{n-12}{6}+s_r})_2\}, \{0_1, (c_{\frac{n-12}{6}+t_r})_2, (d_{\frac{n-12}{6}+t_r})_2\}$$

occur in (vii) because $\{r, s_r, t_r\} \subseteq \{2, 3, \ldots, \frac{n}{2}-7, \frac{n}{2}-5\} \subseteq \{2, 3, \ldots, \frac{n}{2}-3\}$ and they satisfy the projection property; When $n \equiv 6 \pmod{24}$, take (s_r, t_r) ($r = 1, 2, \ldots, \frac{n-6}{6}$) to be an $(A, \frac{n-6}{6})$-system. Then, for each $r = 2, 3, \ldots, \frac{f-7}{6} + 1$, the three triples

$$\{0_1, (c_r)_2, (d_r)_2\}, \{0_1, (c_{\frac{n-6}{6}+s_r})_2, (d_{\frac{n-6}{6}+s_r})_2\}, \{0_1, (c_{\frac{n-6}{6}+t_r})_2, (d_{\frac{n-6}{6}+t_r})_2\}$$

are among those in (vii) because $\{r, s_r, t_r\} \subseteq \{2, 3, \ldots, \frac{n}{2} - 3\}$ and they satisfy the projection property. □

Lemma 3.16 *Let $v \equiv 1 \pmod{12}$, $f \equiv 1 \pmod{12}$. If $f \geq 13$ and $v \geq 3f + 10$, then there exists an $STS_\pi(v)$.*

Proof. Now, $n \equiv 0 \pmod 6$, $f - 7 \equiv 0 \pmod 6$. The assumption also implies $n \geq f + 5 \geq 18$ and, consequently, $\frac{n-6}{6} - 1 \geq \frac{f-7}{6}$. Furthermore, when $n \equiv 6 \pmod{24}$, $n \geq f + 5 \geq 30$; When $n \equiv 0 \pmod{24}$, the conditions $v \equiv 1 \pmod{12}$, $f \equiv 1 \pmod{12}$ and $v \geq 3f + 10$ imply that $v \geq 3f + 22$ and therefore $\frac{n-12}{6} - 1 \geq \frac{f-7}{6}$. So the constructions in the proof of Lemma 3.15 can be adapted to here with only the trivial case in Case 2 where $n \equiv 0 \pmod{24}$ and $v = 3f + 10$ omitted. □

4 MAIN RESULTS

In this section we establish the existence conditions for $(f, 2)$−rotational Steiner triple systems with $f > 1$. The following lemma shows the conditions which an $(f, 2)$-rotational Steiner triple system of order v has to satisfy. We do not require $f > 1$ here.

Lemma 4.1 *If there is an $(f,2)-$rotational Steiner triple system of order v, then* (i) $v \equiv 1,3$ (*mod* 6), (ii) $f \equiv 1,3$ (*mod* 6), *and* (iii) *if* $v - f \equiv 2$ (*mod* 4) *then* $v \geq 3f$; *if* $v - f \equiv 0$ (*mod* 4) *then* $v = f$ *or* $v \geq 3f - 2$.

Proof. Suppose an $(f,2)$-rotational STS(v), $(V, \mathcal{T})$, exists. Then the set of fixed points forms a subsystem. So (i) and (ii) hold trivially.

We prove (iii) by considering the number of base triples involving one fixed point. There are only two types of these triples, Type 4 and Type 5. Furthermore, each fixed point is in one of the base triples of these types since each element of $Z_n \times \{1,2\}$ has to occur once with the fixed point in a triple of $\mathcal{T}$. Suppose first $v - f \equiv 2$ (mod 4). Then n is odd and the base triples of Type 4 do not exist. Consequently, there have to be f distinct base triples of Type 5. Since each of such base triples covers only one mixed difference, hence the number n of the mixed differences is at least f. We then obtain $v = f + 2n \geq 3f$. Now suppose $v - f \equiv 0$(mod 4). Then n is even. In this case, if $v > f$ then there have to be the base triples of Type 4. Moreover, if some fixed point appears in one of the two base triples of Type 4 then so does it in the other. This leaves $f - 1$ fixed points which have to fall in $f - 1$ distinct base triples of Type 5. Therefore the number of distinct mixed differences is at least $f - 1$, which then implies $v \geq 3f - 2$. □

We now deal with the exceptional cases:

(**1**) $v = 3f - 2$ when $f \equiv 1 \pmod{12}$ and $f > 1$,

(**2**) $v = 21$ when $f = 3$,

(**3**) $v = 3f - 2$ or $v = 3f + 10$ when $f \equiv 15 \pmod{24}$,

(**4**) $v = 3f + 4$ when $f \equiv 3 \pmod{12}$,

(**5**) $v = 25$ when $f = 7$,

(**6**) $v = 3f - 2$ when $f \equiv 7 \pmod{12}$,

(**7**) $v = 27$ when $f = 9$,

(**8**) $v = 3f - 2$ when $f \equiv 21 \pmod{24}$, $v = 3f + 10$ when $f \equiv 9 \pmod{24}$,

(**9**) $v = 3f + 4$ when $f \equiv 9 \pmod{12}$.

Lemma 4.2 *No $(f,2)$–rotational Steiner triple system of order v exists whenever v, f fall in any of the above cases.*

Proof. Consider case (2). In this case, $n = 9$ and therefore there are 4 nonzero pure (i,i)-differences, for each $i = 1, 2$, and 9 mixed differences. If there is a (3, 2)-rotational STS(21), then 3 of the mixed differences have to fall in base triples of Type 5. This leaves 6 mixed differences which have to occur in base triples of Type 3. If exactly one pair of these 6 mixed differences fall in a base triple of Type 3 which also covers an (i,i)-difference, then it forces a base triple of Type 1 or Type 2 to cover exactly two $(i+1, i+1)$-differences, which is impossible. So the only possibility is that all the 6 mixed differences are covered by 3 base triples of Type 3 each of which contains exactly one (i,i)-difference. But then all four $(i+1, i+1)$-differences fall in base triples of Type 1 and Type 2, so we have a cyclic STS(9) on $Z_9 \times \{i+1\}$, which is again impossible. Similarly, the nonexistence of a cyclic STS(9) also implies the nonexistence of an $(f,2)$–rotational STS(v) in cases (5) and (7).

Consider case (4). We have $n = f + 2 \equiv 5 \pmod{12}$. There is no base triple of Type 2. There are altogether $(f+1)/2$ nonzero pure (i,i)-differences and $f+2$ mixed differences. Suppose there is an $(f,2)$-rotational STS(v) in this case. Since there are no base triples of Type 4, each fixed point has to fall in a base triple of Type 5. This leaves two mixed differences which have to fall in a base triple, say $\{0_i, x_i, y_{i+1}\}$, of Type 3. This base triple covers only one (i,i)-difference so all other nonzero pure (i,i)-differences are forced to occur in base triples of Type 1. But this is impossible since each base triple of Type 1 covers exactly three (i,i)-differences and $\frac{f+1}{2} - 1 \equiv 1 \pmod 6$. A similar counting of differences demonstrates the nonexistence of an $(f,2)$-rotational STS(v) in case (9).

Now consider case (1). Obviously $n = f - 1$. Suppose there is an $(f,2)$-rotational STS(v) in this case. Since $f - 1$ is even, there has to be a fixed point ∞_j which is used to form the two base triples of Type 4. Consequently, all other fixed points have to fall in $f - 1$ base triples of Type 5 and all nonzero pure (i,i)-differences (for each $i \in \{1,2\}$), except for $(\frac{n}{2})_{ii}$, have to fall in base triples of Type 1 or Type 2. This forces a 1-rotational STS$(n+1)$ on $(Z_n \times \{i\}) \bigcup \{\infty_j\}$ for each $i \in \{1,2\}$, which is impossible since $n + 1 \equiv 1, 13 \pmod{24}$. The nonexistence of an $(f,2)$-rotational STS(v) in cases (3), (6) and (8) follows in a similar manner. □

Theorem 4.3 *Assume $f > 1$. Then there exists an $(f,2)$-rotational Steiner triple system of order v if and only if*

(i) *$v \equiv 1, 3$ (mod 6),*

(ii) *$f \equiv 1, 3$ (mod 6), and*

(iii) *if $v - f \equiv 2$ (mod 4) then $v \geq 3f$; if $v - f \equiv 0$ (mod 4) then $v = f$ or $v \geq 3f - 2$,*

with the exceptions of $v = 3f - 2$ when $f \equiv 1$ (mod 12), $v = 21$ when $f = 3$, $v = 3f - 2$ or $v = 3f + 10$ when $f \equiv 15 \pmod{24}$, $v = 3f + 4$ when $f \equiv 3$ (mod 12), $v = 25$ when $f = 7$, $v = 3f - 2$ when $f \equiv 7$ (mod 12), $v = 27$ when $f = 9$, $v = 3f - 2$ when $f \equiv 21 \pmod{24}$, $v = 3f + 10$ when $f \equiv 9 \pmod{24}$, $v = 3f + 4$ when $f \equiv 9$ (mod 12).

Proof. The existence of an $(f, 2)$-rotational STS(v) is immediate when $v = f$. We assume $v > f > 1$.

Obviously the necessary conditions (as in Lemma 4.1) $v \geq 3f$ when $v - f \equiv 2 \pmod{4}$ and $v \geq 3f - 2$ when $v - f \equiv 0 \pmod{4}$ further imply $v \geq 3f + 4$, when $v \equiv 1 \pmod{12}$, $f \equiv 3, 7 \pmod{12}$ and when $v \equiv 7 \pmod{12}$, $f \equiv 1, 9 \pmod{12}$; and $v \geq 3f + 6$, when $v \equiv 3 \pmod{12}$, $f \equiv 3, 7 \pmod{12}$ and when $v \equiv 9 \pmod{12}$, $f \equiv 1, 9 \pmod{12}$. The nonexistence for the exceptional cases is proved in Lemma 4.2. For each of the remaining cases, an $(f, 2)$-rotational STS(v) has been constructed in Lemma 3.1–3.16. We do not require $f > 1$ in Lemma 3.4 and Lemma 3.6, and therefore the existence of a (1, 2)-rotational (namely, a 2-rotational) STS(v), when $v \equiv 3$ or $7 \pmod{12}$, is established as a special case from these two lemmas. □

Acknowledgments

I express my sincere thanks to my supervisor, Dr. Charles J. Colbourn, for his many valuable suggestions about the research at its early stages, and in the preparation of this paper.

REFERENCES

[1] C. J. Cho, Rotational Steiner triple systems, *Discrete Math.* **42** (1982), 153-159.

[2] C. J. Colbourn and A. Rosa, *Triple Systems*, Oxford University Press, to appear.

[3] J. Doyen, A note on reverse Steiner triple systems, *Discrete Math.* **1** (1971-72), 315-319.

[4] J. Doyen and R. M. Wilson, Embeddings of Steiner triple systems, *Discrete Math.* **5** (1973), 229-239.

[5] R. B. Gardner, Steiner triple systems with near-rotational automorphisms, *J. Combin. Theory, series A* **61** (1992), 322-327.

[6] A. Hartman and D. G. Hoffman, Steiner triple systems with an involution, *Europ. J. Combin.* **8** (1987), 371-378.

[7] E. S. O'Keefe, Verification of a conjecture of Th. Skolem, *Math Scand.* **9** (1961), 80-82.

[8] R. Peltesohn, Eine Lösung der beiden Heffterschen Differenzenprobleme, *Compositio Math.* **6** (1939), 251-257.

[9] K. T. Phelps and A. Rosa, Steiner triple systems with rotational automorphisms, *Discrete Math.* **33** (1981), 57-66.

[10] A. Rosa, Poznámka o cyklických Steinerových systémoch trojíc, *Math.-Fyz. Čas* **16** (1966), 285-290.

[11] A. Rosa, On reverse Steiner triple systems, *Discrete Math.* **2** (1972), 61-71.

[12] Th. Skolem, On certain distributions of integers in pairs with given differences, *Math. Scand.* **5** (1957), 57-68.

[13] Th. Skolem, Some remarks on the triple systems of Steiner, *Math. Scand.* **6** (1958), 273-280.

[14] L. Teirlinck, The existence of reverse Steiner triple systems, *Discrete Math.* **6** (1973), 299-300.

A SIMPLE POLYNOMIAL TIME ALGORITHM FOR A CONVEX HULL PROBLEM EQUIVALENT TO LINEAR PROGRAMMING

Bahman Kalantari

Department of Computer Science
Rutgers University, New Brunswick, NJ 08903, USA

ABSTRACT

Over the rationals, the general linear programming problem is equivalent to the convex hull problem of determining if a given $m \times n$ matrix H has a nontrivial nonnegative zero. We give a polynomial time algorithm that either finds a nontrivial nonnegative zero of H, or it obtains a hyperplane separating the column vectors of H from the origin. In particular, the algorithm provides an alternate proof of a strengthened version of Gordan's duality theorem, previously proved by the author. The algorithm which is motivated by this duality theorem is analogous to Karmarkar's algorithm but its analysis is much simpler.

1 INTRODUCTION

Given an $m \times n$ matrix H of real coefficients, define *Homogeneous Feasibility Problem* (HFP), as the problem of testing the nonemptiness of

$$U(H) = \{x \in \Re^n : Hx = 0, \quad x \geq 0, \quad x \neq 0\},$$

and *Separation Problem* (SP), as the problem of testing the nonemptiness of

$$U'(H) = \{y \in \Re^m : H^T y < 0\}.$$

According to Gordan's Duality Theorem (GDT), see Dantzig [1],

$$U(H) = \emptyset \quad \Longleftrightarrow \quad U'(H) \neq \emptyset. \qquad (GDT)$$

C. J. Colbourn and E. S. Mahmoodian (eds.), Combinatorics Advances, 207–216.

Geometrical interpretation of GDT is simple: $U(H) \neq \emptyset$ if and only if the origin in $\Re^m$ lies in the convex hull of the column vectors of H, and $U'(H) \neq \emptyset$ if and only if the origin can be separated from these vectors via a hyperplane.

From the duality theory in linear programming, see Dantzig [1], it is well-known that the general linear programming problem is equivalent to the problem of testing the feasibility of $\Omega = \{x : Ax = b, x \geq 0\}$, where A is an $m \times n$ matrix. The recessive cone of Ω is the set $\{x : Ax = 0, \quad x \neq 0, \quad x \geq 0\}$, which for $H = A$, coincides with $U(H)$ defined in HFP. If $U(A)$ is empty, it is easy to see that $\Omega \neq \emptyset$ if and only if $U(H) \neq \emptyset$, where $H = [A, -b]$. Suppose $U(A)$ is empty. Then, given that A and b have rational entries with binary input length L, it is well-known that $\Omega \neq \emptyset$ if and only if $\Omega' = \{x' = (x, s) : Ax = b, \quad e^T x + s = M, \quad x' \geq 0\} \neq \emptyset$, where $e = (1, \ldots, 1)^T$, $s \in \Re$ the slack variable, and $M = O(2^L)$ a computable constant. Since the recessive cone of Ω' is empty, and its size is $O(L)$, we can conclude: Over the rationals, a general linear programming problem of binary input length L is reducible to an equivalent HFP of binary input length $O(L)$.

Karmarkar's linear programming algorithm [5], which in fact can be viewed as an algorithm designed for HFP, is an $O(nL)$ iteration algorithm where each iteration consists of solving a system of linear equations in $O(n^3)$ arithmetic operations. The first $O(\sqrt{n}L)$ iteration linear programming path-following Newton method was obtained by Renegar [10]. Karmarkar's rank-one updating scheme which results in $O(n^{2.5})$ amortized complexity per iteration can also be used within path-following algorithms, e.g. see Vaidya [11], so that a linear programming problem is solvable in $O(n^3 L)$ arithmetic operations. A simple $O(\sqrt{n}L)$ iteration algorithm for HFP was given by Khachiyan and Kalantari [8]. The aim of the present paper is to give an extremely simple $O(nL)$ iteration algorithm for HFP. Although the algorithm is analogous to Karmarkar's algorithm, its analysis is much simpler. Furthermore, the algorithm gives an alternate proof of a strengthened version of GDT, previously proved in Kalantari [2], [3]. However, this duality result motivates the algorithm. The algorithm first converts HFP into a problem equivalent to Karmarkar's canonical linear program. In the remaining of this section we introduce the latter problem, show its relationship to HFP, and state the strengthened GDT. Section 2, presents an algorithmic proof of this duality theorem.

Let A be an $m \times n$ matrix and $c \in \Re^n$. Assume that there exists $d^0 \in \Re^n$, $d^0 > 0$ satisfying $Ad^0 = 0$, $c^T d^0 > 0$. Consider the problem of testing the nonemptiness of

$$V(A, c) = \{x \in \Re^n : Ax = 0, \quad c^T x = 0, \quad x \geq 0, \quad x \neq 0\}.$$

We shall refer to the above problem as KCLP, since it is equivalent to Karmarkar's canonical linear programming problem. We claim that KCLP is equivalent to HFP. Clearly, $V(A,c) = U([A^T, c]^T)$. Conversely, given a matrix H, it is easy to see that $U(H) \neq \emptyset$ if and only if $V(A,c) \neq \emptyset$, where $A = [H, -Hx^0]$, $c = (0, \cdots, 0, 1)^T$, x^0 an arbitrary positive point, and $d^0 = (x^0, 1)$. In particular, x^0 and hence d^0 can be taken to be the vector of ones.

If in KCLP, $V(A,c) = \emptyset$, then from GDT there exists $y \in \Re^m$, $z \in \Re$ such that $A^T y + zc < 0$. Multiplying this inequality by d^{0^T}, we get $d^{0^T} A^T y + zc^T d^0 < 0$. Since $Ad^0 = 0$, and $c^T d^0 > 0$, we conclude that $z < 0$. Thus, for the matrix $H = [A^T, c]^T$, GDT can be restated as

$$V(A,c) = \emptyset \quad \Longleftrightarrow \quad V'(A,c) = \{y \in \Re^m : A^T y < c\} \neq \emptyset.$$

Suppose we are given y in $V'(A,c)$. Thus, for some $u > 0$ we have

$$c - u = A^T y. \tag{i}$$

Let $d \in \Re^n$ be any positive vector satisfying $Ad = 0$. Let $\lambda = Du$, where $D = \text{diag}(d)$ is the diagonal matrix with diagonal entries $d_1, \ldots, d_n$. Then (i) can be written as

$$c - D^{-1}\lambda = A^T y. \tag{ii}$$

From (ii), d can be interpreted as the (unique) constrained stationary point of the convex function

$$f(x) = c^T x - \sum_{i=1}^{n} \lambda_i \ln x_i,$$

with respect to the constraints $Ax = 0$, with y as the corresponding vector of Lagrange multipliers. Multiplying (ii) by D, we get

$$Dc - \lambda = DA^T y. \tag{iii}$$

Equivalently, from (iii), the vector $e = (1, \ldots, 1)^T \in \Re^n$ can be interpreted as the (unique) constrained stationary point of the convex function

$$f_d(x) = c^T Dx - \sum_{i=1}^{n} \lambda_i \ln x_i,$$

with respect to the constraints $ADx = 0$.

With the added assumption that A has rank m, multiplying (iii) by AD and solving for y gives

$$y = (AD^2 A^T)^{-1}(AD^2 c - AD\lambda). \tag{iv}$$

Substituting (*iv*) in (*iii*) gives

$$P_d Dc = P_d \lambda, \qquad (v)$$

where $P_d = I - DA^T(AD^2A^T)^{-1}AD$.

What turns the above observations on Gordan's Theorem into a powerful algorithmic tool is that the role of d and λ in equations (*ii*)-(*v*) can be reversed:

$$V(A,c) = \emptyset \iff \textit{given any positive } \lambda \in \Re^n,$$
$$\textit{there exists a unique } d \in \{x \in \Re^n : Ax = 0, \quad x > 0\} \textit{ satisfying } (ii) - (v). \qquad (SGDT)$$

Furthermore, the vector d in SGDT which is a constrained stationary point of $f(x)$, can also be shown to be a constrained stationary point of Karmarkar potential function $K(x) = c^T x / (\prod_{i=1}^n x_i^{(\lambda_i / e^T \lambda)})$. For $\lambda = e$, the proof of SGDT which is a strengthened version of Gordan's Theorem is given in Kalantari [2] (Theorem 3.5). For arbitrary $\lambda > 0$ and in the general context of homogeneous functions, SGDT is proved in Kalantari [3]. An even more general version of SGDT is given in Kalantari [4]. In particular, since for $\lambda = e$, $ADe = Ad = 0$, (*v*) reduces to $P_d Dc = e$, and the corresponding y in (*iv*) becomes $y = (AD^2A^T)^{-1}AD^2c$. For the purpose of the present paper it is enough to consider the weaker inequality $P_d Dc > 0$, from which the following weaker version of SGDT can be deduced

$$V(A,c) = \emptyset \iff V''(A,c) \neq \emptyset, \qquad (SGDT^*)$$

where

$$V''(A,c) = \{y = (AD^2A^T)^{-1}AD^2C \in \Re^m : \quad Ad = 0, \quad d > 0, \quad A^T y < c\}.$$

Obviously, $V''(A,c)$ is a subset of $V'(A,c)$. However, equality may not hold. For example if $A = [1,-1]$, and $c = [1,1]^T$, then $V''(A,c) = \{0\}$, whereas $V'(A,c)$ is the interval $(-1,1)$, justifying the fact that even SGDT* is stronger than GDT. The next section describes an algorithmic proof of SGDT* which in turn is motivated by it.

2 THE ALGORITHM

For the sake of completeness of this section, while introducing some new notations, we reintroduce the problem assumptions from scratch. Let $W = \{x \in$

$\Re^n : Ax = 0\}$, where A is a matrix of rank m. Assume there exists a positive vector $d^0 \in W$, satisfying $c^T d^0 > 0$, where c is a given vector in $\Re^n$. Given a positive $d \in W$, letting $D = \text{diag}(d)$, define $W_d = \{x \in \Re^n : ADx = 0\}$, $V_d = V(AD, Dc) = \{x \in W_d : c^T Dx = 0, \quad x \geq 0, \quad x \neq 0\}$, and $f_d(x) = c^T Dx - \sum_{i=1}^n \ln x_i$. For $d = e = (1, \ldots, 1)^T$, denote V_e and f_e, by V and f, respectively. We shall describe an algorithm that either finds a point in V or a positive d in W satisfying $P_d Dc > 0$, where $P_d = I - DA^T(AD^2A^T)^{-1}AD$, the projection matrix with respect to W_d.

Let $d > 0$ be a given point in W. Suppose we have

$$\lambda = P_d Dc > 0. \tag{1}$$

Then, for any nonnegative, nontrivial $x \in W_d$, since $P_d x = x$, we have $c^T Dx = c^T DP_d x = \lambda^T x > 0$. Thus, if (1) is satisfied, then V_d is empty and clearly so is V. If (1) is not satisfied we shall show how to obtain a point $x \in W_d$ so that $f(Dx) - f(d) \leq -\delta$, where $\delta > 0.3$. For $x > 0$ we have

$$f(Dx) - f(d) = c^T Dx - \sum_{i=1}^n \ln x_i d_i - c^T d + \sum_{i=1}^n \ln d_i =$$

$$c^T Dx - \sum_{i=1}^n \ln x_i - \sum_{i=1}^n \ln d_i - c^T d + \sum_{i=1}^n \ln d_i = f_d(x) - f_d(e). \tag{2}$$

Let $u \in \Re^n$ be any vector satisfying $\|u\| = (\sum_{i=1}^n u_i^2)^{1/2} = 1$. For $\theta \in (-1, 1)$, we have $\ln(1+\theta) = \sum_{k=1}^\infty (-1)^{k+1}\theta^k/k$. From this Taylor's expansion, and since for $k \geq 2$, $|\sum_{i=1}^n u_i^k| \leq \sum_{i=1}^n u_i^2 \leq 1$; for any $\alpha \in [0, 1)$, we have

$$-\sum_{i=1}^n \ln(1 + \alpha u_i) = \sum_{k=1}^\infty (-1)^k \frac{\alpha^k}{k} \sum_{i=1}^n u_i^k = -\alpha \sum_{i=1}^n u_i + \sum_{k=2}^\infty (-1)^k \frac{\alpha^k}{k} \sum_{i=1}^n u_i^k$$

$$\leq -\alpha \sum_{i=1}^n u_i + \sum_{k=2}^\infty \frac{\alpha^k}{k} = -\alpha \sum_{i=1}^n u_i - \alpha + \sum_{k=1}^\infty \frac{\alpha^k}{k} = -\alpha e^T u - \alpha - \ln(1-\alpha). \tag{3}$$

From (3), and that $\nabla f_d(e) = Dc - e$, we get $f_d(e + \alpha u) - f_d(e) =$

$$C^T D(e + \alpha u) - \sum_{i=1}^n \ln(1 + \alpha u_i) - c^T De \leq \alpha \nabla f_d(e)^T u - \alpha - \ln(1 - \alpha). \tag{4}$$

Suppose $u = -P_d \nabla f_d(e)/\|P_d \nabla f_d(e)\|$, i.e. the projected steepest decent direction of $f_d(x)$ with respect to W_d, at e. Since $P_d e = e$, and $P_d^2 = P_d$, we have

$$\gamma = \nabla f_d(e)^T u = -\|P_d \nabla f_d(e)\| = -\|P_d(Dc - e)\| = -\|P_d Dc - e\|. \tag{5}$$

If (1) is not satisfied, we must have $\gamma \leq -1$. Otherwise, letting $\lambda = P_d D c$, from (5) we get $-1 < \lambda_i - 1 < 1$, for all $i = 1, \ldots, n$, implying that $\lambda > 0$, a contradiction. If $\gamma \leq -1$, by setting $\alpha = 0.5$ in (4) we get

$$f_d(e + 0.5u) - f_d(e) \leq -\delta, \quad \delta = 1 - \ln 2 > 0.3. \tag{6}$$

Based on the above results we now state a simple algorithm that either finds a point in V, or a positive $d \in W$ satisfying (1):

Let $d^k > 0$ be the given current iterate in W. If $c^T d^k \leq 0$, set α to be the solution to the equation $\alpha c^T d^0 + (1-\alpha) c^T d^k = 0$. Then, $v = \alpha d^0 + (1-\alpha) d^k$ is in V, stop. Otherwise, compute $\lambda^k = P_{d^k} D_k c$, where $D_k = diag(d^k)$. If $\lambda^k > 0$, stop with the conclusion that $V = \emptyset$. Otherwise, replace d^k with $d^{k+1} = D_k(e + 0.5u^k)$, where $u^k = -P_{d^k} \nabla f_{d^k}(e) / \|P_{d^k} \nabla f_{d^k}(e)\|$, and repeat.

Assume that the algorithm is executed k times and $c^T d^k > 0$. From the repeated application of (6) and (2) we get

$$f(d^k) \leq f(d^0) - k\delta. \tag{7}$$

Clearly, $f(d^k) \geq \min\{f(td^k) : t \in (0, \infty)\}$. From a simple calculation it can be shown that

$$\min\{f(td^k) : t \in (0,\infty)\} = n - n \ln n + n \ln\left(\frac{c^T d^k}{(\prod_{i=1}^n d_i^k)^{\frac{1}{n}}}\right). \tag{8}$$

Since from the relationship between arithmetic and geometric means,

$$c^T d^k / (\prod_{i=1}^{n} d_i^k)^{\frac{1}{n}}) \geq n c^T d^k / e^T d^k,$$

from (7) and (8) we get

$$\ln \frac{c^T d^k}{e^T d^k} \leq \left(-1 + \frac{f(d^0)}{n}\right) - \frac{\delta k}{n}. \tag{9}$$

Equivalently, setting $\sigma = \exp(-1 + f(d^0)/n)$, and $x^k = d^k / e^T d^k$, we get

$$c^T x^k \leq \sigma \exp(-\frac{\delta k}{n}), \quad x^k \in \Omega_0 = \{x : Ax = 0, \ \ e^T x = 1, \ \ x \geq 0\}. \tag{10}$$

If the algorithm does not halt, from (10) it follows that any accumulation point of x^k's is a point in V. However, for rational inputs A and c of binary input length L, and when d^0 (which for HFP can be taken to be e) also has size $O(L)$, then $O(nL)$ iterations are enough: If $V \neq \emptyset$, in $O(nL)$ iterations we can determine a point u^0 in Ω_0 which if not in V satisfies $0 < c^T u^0 < 2^{-L}$. Finally, u^0 can be rounded to a solution v of Ω_0 satisfying $c^T v \leq 0$, thereby giving rise to a point in V. As with the ellipsoid method, see Khachiyan [6], [7], or Karmarkar's algorithm [5], or path-following Newton methods, e.g. Renegar [10] and Vaidya [11], the above algorithm too can be modified so that its arithmetic operations can be carried out to a precision of $O(L)$ bits. Detailed analysis of the sufficiency of $O(L)$ bit precision and/or the rounding schemes are given in [7], [9], [10], [11]. In the remaining of this section however, we shall describe one possible and efficient way to do the final rounding scheme which is well-suited for the problem at hand.

First we convert u^0, described above, into a basic feasible solution u^1 of $\Omega_1 = \{x : Ax = 0, \;\; e^T x = 1, \;\; c^T x = c^T u^0, \;\; x \geq 0\}$. Since we may assume that $c^T u^0$ has size $O(L)$, say satisfying $2^{-L-1} \leq c^T u^0 < 2^{-L}$, the size of Ω_1 will remain to be $O(L)$. Standard procedures for conversion of u^0 to u^1, using a precision of $O(L)$ bits are well-known, see e.g. [9]. In the Appendix we shall describe one possible and efficient procedure for doing this conversion. Now u^1 is a feasible solution of Ω_0. Since the rank of the matrix of Ω_1 is at most $m+2$, u^1 has at most $m+2$ positive components. If u^1 is not already a vertex of Ω_0, by forcing each of its $m+2$ positive components into zero, we obtain at most $m+2$ feasible vertices $v^1, \ldots, v^{m+2}$ of Ω_0. Since u^1 must lie in the convex hull of these vertices, there must exist $v \in \{v^1, \ldots, v^{m+2}\}$ such that $c^T v \leq c^T u^0$. Assuming without loss of generality that c_i's are integers, from the basis representation of v and Cramer's rule, it is easy to see that $c^T v = (1/q) \sum_{i=1}^{n} c_i p_i$, where p_i's and $q > 0$ are integers less that 2^L. Thus $qc^T v < q 2^{-L} < 1$, implying that $c^T v \leq 0$.

Appendix

Suppose we are given a feasible solution of $Ax = b$, $x \geq 0$, where A is an $m \times n$ matrix and $b \neq 0$. Here we describe an $O(m^2 n)$ algorithm to compute a solution so that the column vectors of A corresponding to the nonzero components of the solution are linearly independent. In particular, if A has rank m the obtained solution is a basic feasible solution.

Since $b \neq 0$, we may assume that no column of A is zero. At the beginning of each iteration we have a solution $x \geq 0$ of $Ax = b$, and a submatrix A' of A

consisting of the columns of A corresponding to $x_j > 0$. We also have an $m \times r$ working submatrix B of A', of rank r. Initially, B consists of a single column of A'. Given B and A', we pick a column, a, of A' other than those of B, and test if it is linearly independent of the columns of B. We also assume that we have computed the inverse of $M = B^T B$. It is easy to see that a is not linearly independent of columns of B if and only if $a = B(B^T B)^{-1} B^T a$.

Case (I): a is linearly independent of the columns of B. We obtain the inverse of

$$M' = \begin{bmatrix} a^T \\ B^T \end{bmatrix} [a, B] = \begin{pmatrix} a^T a & a^T B \\ B^T a & B^T B \end{pmatrix} = \begin{pmatrix} c & u^T \\ u & M \end{pmatrix}.$$

From the well-known formula for the inverse of partitioned matrices we have

$$M'^{-1} = \begin{pmatrix} d & v^T \\ v & N \end{pmatrix},$$

where $d = (c - u^T M u)^{-1}$, $v = -dM^{-1}u$, and $N = M^{-1} + \frac{1}{d} v v^T$. Thus, M'^{-1} can be computed in $O(m^2)$ operations.

Case (II): a is not linearly independent of the columns of B. Without loss of generality assume that $B = [a^1, \ldots, a^r]$, and $a = a^{r+1}$. Thus, $a^{r+1} = B(B^T B)^{-1} B^T a^{r+1}$, and letting $(\alpha_1, \ldots, \alpha_r)^T = (B^T B)^{-1} B^T a^{r+1}$, and $\alpha_{r+1} = -1$, we have $\sum_{j=1}^{r+1} \alpha_j a^j = 0$. Thus, for any real number θ we have $\sum_{j=1}^{r+1} x_j a^j = \sum_{j=1}^{r+1} (x_j - \theta \alpha_j) a^j$. In particular if θ is taken to be $\min\{\theta : x_j - \theta \alpha_j \geq 0, \quad j = 1, \ldots, r+1\}$, we obtain a new solution $x' \geq 0$, setting $k \geq 1$ positive components of the previous solution to zero. The components which become zero correspond to k columns of B. We delete these k columns from B and update $(B^T B)^{-1}$ for the new B. Suppose $k = 1$, i.e. only one column of B is deleted. Except for a permutation of the columns of old B, we may assume that the old $B^T B$ is as the matrix M' given above. Given that M'^{-1} is available, $M^{-1} = N - \frac{1}{d} v v^T$ can be computed in $O(m^2)$ operations. If $k > 1$, the above scheme can be repeated one at a time.

In either case, having updated the working matrix B and A', the above scheme is repeated as many times as needed. The total complexity of this procedure is $O(m^2 n)$. This is because each column of A initially with a positive x_j, enters the working matrix B at most once. Once such a column is added to B, it gets deleted from it at most once. Thus the work per each such column of A is $O(m^2)$. Assuming that A and b are rationals with binary input length L, all the operations can be carried out with precision of $O(L)$ bits.

Acknowledgments

The author thanks the Institute for Studies in Theoretical Physics and Mathematics, Tehran for hospitality during his sabbatical leave there. This research was supported in part by the National Science Foundation under Grant No. CCR-9208371.

REFERENCES

[1] G.B. Dantzig, *Linear Programming and Extensions*, Princeton University Press, Princeton, NJ, 1963 .

[2] B. Kalantari, Canonical problems for quadratic programming and projective methods for their solution, Proceedings of AMS conference "Mathematical problems arising from linear programming", 1988. In: *Contemporary Mathematics*, Volume 114 (1990), 243-263.

[3] B. Kalantari, Derivation of a generalized and strengthened Gordan theorem from generalized Karmarkar potential and logarithmic barrier functions, Technical Report LCSR-TR-121, Department of Computer Science, Rutgers University, New Brunswick, NJ, 1989.

[4] B. Kalantari, A theorem of the alternative for multihomogeneous functions and its relationship to diagonal scaling of matrices, Technical Report LCSR-TR-202, Department of Computer Science, Rutgers University, New Brunswick, NJ, 1993. To appear in *Linear Algebra and its Applications*.

[5] N. Karmarkar, A new polynomial time algorithm for linear programming, *Combinatorica*, 4 (1984), 373-395.

[6] L. Khachiyan, A polynomial algorithm in linear programming, *Soviet Mathematics Doklady*, 20 (1979), 191-196.

[7] L. Khachiyan, Polynomial algorithms in linear programming. *USSR Computational Mathematics and Mathematical Physics*, 20 (1980), 53-72.

[8] L. Khachiyan and B. Kalantari, Diagonal matrix scaling and linear programming, *SIAM J. Optimization*, 2 (1992), 668-672.

[9] C.H. Papadimitriou, and K. Steiglitz, *Combinatorial Optimization Algorithms and Complexity*, Prentice-Hall, Inc., Englewood Cliffs, NJ, 1982.

[10] J. Renegar, A polynomial-time algorithm based on Newton's method for linear programming, *Mathematical Programming*, 40 (1988), 59-93.

[11] P.M. Vaidya, An algorithm for linear programming which requires $O(((m+n)n^2 + (m+n)^{1.5}n)L)$ arithmetic operations, *Mathematical Programming*, 47 (1990), 175-201.

A LINEAR-TIME ALGORITHM FOR MINIMUM COST FLOW ON UNDIRECTED ONE-TREES

Bahman Kalantari
*** Iraj Kalantari**

Department of Computer Science,
Rutgers University, New Brunswick, NJ 08903, USA

** Department of Mathematics*
Western Illinois University, Macomb, IL 61455, USA

ABSTRACT

We give an $O(n)$-time algorithm for the minimum cost flow problem over an undirected one-tree with n vertices. A one-tree is a spanning tree with one additional edge.

1 INTRODUCTION

Let $G = (V, E)$ be an undirected graph with a real number b_i (supply or demand) associated with each vertex $v_i \in V$, $i = 1, \cdots, n = |V|$, satisfying $\sum_{i=1}^{n} b_i = 0$. For each edge $e \in E$, let $c_e \geq 0$ be the unit shipping cost of the edge. The minimum cost flow problem (MCFP) is to ship the supplies through G satisfying the demands at minimal cost. More generally, minimum cost flow problem is considered over directed graphs which may also be *capacitated*, i.e. the flow across each arc is restricted to be within given upper and lower bounds. In the *uncapacitated* case, the upper and lower bounds for the arcs are infinity and zero, respectively. The first polynomial-time algorithm for the problem was given by Edmonds and Karp [5]. Other polynomial and strongly polynomial-time algorithms have been discovered by Tardos [13], Galil and Tardos [6], Goldberg and Tarjan [7, 8], Orlin [12], and others. For a network simplex algorithm see Grigoriadis [9]. In this paper, we give an $O(n)$-time algorithm for the special case where G is an undirected, uncapacitated, one-tree. A one-tree is a spanning tree with one additional edge. One-trees where defined by Held and Karp [10] in the context of the minimum Hamiltonian cycle problem. Our algorithm is motivated by, and generalizes, the algorithm for bipartite matching on a line or a circle due to Werman, Peleg, Melter, and Kong [14].

C. J. Colbourn and E. S. Mahmoodian (eds.), Combinatorics Advances, 217–223.

In Section 2 we first give a straight-forward linear-time algorithm for MCFP over spanning trees. Using this algorithm, we reduce the optimization problem over a one-tree to an equivalent optimization over a cycle. To arrive at an $O(n)$-time algorithm for MCFP over a cycle, we employ results from linear programming, least square absolute deviations, and the existing linear-time median algorithm.

2 THE ALGORITHM

Throughout the rest of this paper, we assume that G is an undirected, uncapacitated graph. Let $V(G)$ be the value of optimal flow cost over G. Consider the case where G is a spanning tree on the vertices $v_1, \cdots, v_n$. Assume that v_i is a vertex of degree one. Let v_j be the (unique) neighbor of v_i. Let G' be the graph obtained from G by removing the vertex v_i, the edge (v_i, v_j), and by replacing b_j by $b_j + b_i$. Let $V(G')$ be the optimal shipping cost over G'. Let c_e be the unit shipping cost of the edge $e = (v_i, v_j)$. We have

Theorem 2.1. $V(G) = V(G') + c_e|b_i|$.

Proof. If v_i is a source node ($b_i > 0$), b_i units must be transferred to v_j at the cost of $c_e b_i$. Otherwise, v_i receives b_i units which must pass through v_j at the partial cost of $-c_e b_i$. □

The above theorem suggest a linear-time algorithm for computing $V(G)$ when G is a spanning tree: Initialize $V(G) = 0$. While there exists a vertex v_i of degree one in G, modify G by removing v_i, the edge (v_i, v_j), replacing b_j by $b_i + b_j$, and updating $V(G)$ by $V(G) + c_e|b_i|$. The final value of $V(G)$ is the optimal flow cost. The data structure to realize this algorithm may be constructed in $O(n)$ space by a depth first traversal of the tree starting with an arbitrary vertex of degree one as the root vertex.

Now assume that G is a one-tree. As in the previous case, if there is a vertex of degree one, we remove this vertex, its incident edge, and adjust the supply (or demand) of its neighbor. We repeat this while updating the partial cost until no vertex of degree one exists. Since a one-tree contains a unique cycle, continuing the above process we eventually are left with a cycle. Thus without loss of generality we assume that G is the cycle given by the edges (v_1, v_2), $(v_2, v_3), \cdots, (v_{n-1}, v_n)$, (v_n, v_1), with the corresponding unit shipping cost of $c_1, \cdots, c_n$, respectively.

From the theory of linear programming, it is well known that when the underlying graph is directed, the basic solutions of the equivalent linear programming formulation correspond to spanning trees of G (see for example Chvatal [3]). An undirected graph may be converted into a directed one by replacing each edge with two arcs with opposite orientation. Hence, for undirected graphs as well, we may conclude that the optimal flow cost is attained over a spanning tree. Thus we have

Theorem 2.2. Over a one-tree, there exists an optimal solution for which the flow value is zero on at least one edge. □

This immediately suggests an $O(n^2)$-time algorithm for MCFP over a one-trees: Remove each edge of the cycle, apply the spanning tree algorithm described earlier, and among the n solutions that result, select the one with the least cost. However, the complexity of this algorithm is not optimal. In what follows we first describe an $O(n \log n)$-time algorithm. Finally, we give a linear-time algorithm.

For each $i = 1, \cdots, n$, define $w_i = \sum_{j=1}^{i} b_j$ and note that $w_n = 0$. Suppose there is an optimal solution with zero flow on the edge (v_n, v_1). From Theorem 2.1, it is easy to see that the optimal cost $V(G)$ is given by

$$V(G) = \sum_{i=1}^{n} |w_i| c_i.$$

Similarly, if for an optimal solution the flow value is zero on (v_j, v_{j+1}), for some $j \neq n$, the optimal cost is given by

$$V(G) = \sum_{i=1}^{n} |w_i - w_j| c_i.$$

Thus, if $f(w) = \sum_{i=1}^{n} |w_i - w| c_i$ and $W = \{w_1, \cdots, w_n\}$, we have

Theorem 2.3. $V(G) = \min\{f(w) : w \in W\}$. □

We now consider the optimization of f. Let π be the permutation of $1, \cdots, n$ which sorts the w_i's, i.e. $w_{\pi(1)} \leq w_{\pi(2)} \leq \cdots \leq w_{\pi(n)}$. The next theorem characterizes the point of W at which f is minimized.

Theorem 2.4. Let $k = \min\{j : \sum_{i=1}^{j} c_{\pi(i)} \geq \frac{1}{2} \sum_{i=1}^{n} c_i\}$. Then, $w^* = w_{\pi(k)}$ is the minimizer of $f(w)$.

Proof. Consider $f(w)$ for $w \in \Re$. Clearly, $f(w)$ is piecewise linear and since $c_i \geq 0$ for all $i = 1, \cdots, n$, it is convex. It is differentiable everywhere except for $w \in W$, and $\min\{f(w) : w \in W\} = \min\{f(w) : w \in \Re\}$. If $w < w_{\pi(1)}$, we have $f'(w) = -\sum_{i=1}^{n} c_i < 0$. If $w > w_{\pi(n)}$, $f'(w) = \sum_{i=1}^{n} c_i > 0$. It is easy to see that the minimum value for f occurs at the smallest point of W for which the right-hand derivative of f is nonnegative. Suppose $w_{\pi(j)} \leq w < w_{\pi(j+1)}$, for some $j \in \{1, \cdots, n-1\}$. Then,

$$f(w) = \sum_{i=1}^{j}(-w_{\pi(i)} + w)c_{\pi(i)} + \sum_{i=j+1}^{n}(w_{\pi(i)} - w)c_{\pi(i)}.$$

The right-hand derivative at such a w is given by

$$f'_+(w) = \sum_{i=1}^{j} c_{\pi(i)} - \sum_{i=j+1}^{n} c_{\pi(i)}.$$

If we add and subtract the quantity $\sum_{i=1}^{j} c_{\pi(i)}$ to the above equation and denote $\sum_{i=1}^{n} c_{\pi(i)}$ by σ we get

$$f'_+(w) = -\sigma + 2\sum_{i=1}^{j} c_{\pi(i)}.$$

The proof now follows from the previous analysis and the above equation. □

Once $w_{\pi(k)}$, known as the *weighted median*, is obtained, we remove the corresponding edge and apply the tree algorithm. The weighted median may be obtained by sorting the w_i's in $O(n \log n)$. In the context of least absolute-deviation curve-fitting, this complexity for computing the weighted median together with the above theorem is proved in Bloomfield and Steiger [1]. However, here we give a linear-time algorithm for computing the weighted median, and hence for MCFP over a one-tree. Before describing the linear-time algorithm we consider a special case considered in Werman et al. [14]. If b_i's are integers, we could sort the w_i's by bucket sorting in $O(\delta)$-time, where $\delta = \sum_{i=1}^{n} |b_i|$. In particular, bipartite matching of two sets of point of equal cardinality on a circle considered by Werman et al. [14], corresponds to the case where n is even, b_i is equal to 1 for all elements of one set, and -1 for all elements of the other set. Thus, in this special case since $O(\delta) = O(n)$, we have

Corollary 2.1. If for all $i = 1, \cdots, n$, b_i is either 1 or -1, then MCFP is solvable in $O(n)$-time. □

Finally, we show that more generally an $O(n)$-time algorithm is possible. We will use an approach similar to that of Megiddo [11] and Dyer [4], where we first exclude from consideration half of the points in W, next one-forth of the points, etc.

In the first iteration, using the linear-time median algorithm of Blum, Floyd, Pratt, Rivest, and Tarjan [2], in $O(n)$-time we compute the median m_0 of $W_0 = W$, and $w_{i_0} = \max\{w \in W_0 : w \leq m_0\}$. For a given $w \in W$, if we let $I(w) = \{i : w_i \leq w\}$, we have

$$f'_+(w) = -\sigma + \sum_{i \in I(w)} 2c_i.$$

Thus, $f'_+(w_{i_0})$ can be computed in $O(n)$-time. If $f'_+(w_{i_0}) \geq 0$, we define $W_1 = \{w \in W : w < w_{i_1}\}$. Otherwise, $W_1 = \{w \in W : w > w_{i_0}\}$. In the next iteration, in $O(\frac{1}{2}n)$-time we compute the median m_1 of W_1, and $w_{i_1} = \max\{w \in W_1 : w \leq m_1\}$. In general given $w, w' \in W$, we may write

$$f'_+(w') = f'_+(w) + \sum_{i \in I(w')\setminus I(w)} 2c_i - \sum_{i \in I(w)\setminus I(w')} 2c_i.$$

Suppose in the above equation we choose $w = w_{i_0}$ and $w' = w_{i_1}$. Since the cardinality of $I(w_{i_1})\setminus I(w_{i_0})$ (similarly $I(w_{i_0})\setminus I(w_{i_1})$) is at most $O(\frac{1}{2}n)$, we conclude that given $f'_+(w_{i_0})$, $f'_+(w_{i_1})$ can be computed in $O(\frac{1}{2}n)$-time. If $f'_+(w_{i_1}) \geq 0$, we define $W_2 = \{w \in W_1 : w < w_{i_1}\}$. Otherwise, $W_2 = \{w \in W_1 : w > w_{i_1}\}$. More generally, given W_j, m_j, w_{i_j}, and $f'_+(w_{i_j})$, we can compute W_{j+1}, m_{j+1}, $w_{i_{j+1}}$, and $f'_+(w_{i_{j+1}})$, in $O(n/2^j)$-time. Hence the total time in computing w^*, the minimizer of $f(w)$ is given by

$$O(n + \frac{n}{2} + \frac{n}{4} + \frac{n}{8} \cdots) = O(n).$$

To summarize, we have shown

Theorem 2.5. The MCFP over undirected one-trees is solvable in $O(n)$-time. □

Remarks. We note that our algorithm for MCFP over a one-tree insists on treating the graph as an undirected one as opposed to its conversion into an equivalent directed graph. While we have only considered the cases where G is either a tree or a one-tree, our algorithm can be modified to solve some other cases as well. For example when G is the union of two cycles, say, C_1 and C_2 with a single common vertex, say v_1. For $i = 1,\ 2$, let V_i be the set of vertices in

C_i excluding v_1. Let $\beta_1 = \sum_{\{i\ :\ v_i \in V_1\}} b_i$, and $\beta_2 = \sum_{\{i\ :\ v_i \in V_2\}} b_i$. We compute $V(C_1)$, the minimum cost flow over C_1, where we replace b_1 with $-\beta_1$. Next we compute $V(C_2)$, the minimum cost flow over C_2, where we replace b_1 with $-\beta_2$. It is easy to see that the optimal cost over G is $V(C_1) + V(C_2)$.

It is also worth mentioning that the algorithm of this paper gives rise to a primal simplex algorithm for solving MCFP over an arbitrary undirected graph as follows: Select an arbitrary spanning tree T of G. Solve the problem over T. Define an edge to be nonbasic if and only if it belongs to $G-T$. Select a nonbasic edge e and solve MCFP (using the algorithm of this paper) over the one-tree $T \cup \{e\}$. If the cost improves, replace T by the spanning tree $\{T \cup \{e\}\} - \{e'\}$, where e' is an edge in $T \cup \{e\}$ on which the new flow is zero. This results in an improved spanning tree and the process is repeated. Otherwise, select another nonbasic edge in $G - T$ and repeat the above. If the current solution does not improve for any nonbasic edge, it must be optimal. Clearly, in the absence of degenerate pivots, this gives a finite algorithm. Such an algorithm would most likely have a better performance than the primal simplex algorithm which works on the directed graph G', obtained by the duplication of the edges in G. This is because the number of feasible undirected spanning trees of G (possibly exponentially many) are far fewer than the number of feasible directed trees of G'. This algorithm however, is most likely non-polynomial. An interesting problem is whether there exists polynomial extensions of the algorithm of this paper to general undirected graphs. Needless to say that for such an algorithm to be of interest, its complexity must be faster than the existing polynomial algorithms which solve the equivalent directed formulation.

Acknowledgments

The first author thanks the Institute for Studies in Theoretical Physics and Mathematics, Tehran for hospitality during his sabbatical leave there. This research was supported in part by the National Science Foundation under Grant No. CCR-9208371.

REFERENCES

[1] P. Bloomfield and W.L. Steiger, *Least Absolute Deviation, Theory, Application, and Algorithms* (Birkhauser, 1983).

[2] M. Blum, R.W. Floyd, V.R. Pratt, R.L. Rivest, and R.E. Tarjan, Time bounds for selection, *Computer and System Sciences* 7 (1972) 448-461.

[3] V. Chvatal, *Linear Programming* (W.H. Freeman and Co., 1983).

[4] M.E. Dyer, Linear time algorithms for two- and three-variable linear programs, *SIAM J. Comput.* 13 (1984) 31-45.

[5] J. Edmonds and R.M. Karp, Theoretical improvements in algorithmic efficiency for network flow problem, *J. ACM* 19 (1972) 248-264.

[6] Z. Galil and E. Tardos, An $O(n^2(m + n\log n)\log n)$ min-cost flow algorithm, *Proc. 27th Annual Sympos. of Found. of Comp. Sci.* (1986) 1-9.

[7] A.V. Goldberg and R.E. Tarjan, Solving minimum cost flow problem by successive approximation, *Proc. 19th ACM Sympos. on the Theory of Computing* (1987) 7-18.

[8] A.V. Goldberg and R.E. Tarjan, Finding minimum-cost circulation by canceling negative cycles, *Proc. 20th ACM Sympos. on the Theory of Computing* (1988) 388-397.

[9] M.D. Grigoriadis, An efficient implementation of the network simplex method, *Mathematical Programming Study* 26 (1986) 83-111.

[10] M. Held and R. Karp, The traveling salesman problem and minimum spanning trees, *Operations Research* 18 (1970) 1138-1162.

[11] N. Megiddo, Linear-time algorithms for linear programming in R^3 and related problems, *SIAM J. Comput.* 12 (1983) 759-776.

[12] J. Orlin, A faster strongly polynomial minimum cost flow algorithm, *Proc. 20th ACM Sympos. on the Theory of Computing* (1988) 377-387.

[13] E. Tardos, A strongly polynomial minimum cost circulation algorithm, *Combinatorica*, 5 (1985) 247-255.

[14] M. Werman, S. Peleg, R. Melter, and T.Y. Kong, Bipartite graph matching for points on a line or a circle, *Journal of Algorithms* 7 (1986) 277-284.

AN ASYMPTOTIC EXISTENCE RESULT FOR ORTHOGONAL DESIGNS

Hadi Kharaghani

Dept. of Mathematics & Computer Science
University of Lethbridge
Lethbridge, AB T1K 3M4, Canada
and
Institute for Studies in Theoretical
Physics and Mathematics
Tehran, Iran

ABSTRACT

Orthogonal matrices with indeterminate entries are called orthogonal designs. There is a strong relationship between orthogonal designs and quadratic forms. Orthogonal designs are used to construct Hadamard matrices and, more generally, weighing matrices. Despite the importance of orthogonal designs, not much is known about their existence or construction. We use a powerful new constructive technique to find a kind of asymptotic existence result for orthogonal designs.

1 INTRODUCTION

In a recent paper, Craigen [1] proved the existence of signed group Hadamard matrices. As a consequence he provided a new asymptotic existence theorem for Hadamard matrices. Using a conventional method, Craigen, Holzmann and Kharaghani [2] subsequently showed an asymptotic existence result for complex Hadamard matrices, improving Craigen's result. They were also able to "transfer" an orthogonal design of order 2^α to a complex orthogonal design of order $2^{\alpha+1}q$, for certain odd integer q. In this paper, we further improve the results of [2]. As a consequence, we see that if $N(q)$ is the number of nonzero digits in the binary expansions of the odd integer q, then there is an orthogonal design which in turn provides an Hadamard matrix of order $2^\alpha q$, where $\alpha \leq 2N(q)-1$. A complex orthogonal design of order n and type $(s_1, s_2, \ldots, s_k)$, s_i positive integers, denoted $COD(n; s_1, s_2, \ldots, s_k)$, is an $n \times n$ matrix X, with entries in $\{0, \epsilon_1 x_1, \epsilon_2 x_2, \ldots, \epsilon_k x_k\}$, $\epsilon_i \in \{\pm 1, \pm i\}$, x_is commuting indetermi-

C. J. Colbourn and E. S. Mahmoodian (eds.), Combinatorics Advances, 225–233.

nates, which satisfies $XX^* = (\sum_{i=1}^{k} s_i x_i^2)I_n$. Here X^* is the conjugate transpose of X. When all the ϵ_is are real, such a design is an orthogonal design and is denoted $OD(n; s_1, s_2, \ldots, s_k)$. A pair of matrices X, Y are said to be amicable (antiamicable) if $XY^* = YX^*$ $(XY^* = -YX^*)$. $(a_1, a_2, \ldots, a_n)$ denotes a circulant matrix with first row as shown. α_ℓ means α repeated ℓ times. As usual $+ = 1$, $- = -1$, $\bar{x} = -x$, $*$ is the Hadamard product and $\otimes$ denotes the Kronecker product.

All notations follow that of Geramita and Seberry [3] unless otherwise specified.

2 BASIC RESULTS

We begin with a known result.

Lemma 2.1 *Let X, a $COD(n; s_1, s_2, \ldots, x_k)$, and Y, a $COD(n; t_1, t_2, \ldots, t_\ell)$, be amicable. Let $X = \sum_{i=1}^{k} x_i P_i$, $\quad Y = \sum_{i=1}^{\ell} y_i Q_i$. Then*

1. *P_i and Q_j are $(0, \pm 1, \pm i)$-matrices.*
2. *$P_i * P_j = 0, i \neq j$, $Q_{i'} * Q_{j'} = 0, i' \neq j'$.*
3. *$P_i P_i^* = s_i I_n$, $Q_j Q_j^* = t_j I_n$.*
4. *$P_i P_j^* + P_j P_i^* = 0, i \neq j$, $Q_{i'} Q_{j'}^* + Q_{j'} Q_{i'}^* = 0, i' \neq j'$.*
5. *$P_i Q_j^* = Q_j P_i^*$.*

Proof: See page 213 of [3] for details. □

The following lemma has its origin in rational quadratic forms.

Lemma 2.2 (Warren Wolfe [4]) *For each positive integer n, there is a pair of amicable orthogonal designs $OD(2^n; 1, 1, \ldots, 1)$ in $n + 1$ variables.*

Proof: This follow from Corollary 2 [4, p. 1015] and table (iv) of the same paper by considering four cases $q \equiv i \pmod 4$, $i = 0, 1, 2, 3$. □

Corollary 2.3 *For positive integer n, let $X = \sum_{i=1}^{n+1} x_i P_i$ and $Y = \sum_{i=1}^{n+1} y_i Q_i$ be two amicable $OD(2^n; 1, 1, \ldots, 1)$ in $n+1$-variables. Then P_i, Q_j are $(0, \pm 1)$-matrices satisfying 2–5 of Lemma 2.1*

Let $\mathcal{P} = \{P_1, P_2, \ldots, P_{n+1}\}$, $\mathcal{Q} = \{Q_1, Q_2, \ldots, Q_{n+1}\}$ be the above matrices. Then $\sqrt{-1}P_j = iP_j \in \mathcal{Q}$ and $iQ_j \in \mathcal{P}$ for $j = 1, 2, \ldots, n+1$.

Lemma 2.4 *Let A be a circulant $(0, \pm 1, \pm i)$-matrix of order n such that the matrices $A_1 = \alpha(A + A^*)$ and $A_2 = \alpha i(A - A^*)$ are $(0, \pm 1, \pm i)$-matrices for some α. Then A_1 and A_2 are disjoint, circulant and Hermitian matrices satisfying $A_1^2 + A_2^2 = 4\alpha^2 AA^*$.*

Proof: Straightforward. □

Note that the class of matrices satisfying the conditions of Lemma 2.4 includes the circulant ± 1-matrices, circulant ± 1- quasi-symmetric matrices and many other matrices. For example, for the circulant matrix $A = (0\ X_\ell\ 0_k\ 0_{\ell+k})$, where X_ℓ is any $(\pm 1, \pm i)$ sequence of length ℓ, A_1 and A_2 are $(0, \pm 1, \pm i)$-matrices for $\alpha = 1$. Obviously, a more general result than above is correct, but we choose to mention only what we use in this paper.

Lemma 2.5 *Let $\{A_1, A_2, \ldots, A_k\}$ be a set of commuting Hermitian matrices and $\{P_1, P_2, \ldots, P_k\}$ a set of mutually antiamicable matrices. Let $H = \sum_{i=1}^{k} A_i \otimes P_i$. Then $HH^* = \sum_{i=1}^{k} A_i^2 \otimes P_i P_i^*$.*

Furthermore, if the A_is are mutually disjoint circulant $(0, \pm 1)((0, \pm 1, \pm i))$-matrices and the P_is are $(0, \pm 1)((0, \pm 1, \pm i))$-matrices, then H is a block circulant $(0, \pm 1)((0, \pm 1, \pm i))$-matrix.

Proof:

$$
\begin{aligned}
HH^* &= \Big(\sum_{i=1}^{k} A_i \otimes P_i\Big)\Big(\sum_{i=1}^{k} A_i^* \otimes P_i^*\Big) \\
&= \sum_{i,j} A_i A_j \otimes P_i P_j^*
\end{aligned}
$$

$$\begin{aligned}
&= \sum_{i<j} A_i A_j \otimes (P_i P_j^* + P_j P_i^*) + \sum_{i=1}^{k} A_i^2 \otimes P_i P_i^j \\
&= \sum_{i=1}^{k} A_i^2 \otimes P_i P_i^*.
\end{aligned}$$

The rest is clear. □

3 MAIN RESULTS

For the remaining sections of the paper, for the positive integer q, $N(q) = N$ is the number of nonzero digits in the binary expansion of q.

Theorem 3.1 *Let* $q = 1 \pmod 4$ *and* $n = 2^{N(q)-2}$. *If there is an*

$$OD(n; s_1, s_2, \ldots, s_k),$$

then there is an $OD(2nq; 2qs_1, 2qs_2, \ldots, 2qs_k)$.

Proof: Let $q = 1 + \sum_{i=1}^{N-1} 2^{\alpha_i}$, where $2 \leq \alpha_1 < \alpha_3 < \cdots < \alpha_{N-1}$. For $1 \leq i \leq N-1$, let X_i, Y_i be a pair of Golay sequences of length $\ell_i = 2^{\alpha_i - 1}$. Let $M_0 = (+, 0_{q-1})$, $M_1 = (0, X_1, 0_{q-2\ell_1-1}, Y_1)$, $M_2 = (0, 0_{\ell_1}, X_2, 0_{q-2(\ell_1+\ell_2)-1}, Y_2, 0_{\ell_1})$, $\ldots$, $M_{N-1} = (0, 0_{\ell_1+\ell_2+\cdots+\ell_{N-2}}, X_{N-1}, Y_{N-1}, 0_{\ell_1+\ell_2+\cdots+\ell_{N-2}})$
$L_1 = (0, X_1, 0_{q-2\ell_1-1}, \bar{Y}_1)$, $L_2 = (0, 0_{\ell_1}, X_2, 0_{q-2(\ell_1+\ell_2)-1}, \bar{Y}_2, 0_{\ell_1})$, $\ldots$, $L_{N-1} = (0, 0_{\ell_1+\ell_2+\cdots+\ell_{N-2}}, X_{N-1}, \bar{Y}_{N-1}, 0_{\ell_1+\ell_2+\cdots+\ell_{N-2}})$.

Let $A_0 = B_0 = M_0$ and for $1 \leq i \leq N-1$, let $A_{2i-1}, A_{2i}(B_{2i-1}, B_{2i})$ be the matrices of Lemma 2.4 corresponding to $M_i(L_i)$ for $\alpha = \frac{1}{2}$. Then

$$\begin{aligned}
\sum_{i=0}^{2N-2} (A_i^2 + B_i^2) &= 2I_q + \sum_{i=1}^{N-1} (L_i L_i^* + M_i M_i^*) \\
&= 2I_q + \sum_{i=1}^{N-1} 2(X_i X_i^* + Y_i Y_i^*) = 2I_q + \sum_{i=1}^{N-1} 4 \cdot 2^{\alpha_i - 1} \\
&= 2qI_q.
\end{aligned}$$

Let $n = 2N - 2$ in the Corollary 2.3 and let

$$\mathcal{P} = \{P_1, P_2, \ldots, P_N, iQ_1, iQ_2, \ldots, iQ_{N-1}\},$$

$$\mathcal{Q} = \{Q_N, Q_{N+1}, \ldots, Q_{2N-1}, iP_{N+1}, iP_{N+2}, \ldots, iP_{2N-1}\}.$$

Let O be any $OD(n; s_1, s_2, \ldots, s_k)$. Let $C = A_0 \otimes P_1 O + A_1 \otimes P_2 O + A_3 \otimes P_3 O + \cdots + A_{2N-3} \otimes P_N O + A_2 \otimes iQ_1 O + A_4 \otimes iQ_2 O + \cdots + A_{2N-2} \otimes iQ_{N-1} O$, $D = B_0 \otimes Q_N O + B_1 \otimes Q_{N+1} O + B_3 \otimes Q_{N+2} O + \cdots + B_{2N-3} \otimes Q_{2N-1} O + B_2 \otimes iP_{N+1} O + B_4 \otimes iP_{N+2} O + \cdots + B_{2N-2} \otimes iP_{2N-1})$. Since for each $1 \leq i \leq N-1$, A_{2i} and B_{2i} are $(0, \pm 1)$-matrices, C and D have all their entries ± 1. Furthermore, C and D are two amicable matrices satisfying

$$\begin{aligned} CC^t + DD^t &= \left(\sum_{i=0}^{2N-2} A_i^2\right) \otimes OO^* + \left(\sum_{i=0}^{2N-2} B_i^2\right) \otimes OO^* \\ &= \left(\sum_{i=0}^{2N-2} (A_i^2 + B_i^2)\right) \otimes OO^* \\ &= 2qI_q \otimes OO^*. \end{aligned}$$

Let $X = \begin{pmatrix} C & D \\ D & -C \end{pmatrix}$, then X is an $OD(2^{2N-1}q; 2qs_1, 2qs_2, \ldots, 2qs_k)$. □

Theorem 3.2 *Let* $q = 1 + \sum_{i=1}^{N-1} 2^{\alpha_i}$, $2 \leq \alpha_1 < \alpha_2 < \cdots < \alpha_{N-1}$, $n = 2^{2N-2}$. *There is an* $OD(2nq; n, n, 2^{\alpha_1} \cdot n, 2^{\alpha_1} \cdot n, 2^{\alpha_2} \cdot n, 2^{\alpha_2} \cdot n, \ldots, 2^{\alpha_{N-1}} \cdot n, 2^{\alpha_{N-1}} \cdot n)$ *in* $2N$ *-variables.*

Proof: Let H be any Hadamard matrix of order n, A_i, B_i, $0 \leq i \leq 2N-2$ and $\mathcal{P}$, $\mathcal{Q}$ as in the proof of Theorem 3.1. Let $C = A_0 \otimes x_1 P_1 H + A_1 \otimes x_2 P_2 H + A_3 \otimes x_3 P_3 H + \cdots + A_{2N-3} \otimes x_N P_N H + A_2 \otimes i x_{N+1} Q_1 H + \cdots + A_{2N-2} \otimes x_{2N-1} Q_{N-1} H$, $D = B_0 x_{2N} Q_N H + B_1 \otimes x_{N+1} Q_{N+1} H + B_2 \otimes x_{N+2} Q_{N+2} H + \cdots + B_{2N-3} \otimes x_{2N-1} Q_{2N-1} H + B_2 \otimes i x_2 P_{N+1} H + \cdots + B_{2N-2} \otimes i x_N P_{2N-1} H$. It can be seen now that C, D are two block circulant real matrices satisfying

$$\begin{aligned} CC^t + DD^t &= A_0^2 \otimes x_1^2 HH^t + B_0^2 \otimes x_{2N}^2 HH^t + \sum_{i=1}^{N-1} (A_{2i-1}^2 + B_{2i}^2) \otimes x_{2i}^2 HH^t \\ &\quad + \sum_{i=1}^{N-1} (A_{2i}^2 + B_{2i-1}^2) \otimes x_{N+i}^2 HH^t \end{aligned}$$

$$= I_q \otimes nx_1^2 I_n + I_q \otimes nx_{2N}^2 I_n + \sum_{i=1}^{N-1} 2^{\alpha_i} I_q \otimes nx_{i+1}^2 I_n$$
$$+ \sum_{i=1}^{N-1} 2^{\alpha_i} I_q \otimes nx_{i+N} I_n.$$

Hence the matrix $\begin{pmatrix} C & D \\ -D & C \end{pmatrix}$ is an $OD(2nq; n, n, 2^{\alpha_1}\cdot n, 2^{\alpha_1}\cdot n, 2^{\alpha_2}\cdot n, 2^{\alpha_2}\cdot n, \ldots, 2^{\alpha_{N-1}}\cdot n, 2^{\alpha_{N-1}}\cdot n)$ in $2N$ variables. □

Example 3.3 *Let $q = 5 = 1 + 2^2$ in Theorem 3.1. Let $X_1 = (++)$, $Y = (+-)$ be a Golay sequence of length 2. Then $M_0 = (+0000)$, $M_1 = (0+++-)$, $L_1 = (0++-+)$, $A_0 = B_0 = (+0000)$, $A_1 = (00++0)$, $A_2 = (0i00\bar{i})$, $B_1 = (0+00+)$ and $B_2 = (00i\bar{i}0)$.*

Let $P_1 = \begin{pmatrix} 0 & + & 0 & 0 \\ - & 0 & 0 & 0 \\ 0 & 0 & 0 & + \\ 0 & 0 & - & 0 \end{pmatrix}$, $P_2 = \begin{pmatrix} 0 & 0 & + & 0 \\ 0 & 0 & 0 & - \\ - & 0 & 0 & 0 \\ 0 & + & 0 & 0 \end{pmatrix}$, $P_3 = \begin{pmatrix} 0 & 0 & 0 & + \\ 0 & 0 & + & 0 \\ 0 & - & 0 & 0 \\ - & 0 & 0 & 0 \end{pmatrix}$,

$Q_1 = \begin{pmatrix} 0 & + & 0 & 0 \\ - & 0 & 0 & 0 \\ 0 & 0 & 0 & - \\ 0 & 0 & + & 0 \end{pmatrix}$, $Q_2 = \begin{pmatrix} 0 & 0 & + & 0 \\ 0 & 0 & 0 & + \\ - & 0 & 0 & 0 \\ 0 & - & 0 & 0 \end{pmatrix}$, *and* $Q_3 = \begin{pmatrix} 0 & 0 & 0 & + \\ 0 & 0 & - & 0 \\ 0 & + & 0 & 0 \\ - & 0 & 0 & 0 \end{pmatrix}$.

Let $O = \begin{pmatrix} a & b & c & d \\ \bar{b} & a & d & \bar{c} \\ \bar{c} & \bar{d} & a & b \\ \bar{d} & c & \bar{b} & a \end{pmatrix}$.

Then $C = A_0 \otimes P_1 O + A_1 \otimes P_2 O + A_2 \otimes iQO = (P_1 O, \overline{Q_1 O}, P_2 O, P_2 O, Q_1 O)$ $D = B_0 \otimes Q_2 O + B_1 \otimes Q_3 O + B_3 \otimes iP_3 O = (Q_2 O, Q_3 O, \overline{P_3 O}, P_3 O, Q_3 O)$. Now, $X = \begin{pmatrix} C & D \\ D & -C \end{pmatrix}$ is an $OD(40; 10, 10, 10, 10)$. If we use monomials of order 8 and $OD(8; 1, \ldots, 1)$ in 8 variables, we will get an $OD(80; 10, \ldots, 10)$ in 8 variables. Applying the construction method for Theorem 3.2, we have

$$\begin{aligned} C &= A_0 \otimes x_1 P_1 H + A_2 \otimes x_2 P_2 H + A_2 \otimes i x_3 Q_1 H \\ &= (x_1 P_1 H, \overline{x_3 Q_1 H}, x_2 P_2 H, x_2 P_2 H, x_3 Q_1 H) \\ D &= B_0 \otimes x_4 Q_2 H + B_1 \otimes x_3 Q_3 H + B_2 \otimes i x_2 P_3 H \\ &= (x_4 Q_2 H, x_3 Q_3 H, \overline{x_2 P_3 H}, x_2 P_3 H, x_3 Q_3 H) \end{aligned}$$

Here H is any Hadamard matrix of order 4.

Let $Y = \begin{pmatrix} C & D \\ D & -C \end{pmatrix}$, *then* Y *is an* $OD(40; 4, 4, 16, 16)$.

Remark: By using monomials of order 8 and minor modification, we can get an $OD(80; 4, 4, 4, 4, 32, 32)$. Such an OD is not obtained from $OD(40; 4, 4, 16, 16)$ by a trivial method.

We now consider the odd integers q, $q \equiv 3 \pmod 4$.

Theorem 3.4 *Let* $q \equiv 3 \pmod 4$ *and* $n = 2^{N(q)-3}$. *If there is a*

$$COD(n; s_1, s_2, \ldots, s_k),$$

then there is a $COD(2nq; 2qs_1, \ldots, 2qs_k)$.

Proof: Let $q = 1 + \sum_{i+1}^{N-1} 2^{\alpha_i}$, where $1 = \alpha_1 < \alpha_2 < \alpha_3 < \cdots < \alpha_{N-1}$. For $1 \le i \le N-1$, let X_i, Y_i be a pair of Golay sequences of length $\ell_i = 2^{\alpha_i - 1}$.

Let $M_0 = (+, 0_{q-1})$, $M_1 = (0, +, 0_{q-3}, +)$, $L_1 = (0, +, 0_{q-3}, -)$, $M_2, \ldots, M_{N-1}$, $L_2, \ldots, L_{N-1}$ as in the proof of Theorem 3.1. Let $A_0 = B_0 = M_0$, $A_1 = M_1$, $B_1 = iL_1$ and for $2 \le i \le N-1$, let A_{2i-1}, A_{2i}, B_{2i-1}, B_{2i} be the matrices in the proof of Theorem 3.1. The rest of the proof is similar to the proof of Theorem 3.1. See also [2]. □

Theorem 3.5 *Let* $q = 1 + \sum_{i=1}^{N-1} 2^{\alpha_i}$, $1 = \alpha_1 < \alpha_2 < \cdots < \alpha_{N-1}$, $n = 2^{2N-2}$. *There is a* $COD(2nq; n, n, 4n, 2^{\alpha_2} \cdot n, 2^{\alpha_2} \cdot n, \ldots, 2^{\alpha_{N-1}} \cdot n, 2^{\alpha_{N-1}} \cdot n)$ *in* $2N-1$ *variables.*

Proof: For $1 \le i \le N-1$, let X_i, Y_i be a pair of Golay sequences of length $2^{\alpha_i - 1}$. Let $M_0, M_1, L_1, A_1, A_3, A_4, \ldots, A_{2N-2}, B_1, B_3, B_4, \ldots, B_{2N-2}$ as in the proof of Theorem 3.4. Now follow the construction of Theorem 3.2. □

Example 3.6 *Let* $q = 3 = 1 + 2$ *in Theorem 3.5. Then* $M_0 = (+, 0, 0)$, $M_1 = (0, +, +)$, $L_1 = (0, +, -)$. *Thus* $A_0 = B_0 = (+, 0, 0)$, $A_1 = (0, +, +)$, $B_1 = (0, i, \bar{i})$.

Let $P_1 = \begin{pmatrix} + & 0 \\ 0 & + \end{pmatrix}$, $P_2 = \begin{pmatrix} 0 & + \\ - & 0 \end{pmatrix}$, $Q_1 = \begin{pmatrix} + & 0 \\ 0 & - \end{pmatrix}$, $Q_2 = \begin{pmatrix} 0 & + \\ + & 0 \end{pmatrix}$, *and* $O = \begin{pmatrix} a & b \\ b & \bar{a} \end{pmatrix}$. *Then* $C = A_0 \otimes P_1O + A_1 \otimes P_2O = (P_1O, P_2O, P_2O)$ $D = B_0 \otimes Q_1O + B_1 \otimes Q_2O = (Q_1O, iQ_2O, \bar{i}Q_2O)$. *The matrix* $\begin{pmatrix} C & D \\ D & -C \end{pmatrix}$ *is a* $COD(12; 6, 6)$. *Using* $A_0, B_0, A_1, B_1, P_1, P_2, Q_1, Q_2$ *and any Hadamard matrix of order 2 in Theorem 3.5, we get*

$$C = A_0 \otimes x_1P_1H + A_1 \otimes x_2P_2H = (x_1P_1H, x_2P_2H, x_2P_2H)$$

$$D = B_0 \otimes x_3Q_1H + B_1 \otimes x_2Q_2H = (x_3Q_1H, ix_2Q_2H, \bar{i}x_2Q_2H).$$

Now the matrix $X = \begin{pmatrix} C & D \\ D & -C \end{pmatrix}$ *is a* $COD(12; 2, 2, 8)$.

Remarks:

1. Theorem 3.1 extends the main results of [2].
2. All the orthogonal designs obtained in Theorems 3.2 and 3.5 are new.
3. A repeated application of Theorem 3.1 (and all other theorems of this section) provides many new orthogonal designs. For example, by repeating the method of Example 3.3, we get $OD(10^p \cdot 4; 10^p, 10^p, 10^p, 10^p)$ for any positive integer p.
4. We have used only Golay sequences of length 2^α in our results. Using Golay sequences of other lengths or Base sequences will result in many more orthogonal designs.
5. Our result improves the asymptotic existence result for Hadamard matrices of [2] at least by a power of 2.

Acknowledgment: This work was completed while the author was visiting the institute for the Theoretical Physics and Mathematics (IPM) in Iran. The author wishes to thank the combinatorics group for their hospitality. The work is supported by an NSERC grant.

REFERENCES

[1] R. Craigen, "Signed groups, sequences and asymptotic existence of Hadamard matrices", to appear in JCT.

[2] R. Craigen, W.H. Holzmann, and H. Kharaghani, "On the asymptotic existence of Complex Hadamard matrices", preprint 1994.

[3] A. Geramita and J. Seberry, *Quadratic Forms, Orthogonal Designs, and Hadamard Matrices*, Vol. 45 of Lecture Notes in Pure and Applied Mathematics, Marcel Dekker Inc., New York and Basel, 1979.

[4] W. Wolfe, "Amicable orthogonal designs - existence", *Can. J. Math*, 28 (1976), pp. 1006-1020.

DECOMPOSITION OF COMPLETE TRIPARTITE GRAPHS INTO 5-CYCLES

E.S. Mahmoodian and Maryam Mirzakhani

Department of Mathematical Sciences
Sharif University of Technology
P.O. Box 11365-9415
Tehran, Iran
emahmood@irearn.bitnet

ABSTRACT

A result of Sotteau on the necessary and sufficient conditions for decomposing the complete bipartite graphs into even cycles has been shown in many occasions, that it is a very important tool in the theory of graph decomposition into *even* cycles. In order to have similar tools in the case of odd cycle decomposition, obviously bipartite graphs are not suitable to be considered. Searching for such tools, we have considered decomposition of complete tripartite graphs, $K_{r,s,t}$, into 5-cycles. There are some necessary conditions that we have shown their sufficiency in the case of $r = t$, and some other cases. Our conjecture is that these conditions are always sufficient.

1 INTRODUCTION

A theorem of Sotteau [2] states that,

> The necessary and sufficient conditions in order that $K_{m,n}$ admits a decomposition into $2k$-cycles are:
>
> (i) $m \geq k, \quad n \geq k$;
>
> (ii) m and n both are even;
>
> (iii) $2k \mid mn$.

This theorem is a very useful tool in recursive construction in cycle decomposition problems. But it may be used only when cycles are of even length. So

C. J. Colbourn and E. S. Mahmoodian (eds.), Combinatorics Advances, 235–241.

if we want to work with odd cycles we should look at other alternatives, say complete tripartite graphs, $K_{r,s,t}$.

Decomposition of $K_{r,s,t}$ into triangles is easy:

Proposition 1.1 *Necessary and sufficient conditions for $K_{r,s,t}$ to be decomposed into 3-cycles is that $r = s = t$.*

Proof. Necessity of the conditions is trivial and to show the sufficiency one may use a latin square of size r. ■

The next case is 5-*cycle* decomposition.

Example 1. The complete tripartite graph $K_{4,2,2}$ may be decomposed into 5-cycles. Suppose the sets $A = \{a_1, a_2, a_3, a_4\}$, $B = \{b_1, b_2\}$, and $C = \{c_1, c_2\}$ are three parts of this graph. Then a decomposition may be done as follows:

$$(b_1, a_1, c_2, a_3, c_1), (b_2, a_1, c_1, a_4, c_2), (c_2, a_2, b_2, a_3, b_1), \text{ and } (c_1, a_2, b_1, a_4, b_2).$$

2 NECESSARY CONDITIONS

To decompose $K_{r,s,t}$ into 5-cycles some immediate necessary conditions follow.

Theorem 2.1 *Let $r \leq s \leq t$. If there exists a decomposition of $K_{r,s,t}$ into 5-cycles, then the following conditions are satisfied:*

(i) *r, s, and t all are even or all are odd;*

(ii) $5 \mid rs + rt + st$*;*

(iii) $t \leq 4rs/(r + s)$.

Proof. Each cycle uses two edges from each vertex, thus the degree of each vertex must be even. In other words $r + s$, $r + t$, and $s + t$ all are even. Thus (i) follows. To see (ii), note that $rs + rt + st$ is the total number of edges in $K_{r,s,t}$.

For (iii), we note that each 5-cycle uses *at least one* edge and *at most three* edges from between any two parts of $K_{r,s,t}$. This results in six inequalities:

$$2st/3(s+t) \leq r \leq 4st/(s+t),$$
$$2rt/3(r+t) \leq s \leq 4rt/(r+t),$$
$$2rs/3(r+s) \leq t \leq 4rs/(r+s).$$

Since $r \leq s \leq t$, the inequality $t \leq 4rs/(r+s)$ implies the other ones. ∎

Conjecture. The three conditions given in Theorem 2.1 are also sufficient.

We have proved this conjecture in the case when two parts have equal number of vertices and satisfy the necessary conditions with one exception of $K_{5x,5x,z}$ (z not multiple of 5). Also we have some results for the remaining cases. For some small cases it has been shown to be true.

3 AN APPLICATION

Before going further we show that how the decomposition of $K_{r,s,t}$ can be a *tool* for 5-cycle decompositions. In 1966, Rosa [1] proved that

> A complete graph on n vertices ,K_n, admits a 5-cycle decomposition if and only if $n \equiv 1$ or $5 \pmod{10}$.

We may prove the sufficiency of conditions in Rosa's theorem by applying the decomposition of $K_{r,s,t}$ into 5-cycles. (Necessity of the conditions is trivial as usual).

For n we have two cases:

(i) $n = 10l + 1$,
(ii) $n = 10l + 5$.

(i) In this case we let $r = 10l' + 5$ and $s = 10l'' + 1$ where:

$$l' = \lfloor l/3 \rfloor \quad \text{and} \quad l'' = l - 2\lfloor l/3 \rfloor - 1.$$

Then $n = r + r + s$. Now we proceed by induction, by decomposing K_r, K_r, K_s, and $K_{r,r,s}$ into 5-cycles and "patching" them together to obtain a decomposition for K_n.

(ii) In this case we let $r = 10l' + 5$ and $s = 10l'' + 5$, where l' and l'' are as above. And by a similar method we may decompose K_n.

4 SUFFICIENCY OF CONDITIONS

In this section we prove some useful theorems and some results on the sufficiency of conditions. First an extension theorem:

Theorem 4.1 *If $K_{r,s,t}$ admits a 5-cycle decomposition, then so does $K_{ar,as,at}$ for each positive integer a.*

Proof. We apply Proposition 1.1 to the decomposition of $K_{a,a,a}$ into triangles. Indeed, each part in $K_{ar,as,at}$ may be partitioned into a classes, each with equal number of vertices. Then each class of vertices in each part may be considered as a vertex. This will give a $K_{a,a,a}$ which, in turn, can be decomposed into triangles. Each of these triangles represents a $K_{r,s,t}$. ∎

Corollary 4.2 *$K_{r,r,r}$ admits a 5-cycle decomposition if and only if $5 \mid r$.*

Proof. The necessary conditions of Section 2, in this case, imply $5 \mid r$. To show sufficiency, according to Theorem 4.1, we just need a decomposition of $K_{5,5,5}$. Suppose that the sets $A = \{a_1, a_2, a_3, a_4, a_5\}$, $B = \{b_1, b_2, b_3, b_4, b_5\}$, $C = \{c_1, c_2, c_3, c_4, c_5\}$ are three parts of this graph. Then a decomposition may be done with the base cycles as follows:

$$(a_2, b_5, c_1, b_1, c_5), (b_2, c_5, a_1, c_1, a_5), \text{ and } (c_2, a_5, b_1, a_1, b_5).$$

In other words, we may generate other cycles by adding the indices of the above base cycles (mod 5). ∎

Corollary 4.3 *For every positive integer n, $K_{2n,2n,4n}$ admits a 5-cycle decomposition.*

Proof. In Example 1 of Section 1, we gave a decomposition for $K_{2,2,4}$. ∎

Corollary 4.4 *For every positive integer m, $K_{m,3m,3m}$ admits a 5-cycle decomposition.*

Proof. $K_{1,3,3}$ may be decomposed easily as follows. Suppose that the sets $A = \{a_1\}$, $B = \{b_1, b_2, b_3\}$, $C = \{c_1, c_2, c_3\}$ are three parts of this graph. Then a decomposition may be done with the base cycles as follows:

$$(a_1, b_i, c_i, b_{i+1}, c_{i+2}) \qquad (\text{mod } 3),$$

where, $i = 1, 2, 3$. ■

Next we use the idea of decomposition of $K_{2n,2n,4n}$ and $K_{m,3m,3m}$, which were done by *base cycles*, in the following theorem which is our main theorem.

Theorem 4.5 *Suppose that at least two parts in a complete tripartite graph have the same number of vertices, say $K_{r,r,s}$. And suppose that the triplet (r, r, s) satisfies all three necessary conditions given in Theorem 2.1. Then $K_{r,r,s}$ has a 5-cycle decomposition except possibly when r is a multiple of 5 but s is not.*

Proof. By simple arithmetic from the necessary conditions we may deduce that there exists m and n such that $r = 3m + 2n$ and $s = m + 4n$.

Now we give the following base cycles, assuming that three parts are $A = \{a_1, \ldots, a_s\}$, $B = \{b_1, \ldots, b_r\}$, and $C = \{c_1, \ldots, c_r\}$.

- r is even, which implies that s is also even:

$$(b_{i+2n-j}, a_{m+j}, c_{i+1}, a_{m+2n+2j-i+2\lfloor i/2 \rfloor}, c_i)$$
$$(c_{i+3m+n+j}, a_{m+n+j}, b_{i+1}, a_{m+2n+2j-i+2\lfloor i/2 \rfloor}, b_i)$$
$$(a_j, b_i, c_{i+3j}, b_{i+2}, c_{i+3j+1})$$
$$i = 1, 2, \ldots, r; \; j = 1, 2, \ldots, m.$$

  The indices of b's and c's are all to be computed modulo r.

- r and s are odd:

$$(b_{i+2n-j}, a_{m+j}, c_{i+1}, a_{m+2n+2j-i+2\lfloor i/2 \rfloor}, c_i)$$
$$(b_{r+2n-j}, a_{m+j}, c_{r+1}, a_{m+n+j}, c_r)$$
$$i = 1, 2, \ldots, r-1; \; j = 1, 2, \ldots, n;$$

$$\begin{gathered}(c_{i+1}, a_{m+n+j}, b_{i+n+2-j}, a_{m+2n+2j-i+2\lfloor i/2\rfloor}, b_{i+n+1-j})\\(c_r, a_{m+2n+2j-1}, b_{r+n+1-j}, a_{m+2n+2j}, b_{r+n-j})\\(c_1, a_{m+2n+2j}, b_{r+n+2-j}, a_{m+j+n}, b_{r+n+1-j})\\i = 1, 2, \ldots, r-2;\;\; j = 1, 2, \ldots, n;\end{gathered}$$

$$\begin{gathered}(a_j, b_i, c_{i+3j}, b_{i+2}, c_{i+3j+1})\\i = 1, 2, \ldots, r;\;\; j = 1, 2, \ldots, m.\end{gathered}$$

Again the indices of b's and c's are all to be computed modulo r. ∎

5 SEARCHING FOR A DECOMPOSITION IN OTHER CASES

One may get discouraged and conjecture that the condition $r = t$ is also necessary!

But it is not true. We show this in the following:

Lemma 5.1 *If $K_{a,b,b}$, $K_{a,c,c}$, and $K_{b,c,c}$ each admits a 5-cycle decomposition, then so does $K_{a+c,b+c,b+c}$.*

Proof. Consider the following 2×2 latin square:

	a	c
b	b	c
c	c	b

Using this latin square we may decompose $K_{a+c,b+c,b+c}$ into $K_{b,a,b}$, $K_{b,c,c}$, $K_{c,a,c}$, and $K_{c,c,b}$ each of which, by assumption, admits a 5-cycle decomposition. ∎

Proofs of the following lemmas are also immediate.

Lemma 5.2 *If $K_{a,b,b}$ and $K_{c,b,b}$ each admits a 5-cycle decomposition, then so does $K_{a+c,2b,2b}$.*

Lemma 5.3 *If $K_{a,b,b}$ and $K_{b,a,a}$ each admits a 5-cycle decomposition, then so does $K_{a+b,2a,2b}$.*

Example 2. $K_{30,20,40}$ admits a 5-cycle decomposition.

Proof. By Corollary 4.3, $K_{20,10,10}$ can be decomposed.

In the following we give two more decompositions:

- $K_{3,5,5}$, where its parts are $A = \{a, b, c\}$, $B = \{0, 1, 2, 3, 4\}$, and $C = \{5, 6, 7, 8, 9\}$:

  $a0615$ $a1726$ $a2837$ $a3948$ $a4509$ $b5291$ $c5364$ $c6b70$ $b9c80$ $c3b47$ $b2c18$.

- $K_{4,10,10}$, where its parts are $A = \{a, b, c, d\}$, $B = \{0, 1, 2, 3, 4, 5, 6, 7, 8, 9\}$, and $C = \{\bar{0}, \bar{1}, \bar{2}, \bar{3}, \bar{4}, \bar{5}, \bar{6}, \bar{7}, \bar{8}, \bar{9}\}$:

  $a1\bar{1}2\bar{2}$ $a2\bar{3}1\bar{4}$ $a3\bar{1}4\bar{3}$ $a4\bar{2}1\bar{5}$ $a5\bar{2}6\bar{1}$ $a6\bar{3}3\bar{6}$ $a7\bar{7}6\bar{0}$ $a8\bar{6}0\bar{8}$ $b1\bar{6}2\bar{4}$ $b2\bar{5}3\bar{7}$ $b3\bar{4}0\bar{5}$ $b4\bar{6}5\bar{3}$ $b5\bar{5}6\bar{6}$ $b6\bar{4}7\bar{8}$ $b7\bar{2}9\bar{1}$ $b8\bar{1}0\bar{2}$ $c1\bar{7}0\bar{9}$ $c\bar{0}1\bar{8}6$ $c3\bar{9}9\bar{3}$ $c\bar{1}5\bar{7}9$ $c\bar{6}9\bar{5}4$ $c2\bar{7}8\bar{8}$ $c5\bar{8}3\bar{2}$ $d1\bar{9}4\bar{8}$ $d3\bar{0}9\bar{4}$ $d\bar{0}5\bar{9}6$ $d7\bar{0}4\bar{7}$ $d2\bar{0}8\bar{3}$ $d\bar{6}7\bar{5}8$ $d\bar{1}7\bar{3}0$ $d\bar{9}2\bar{8}9$ $d5\bar{4}8\bar{2}$ $\bar{0}b9a0$ $\bar{9}b0c8$ $7c\bar{7}a\bar{9}$ $4d\bar{5}c\bar{4}$.

Thus, $K_{6,10,10}$ can be decomposed by Corollary 4.3. And $K_{10,20,20}$ may be decomposed by Lemma 5.2 and using $K_{6,10,10}$ and $K_{4,10,10}$. Therefore, $K_{30,20,40}$ admits a decomposition by Lemma 5.3 and using $K_{10,20,20}$ and $K_{20,10,10}$. ∎

REFERENCES

[1] A. Rosa. O cyklických rozkladoch kompletného grafu na nepárnouholníky. *Čas. Pěst. Mat.*, 91:53–63, 1966.

[2] D. Sotteau. Decomposition of $K(m, n)$ into cycles (circuits) of length $2k$. *J. of Combinatorial Theory B*, 30:75–81, 1981.

THE NSM OF A GRAPH

Dara Moazzami

Center for Theoretical Physics and Mathematics(AEOI)
and Shahid Beheshti University
Tehran, Iran

ABSTRACT

The purpose of this paper is to introduce a new invariant for measures of stability in networks. Since many network properties are actually properties of the underlying graph, we restrict this discussion to undirected graphs. We prove a number of basic results about this new parameter, including several relating it to other parameters of a graph, operations on graph, and Hamiltonian properties.

1 THE NEW STABILITY MEASURE OF A GRAPH (NSM)

The NSM of a graph G is defined as $NSM(G) = \min\{\frac{|A|\cdot\tau(G-A)}{\omega(G-A)}\}$, where the minimum is taken over all cut-sets A of G. A *cutset* of a connected graph is a collection of vertices whose removal results in a disconnected graph. For G, $\tau(G)$ is the number of vertices in a largest component of G and $\omega(G)$ is the number of components of G. We see that $\mathrm{NSM}(K_p) = p-1$ for every $p \geq 1$. A set $A \subset V(G)$ is said to be a NSM-set of G if $\mathrm{NSM}(G) = \frac{|A|\cdot\tau(G-A)}{\omega(G-A)}$. Also note that if G is disconnected the set A may be empty.

Throughout this paper we will let p and q be the number of vertices and edges, respectively, of G and we use $\beta(G)$ to denote the independence number of G. A graph G with at least 3 vertices is called n-Hamiltonian for $0 \leq n \leq p-3$, if the removal of any set of k vertices from G, $0 \leq k \leq n$, results in a Hamiltonian graph. By the inflation G^* of G we mean the graph whose vertices are all ordered

C. J. Colbourn and E. S. Mahmoodian (eds.), Combinatorics Advances, 243–250.

pairs (u, x), where x is an edge of G and u is an endpoint of x; two vertices of G^* are adjacent if they differ in exactly one coordinate. Any undefined terms can be found in the standard references on graph theory, including Behzad, Chartrand and Lesniak [1].

Proposition 1.1 *If G is a spanning subgraph of H, then $NSM(G) \leq NSM(H)$.*

Proof: If G is a spanning subgraph of H, we have $\omega(G-A) \geq \omega(H-A)$ and $\tau(G-A) \leq \tau(H-A)$ for any subset $A \subset V(G)$. Thus $NSM(G) \leq NSM(H)$.

Proposition 1.2 *For any graph G, $NSM(G) \geq \frac{\kappa(G)}{\beta(G)}$, where $\kappa(G)$ is the connectivity of G.*

Proof: For every cut-set A of G, we know that $|A| \geq \kappa(G)$. Thus $|A| \cdot \tau(G-A) \geq \kappa(G) \cdot 1$. Also we can see that the number of components of $G-A$ is at most $\beta(G)$.

Proposition 1.3 *If G is not complete, then $NSM(G) \leq \frac{p-\beta(G)}{\beta(G)}$.*

Proof: Let X be a largest independent set of vertices in G. Define $A = V(G) - X$, then $|A| = p - \beta(G), \tau(G-A) = 1$, and $\omega(G-A) = \beta(G)$. The result follows.

Proposition 1.4 *If $m \leq n$ then $NSM(K_{m,n}) = \frac{m}{n}$.*

Proof: If $G = K_{m,n}$, with $m \leq n$, then $\kappa(G) = m$ and $\beta(G) = n$. Combining Propositions 1.2 and 1.3 we obtain $\frac{\kappa(G)}{\beta(G)} \leq NSM(G) \leq \frac{m+n-\beta(G)}{\beta(G)}$. Hence $NSM(G) = \frac{m}{n}$.

Theorem 1.5 *For any graph G, $NSM(G) \geq t(G)$, where $t(G)$ is the toughness of a graph.*

Proof: Let $A \subset V(G)$ be a t-set, [5, 7, 8, 9, 11, 12, 13] and $B \subset V(G)$ be an NSM-set. Then $\frac{|B| \cdot \tau(G-B)}{\omega(G-B)} \geq \frac{|B|}{\omega(G-B)} \geq \frac{|A|}{\omega(G-A)}$ and hence we have that $NSM(G) \geq t(G)$.

Corollary 1.6 *For any graph G, $NSM(G^2) > \kappa(G)$.*

Proof: In [3] Chvàtal obtained the result $t(G^2) \geq \kappa(G)$ and by Theorem 1.5, $NSM(G^2) \geq t(G^2)$.

Corollary 1.7 *If m is a positive integer and $n = 2^m$, then $NSM(G^n) \geq \frac{1}{2}n\kappa(G)$.*

Proof: In [3] Chvàtal obtained the result $t(G^n) \geq \frac{1}{2}n\kappa(G)$.

Let us note that inequality $NSM(G^n) \geq \frac{1}{2}n\kappa(G)$ does not hold in general.

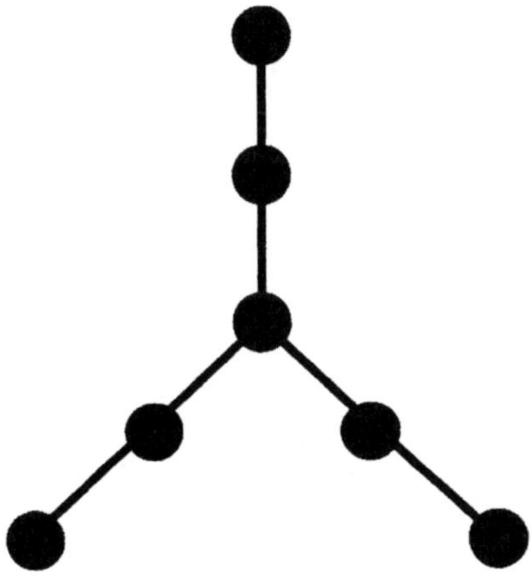

Corollary 1.8 *Let G be an arbitrary graph without isolated vertices and G^* its inflation. If $G \neq K_2$, then $NSM(G^*) \geq \frac{\lambda(G)}{2}$ and $\kappa(G^*) = \lambda(G^*) = \lambda(G)$.*

Proof: In [3] Chvàtal showed that $t(G^*) = \frac{1}{2}\lambda(G)$, and $\kappa(G^*) = \lambda(G^*) = \lambda(G)$.

Corollary 1.9 *Let G be a non-empty graph and let m be the largest integer such that $K_{1,m}$ is an induced sub-graph of G. Then $NSM(G) \geq \frac{\kappa(G)}{m}$.*

Proof: In [9] Goddard and Swart proved that under these conditions we have $t(G) \geq \frac{\kappa(G)}{m}$.

Corollary 1.10 *For any graph G, $NSM(G) \geq \text{bind}(G) - 1$.*

Proof: In [14] Woodall showed that $t(G) \geq \text{bind}(G) - 1$, where $\text{bind}(G)$ is the binding number of G.

Corollary 1.11 *Let G be a graph with $0 < NSM(G)$, and let $\lambda(G) = \lambda$ then $NSM(L(G)) > \frac{\lambda}{2}$.*

Proof: In [13] Pippert showed that $t(L(G)) \geq \frac{\lambda}{2}$, where $L(G)$ is the line graph of G and λ denotes the edge–connectivity.

2 *NSM* AND OPERATIONS ON GRAPHS

Theorem 2.1 *For any nontrivial, noncomplete graph G with p vertices and any vertex v, $NSM(G') \geq NSM(G) - \frac{\tau(G'-A')}{2}$, where $G' = G - v$ and A' be a NSM-set for G'.*

Proof: Let $G' = K_{p-1}$. Thus $NSM(G') = p - 1$, and by Proposition 1.3, $NSM(G) \leq \frac{p-2}{2}$. So $NSM(G') = p - 1 > \frac{p-2}{2} - \frac{1}{2} \geq NSM(G) - \frac{\tau(G'-A')}{2}$. Therefore the theorem holds. Hence $G' \neq K_{p-1}$. Let $|A'| = m$, then $NSM(G') = \frac{m \cdot \tau(G'-A')}{\omega(G'-A')}$. Now define $A = A' \cup \{v\}$. Clearly, A is a cut-set for G and so $NSM(G) \leq \frac{|A| \cdot \tau(G-A)}{\omega(G-A)}$. But $|A| = m + 1$ and $G - A = G' - A'$, so $NSM(G) \leq \frac{(m+1)\tau(G'-A')}{\omega(G'-A')} = \frac{m \cdot \tau(G'-A')}{\omega(G'-A')} + \frac{\tau(G'-A')}{\omega(G'-A')} \leq NSM(G') + \frac{\tau(G'-A')}{2}$.

Theorem 2.2 *If G is a bipartite, r-regular, r-connected graph on p vertices, then $NSM(G) = 1$.*

Proof: From [9] we know that $t(G) \geq 1$ and so by Theorem 1.5 we have $NSM(G) \geq 1$. Let A be one of the partite sets. Then, since G is r-regular, $|A| = \frac{p}{2}$, $\tau(G - A) = 1$ and $\omega(G - A) = \frac{p}{2}$. Therefore $NSM(G) = 1$.

Corollary 2.3 *If G_1 is a bipartite, n-regular, n-connected graph with p_1 vertices and G_2 is a bipartite, m-regular, m-connected graph with p_2 vertices then $NSM(G_1 \times G_2) = 1$.*

Proof: It is well known that $G_1 \times G_2$ is bipartite, $m+n$-regular and $m+n$-connected.

Theorem 2.4 *If $n \geq m$ then*

$$\frac{m+n-2}{2} \leq NSM(K_m \times K_n) \leq \frac{(mn-n)[\frac{n}{m}]}{m}.$$

Corollary 2.5 *For any integer n, $NSM(K_n \times K_n) = n-1$.*

Theorem 2.6 *Let $G = G_1 + G_2$, where $|V(G)| = p$, $|V(G_i)| = p_i$, $\beta(G_i) = \beta_i$, $t(G_i) = t_i$, and $G_i \neq K_{p_i}$, for i = 1, 2. Assume $\beta_1 \leq \beta_2$, then*

$$\min\{\frac{pt_1}{p_1}, \frac{pt_2}{p_2}\} \leq NSM(G) \leq \frac{p-\beta_2}{\beta_2}.$$

Proof: In [13], Pippert proved that $t(G) \geq min\{\frac{pt_1}{p_1}, \frac{pt_2}{p_2}\}$. Thus by Theorem 1.5 we have the bound. From Proposition 1.3 we obtain the upper bound.

3 HAMILTON PROPERTIES OF NSM

Proposition 3.1 *If G is Hamiltonian-connected and $p \geq 3$ then $NSM \geq 1$.*

Proof: In [3], Chvàtal proved that every Hamiltonian graph is 1-tough. Hence $t(G) \geq 1$. But if G is Hamiltonian connected and $p \geq 3$ then G is Hamiltonian. Therefore $NSM(G) \geq 1$.

We can see that a hamiltonian circuit in the inflation G^* of G induces a closed spanning trail in G and vice versa. Hence we have:

Proposition 3.2 *G^* is hamiltonian if and only if G has an eulerian spanning subgraph.*

This proposition and Corollary 1.8 yield:

Corollary 3.3 *Let G be a cubic nonhamiltonian graph with $\lambda(G) = 3$. Then its inflation G^* is a cubic nonhamiltonian graph with $NSM(G^*) \geq \frac{3}{2}$ and $\lambda(G^*) = 3$.*

Indeed, the inflation of a regular graph of degree n is a regular graph of degree n. Moreover, an eulerian spanning subgraph of a cubic graph is necessarily a hamiltonian cycle.

In particular, denoting by G_0 the Petersen graph and setting $G_{k+1} = G_k^*$ we obtain an infinite family $G_1, G_2, \ldots$ of cubic nonhamiltonian $\frac{3}{2}$-tough graphs. The Petersen graph G_0 is not $\frac{3}{2}$-tough; one can show that $t(G_0) = \frac{4}{3}$.

The following theorem, proved by Chvàtal and Erdös [4], enables us to relate Proposition 3.1 to the connectivity and independence number of a graph.

Theorem 3.4 *(Chvàtal and Erdös) If G is k-connected and $k > \beta$, then G is Hamiltonian-connected.*

Thus from Proposition 1.2 we have three possibilities for a graph G:

1. $1 < \frac{\kappa(G)}{\beta(G)} \leq NSM(G)$
2. $\frac{\kappa(G)}{\beta(G)} \leq 1 \leq NSM(G)$
3. $\frac{\kappa(G)}{\beta(G)} \leq NSM(G) < 1.$

By Proposition 3.1, graphs satisfying the third inequality are not Hamiltonian-connected. By Theorem 3.4, graphs satisfying the first inequality are Hamiltonian-connected. The cycle C_p, $p \geq 6$, satisfies the second inequality but is not Hamiltonian-connected while the graph C_p^2, $p \geq 15$ satisfies the second inequality and is Hamiltonian-connected.

In [2] Chartrand, Kapoor and Lick considered some conditions necessary for a graph to be n-Hamiltonian. Let graph G be m-connected. By definition every Hamiltonian graph is 2-connected. Since the removal of any n vertices from an n-Hamiltonian graph G results in a Hamiltonian graph, it follows that G is $(n+2)$-connected.

Theorem 3.5 *If G is n-Hamiltonian then $NSM(G) \geq 1 + \frac{n}{\beta(G)}$.*

Proof: Let A be a cut-set of G. We know that $|A| \geq n+2$. Let A_n be an n-vertex subset of A. Since G is n-Hamiltonian, $G-A_n$ has a Hamiltonian cycle C. The components of $C-A$ are disjoint paths $P_1, \ldots, P_r$. At least two vertices of A lie on C. Let v be one of these vertices. If we start at v and travel around C in a definite direction and return to v, we traverse each P_i exactly once. Let u_i be the next vertex of C encountered after having passed through P_i in the chosen direction. Then $u_1, \ldots, u_r$ are distinct vertices of A. Thus, $|A| \geq r+n$. Also the union of the P_i's include all the vertices of $G-A$. Hence $\omega(G-A) \leq r$. Thus we have $\omega(G-A) \leq r \leq |A|-n$, and so $\frac{|A| \cdot 1}{\omega(G-A)} \geq 1 + \frac{n}{\omega(G-A)}$. Since $\omega(G-A) \leq \beta(G)$ for any A, the theorem follows.

To relate Theorem 3.4 to the connectivity of G, we use a generalization of the following theorem of Chvàtal and Erdős [4]:

Theorem 3.6 *(Chvàtal and Erdős) If G is k-connected and $k \geq \beta$, then G is Hamiltonian.*

Theorem 3.7 *(Molluzzo [11]) If G is k-connected and any integer n, $k-n \geq \beta$, then G is Hamiltonian.*

For such k and n, we have the following three possibilities for a graph G:

1. $1 + \frac{n}{\beta(G)} \leq \frac{\kappa(G)}{\beta(G)} \leq NSM(G)$
2. $\frac{\kappa(G)}{\beta(G)} \leq 1 + \frac{n}{\beta(G)} \leq NSM(G)$
3. $\frac{\kappa(G)}{\beta(G)} \leq NSM(G) < 1 + \frac{n}{\beta(G)}$.

If G satisfies the third inequality it is not n-Hamiltonian, by Theorem 3.5. If G satisfies the first inequality then G is n-Hamiltonian by Theorem 3.7. Define the graph C_p^k for any positive integer k as follows: $V(C_p^k) = \{0, 1, 2, \ldots, p-1\}$ and two vertices i and j are adjacent if and only if $|i-j| \leq k$. The graph C_p^{n+2}, for sufficiently large, satisfies the second inequality and is n-Hamiltonian while the graph $G_{p,2}$, defined below, for p sufficiently large, satisfies the second inequality and is not n-Hamiltonian.

The graph $G_{p,m}$, with $1 \leq m \leq \frac{p-1}{2}$, has p vertices and vertex v which is adjacent to all vertices of the two complete subgraphs, copies of K_m and K_{p-m-1}, in other words we have $G_{p,m} \cong K_1 + (K_m \cup K_{p-m-1})$.

REFERENCES

[1] M. Behzad, G. Chartrand, and L. Lesniak, Graphs and Digraphs 1979, Prindle, Weber and Schmitt.

[2] G. Chartrand, S. F. Kapoor, and D. R. Lick, n-Hamiltonian graphs, **J. Combin. Theory 9** (1970), 305–312.

[3] V. Chvàtal, Tough graphs and Hamiltonian circuits, **Discrete Math. 5** (1973), 215–228.

[4] V. Chvàtal and P. Erdős, A note on Hamiltonian circuits, **Discrete Math. 2** (1972), 111–113.

[5] M. Cozzens, D. Moazzami, and S. Stueckle, The tenacity of a graph, submitted **SIAM J. Discrete Math.**

[6] M. Cozzens, D. Moazzami, and S. Stueckle, Tenacity of the Harary graph, **to appear in J. Combin. Math. and Combin. Comput.**

[7] L. L. Doty, A large class of maximally tough graph, **OR Spektrum 13** (1991), 147–151.

[8] H. Enomoto, B. Jackson, P. Katerinis, and A. Saito, Toughness and the existence of K-factors, **J. Graph Theory 9** (1985), 87–95.

[9] W. D. Goddard and H. C. Swart, On the toughness of a graph, **Quaestiones Math. 13** (1990), 217–232.

[10] D. Mann and S. Stueckle, Tenacity of trees, in preparation.

[11] J. C. Molluzzo, Toughness, Hamiltonian connectedness and n-Hamiltonicity, **Annals N. Y. Acad. Sci. 319, Proceedings of Second International Conf. on Comb. Math.,** New York, 1979 (A. Gewirtz, et al., eds.), 402–404.

[12] T. Nishizeki, 1-tough nonhamiltonian maximal planar graphs, **Discrete. Math. 30** (1980), 305–307.

[13] R. E. Pippert, On the toughness of a graph, in **Graph Theory and its Application, Lect. Notes in Math. 303,** Springer, Berlin, 1972 (Y. Alavi, ed.), 225–233.

[14] Sein Win, On a Connection Between Existence of k-trees and the toughness of a Graph, **Graphs and Combinatorics 5**, 201–205 (1989).

[15] D. R. Woodall, The binding number of a graph and its Anderson number, **J. Combin. Theory Ser. B 15** (1973), 225–255.

ZERO-KNOWLEDGE PROOFS FOR INDEPENDENT SET AND DOMINATING SET PROBLEMS

Javad Mohajeri

Electronic Research Center
Sharif University of Technology
Tehran, Iran

ABSTRACT

Independent set and dominating set problems are two NP-C problems in graph theory. In this paper we propose two zero-knowledge proofs for these problems.

1 INTRODUCTION

Assume that P (The Prover) knows some information, It could be:

1. The Prime factorization of a large integer,
2. A 3-Coloring of a graph,
3. An Independent set of a graph,
4. A Dominating set of a graph,
5. A password or an identification number.

P wants to convince V (Verifier) that he knows some of these information. P can convince V in two ways: Maximum disclosure proof and Minimum disclosure proof.

In a maximum disclosure proof, P simply discloses his information, so that V could do checking herself. And therefore V actually learns the information.

C. J. Colbourn and E. S. Mahmoodian (eds.), Combinatorics Advances, 251–254.

In a minimum disclosure proof, P convinces V that he has the information, in such a way that he dose not reveal even one bit of information, and consequently, dose not help V to determine the information in any way.

In this paper, we propose protocols for Zero-knowledge proofs which are based on independent set and dominating set problems in a graph. These protocols satisfy the following conditions.

1. The prover probably cannot cheat the verifier. If the prover does not know an independent set or a dominating set of the graph, his chance of convincing the verifier that he knows those information is negligible.

2. The verifier cannot cheat the prover. She does not get a slightest hint of the proof, apart from the fact that the prover knows a proof. In particular, the verifier, cannot prove the theorem to anyone else without proving it herself from scratch.

2 A ZERO-KNOWLEDGE PROOF FOR INDEPENDENT SET PROBLEM

Let $G = (V, E)$ be a graph and k a positive integer. Does G contain an independent set of size k or more, i.e., is there a subset I of V such that $|I| \geq k$ and such that no two vertices in I are joined by an edge in E? This problem is called the Independent set problem.

Let P know an independent set in G with $|V| = n$ of size k, and want to convince V that he knows them. We describe a protocol with m rounds. Each round consists of 3 steps and proceeds as follows.

- Step 1: P prepares and represent to V the following locked boxes, B_i, $1 \preceq i \preceq n$. Each of the boxes B_i contains one of the nodes, and each of the boxes $B_{i,j}$ contains either 0 or 1. 1 appears if and only if the following conditions are satisfied.

  - If i_1 and j_1 appear in B_i and B_j respectively, then there is an edge between i_1 and j_1 in G.

  The boxes B_i and $B_{i,j}$ are referred to as node and edge boxes.

- Step 2: V flips a fair coin and tells P the outcome.
- Step 3: (a) If the outcome is "heads", P opens all boxes. (b) If the outcome is "tails", P opens $L = \binom{k}{2}$ edge boxes of ($B_{i,j}$'s).
- Step 4: (a) V verifies that she has got a copy of G. If so she accepts, otherwise rejects. The verification is easy because the opened node boxes reveal the proper node labels, so no problem concerning graph isomorphism has to be settled. (b) V verifies that all of opened edge boxes contain 0, and that all indices of the opened boxes, belong to a subset $S \subseteq V$ of size k, If so she accepts, otherwise rejects.

Note that the boxes have to be reconstructed for each round of protocol. Otherwise, V learns every thing in two rounds by choosing events (a) and (b). If the boxes are always reconstructed, V learns nothing about the independent set. If event (a) happens, she gets just a copy of G. If event (b) happens, she learns, that P knows an independent set in G. P always passes the test if he knows an independent set of size k in G otherwise, he may try to cheat in two ways.

1. He does not put in the boxes a description of G but rather of some other graph an independent set of which he knows. Then he gets caught if event (a) happens.
2. P uses a false independent set of size k. Then he gets caught if event (b) happens. Thus probability of a false prover passing m rounds is 2^{-m}.

3 A ZERO-KNOWLEDGE PROOF FOR DOMINATING SET PROBLEM

Let $G = (V, E)$ be a graph and $k \leq |V|$ a positive integer. Is there a dominating set of size k or less for G, i.e., a subset $D \subseteq V$ with $|D| \leq k$ such that for all $u \in V - D$ there is a $v \in D$ with $\{u, v\} \in E$.

The above NP-C Problem is called the Dominating set Problem. The Protocol used for this Problem is very similar to the protocol of independent set problem, consisting of m rounds, each round contains 4 Steps as following.

- Step 1: P prepares boxes B_i and $B_{i,j}$.

- Step 2: V flips a fair coin and tells P the outcome.
- Step 3: (a) If the outcome is "heads" P opens all boxes. (b) If the outcome is "tails", P opens at least $n-k$ edge boxes B_{i_t,j_t}.
- Step 4: (a) V verifies that she has got a copy of G. (b) V verifies that all of opened edge boxes contain 1 and moreover for $t \neq s$, $i_t \neq i_s$ moreover:

$$\{i_t : 1 \leq t \leq n-k\} \cap \{j_t : 1 \leq t \leq n-k\} = \oslash$$

If so, she accepts otherwise rejects.

Other descriptions are completely similar to those of the previous problem.

REFERENCES

[1] H. Burk and A. Pfitzmann; "Digital Payment systems enabling security and unobservability." Computers and Security 9 (1989), 399–416.

[2] O. Goldreich, S. Micali, and A. Widgerson; "How to prove all NP-statements in Zero-Knowledge, and an methodology of Cryptographic protocol design". Lecture Notes in Computer Science, Vol. 263. Springer, Berlin Heidelberg New York 1987, pp. 171–185.

[3] S. Goldwasser, S. Micali, and C. Rackoff; "The Knowledge Complexity of interactive proof systems." Proceeding of the 17th ACM Symposium on the theory of Computing. 1985, pp. 291–304.

[4] A. Salomaa; "Public-Key Cryptography", Springer-Verlage (1990).

EXPLORING THE SPECTRUM OF VALUES OF PERMANENTS BY SIMULATED ANNEALING

Yaghout Nourani

Ørsted Laboratory, Niels Bohr Institute
University of Copenhagen, Universitetsparken 5
DK-2100 Copenhagen, Denmark

ABSTRACT

This work presents results for computing the values of permanents of fully indecomposable (0,1)-matrices. There exists an interesting and only partially understood spectrum of possible values for the permanent of a (0,1)-matrix with given topological dimension d. The present work is an experimental study of this spectrum using the handle basis representation of directed graphs and the Monte Carlo based optimization method simulated annealing to sample the possible values. The approach represents an importance sampling technique to focus the search for new values. The results from these experiments show that there exist gaps in the spectrum of values of the permanent.

1 THE PERMANENT

The permanent of a matrix was introduced by Cauchy in 1812 [1]. Let $A = (a_{ij})$ be an $n * n$ matrix. The permanent of A, written $Per(A)$, can be calculated in the same way as the determinant, except that the signs of the summands are all positive:

$$Per(A) = \sum_{\sigma} \prod_{i=1}^{n} a_{i\sigma(i)} \tag{18.1}$$

where σ runs over all permutations of $1, ..., n$. At first glance it would appear that this function is easier to compute than the determinant of the same matrix. But it is not true. All known methods grow exponentially with the size of the matrix. In fact, the complexity of computing the permanent of an $n * n$ (0,1)-

C. J. Colbourn and E. S. Mahmoodian (eds.), Combinatorics Advances, 255–265.

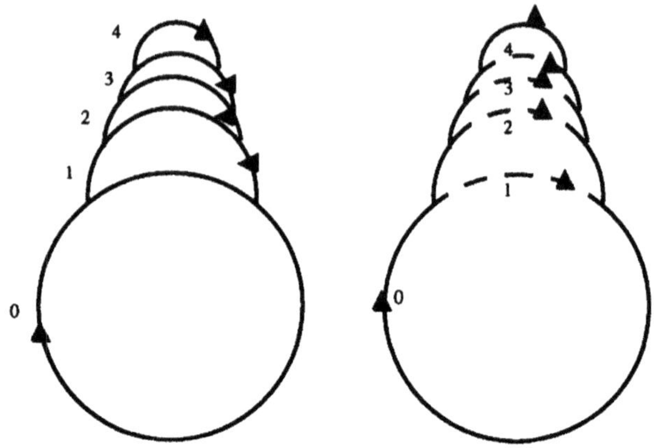

Figure 1 Two different handle basis representation $d = 5$ [7].

matrix is NP-hard [2]—[4]. The direct evaluation of the permanent of an $n * n$ matrix by formula (1) requires $O((n+1)!)$ multiplications. Ryser's formula for computing the permanent of a matrix involves $(n-1)(2n-1)$ multiplications. Nijenhuis and Wilf's implementation of Ryser's method reduces the number of multiplications by a factor of two and the number of additions by a factor of $n/2$ [5]; this is the implementation used in the present work.

There is a one to one correspondence between digraphs and (0,1)-matrices. Assuming that the vertices of a digraph G are numbered $1, ..., n$ in some arbitrary manner, the adjacency matrix representation of G consists of an $n * n$ matrix $A = (a_{ij})$ such that

$$a_{ij} = \begin{cases} 1 & \text{if } i \text{ is connected to } j \\ 0 & \text{otherwise} \end{cases}$$

Conversely, given an $n * n$ (0,1)-matrix A, one can define a corresponding digraph $G(A)$ as follows: The vertices of $G(A)$ are the rows (or columns) of A and there is an edge from vertex i to vertex j if and only if $a_{ij} \neq 0$.

Central to our work is the notion of a handle basis for a finite strongly connected digraph G.

Definition: A handle basis $B = \{H[i], i = 1, \cdots, d-1\}$ of a digraph G expresses G as the edge disjoint union of a sequence of subgraphs, where $H[0]$ is a simple cycle, and $H[i]$, $i > 0$, is a simple path that meets the union of $H[j]$, $j < i$, exactly in its endpoints [6] (see Figure 1).

A path $v_0, \ldots, v_n$ in G of length $n > 0$ is simple if the vertices traversed are distinct. A simple cycle is a simple path except with $v_0 = v_n$. Points which are not endpoints are called interior points of the path.

In other words, a handle is a path which starts from any previous handle and ends on any previous handle but is otherwise disjoint from the digraph of the previous handles. A directed graph admits a handle basis if and only if it is strongly connected. Donald et al. have shown that if B is a handle basis for G, then every point is interior to exactly one handle. In particular, this implies that handles do not cross each other. The number of handles, d, is the topological dimension of the digraph G. This dimension is also known as the cyclomatic number of G [7]–[8].

2 UPPER BOUNDS AND LOWER BOUNDS FOR THE PERMANENT

H. Minc [5] proved that the bounds for the permanent of a nonnegative $n * n$ matrix $A = (a_{ij})$ are

$$\prod_{i=1}^{n}\sum_{t=1}^{i} a'_{it} \leq Per(A) \leq \prod_{i=1}^{n}\sum_{t=1}^{i} a^*_{it} \tag{18.2}$$

where if $A_i=(a_{i1},\ldots,a_{in})$ is the i-th row of A, then $(a^*_{i1},\ldots,a^*_{in})$ denotes the n-tuple A_i arranged in non-increasing order and $(a'_{i1},\ldots,a'_{in})$ denotes the same n-tuple arranged in non-decreasing order. Later Donald et al. [8] proved that

$$Per(A) \leq 1 + min\{\prod(r_i - 1), \prod(c_i - 1)\} \tag{18.3}$$

where A is a fully indecomposable $n * n$ matrix, and c_i and r_i are the column sums and row sums of A, respectively Foregger [9] has obtained the following bound for the permanent of an $n * n$ matrix A with non-negative entries:

$$Per(A) \leq 2^{S(A)-2n} + 1 \tag{18.4}$$

where $S(A)$ is the sum of all the entries of A. Our purpose below is to examine the possible values of the permanent of a d-dimensional strongly connected digraph. It can be shown [8] (see Figure 2) that it is sufficient to restrict our attention to the simplest configurations in which no two handles attach at the same vertices and which have vertices only at points of attachment of handles. Let A be an $n * n$ (0,1)-matrix with unit diagonal. The topological dimension d of the digraph $G(A)$ is then $S(A) - 2n + 1$, where $S(A)$ denotes the sum of entries in A [5]. Foregger's bound can thus be written as:

$$Max[Per(A)] = 2^{d-1} + 1 \tag{18.5}$$

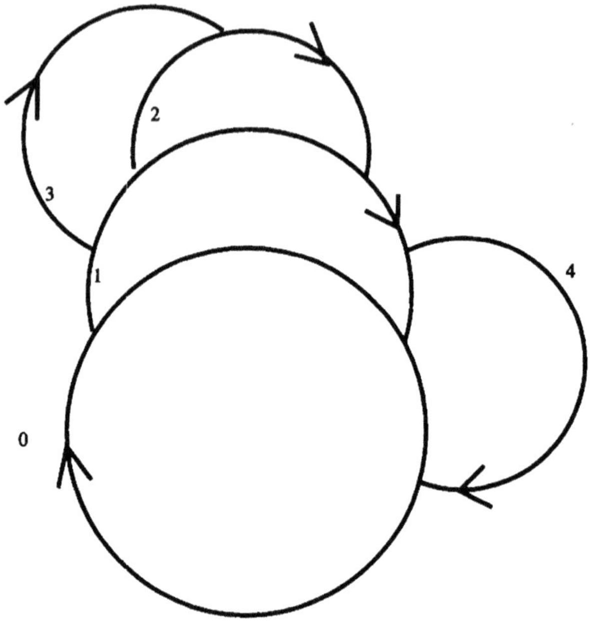

Figure 2 A representation of directed graph, $d = 5$.

Graphically this value is achieved for a digraph $G(A)$ in which all handles start and end on handle number zero ($H[0]$) and are oriented in the same direction (see Fig 3).

3 SIMULATED ANNEALING

Many problems that arise in practice are concerned with finding a best configuration. These kinds of problems are global optimization problems. The subject of combinatorial optimization are often applied to a set of problems that are central to the disciplines of computer science and engineering, e.g. the traveling salesman problem, the hill climbing problem, and VLSI layout. Research in this area aims at developing efficient techniques for finding minimum or maximum values of a function called the cost function. Blind random search and iterative improvement can be used and give acceptable approximations to the global minimum if only limited computer time is available. In order to improve the performance of these algorithms we want to escape from local minima and generate statistically independent configurations. In addition to comparing the

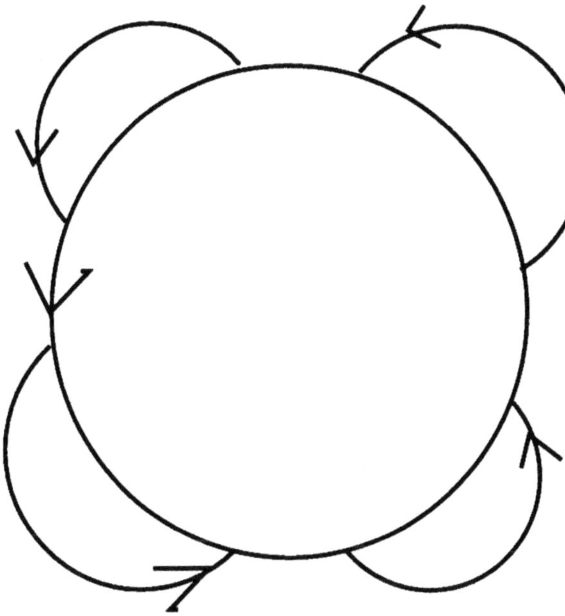

Figure 3 d = 5 and $Per = 2^{d-1} + 1 = 17$.

score of a current configuration with the score of a newly generated one, an acceptance function can be evaluated on the basis of the two scores and a random number. If a true value is returned by th! e acceptance function, the newly Since the acceptance function returns true with a nonzero probability even for moves which increase the cost, this type of algorithm is called probabilistic hill climbing. Simulated annealing algorithms represent one class of hill climbing algorithms.

4 THE METROPOLIS ALGORITHM

Annealing is the process of cooling a physical system in order to obtain states with minimum energy. The cooling rate of annealing is usually slow in order to reduce the number of defects in the crystal structure and to minimize the potential energy stored in the molecular configuration.

Metropolis et al. [10] introduced an algorithm for simulating the state of a physical system at a temperature T. The Metropolis algorithm can be used to generate sequences of configurations of a combinatorial optimization problem in which the cost function plays the role of energy and the control parameter is represented by the temperature. Let $S = \{s\}$ represent the state space. For the problem considered in this work, s is the set of strongly connected d-dimensional digraphs. Let

$$E(s) = -Per(A(s)) \tag{18.6}$$

be the cost function and T the temperature. We assume that for each state s in S there exists a set $N(s)$, with $N(s) \subset S$, known as the move class or the set of neighbors of s. The "meta-graph" structure on an state is defined by the $N(s)$, $s \in S$.

At each step of the Metropolis algorithm a state $s' \in N(s)$ is selected at random. If the energy E of this s' is lower than the energy of s, the acceptance function returns a true value. In this case the move is accepted, and the new state becomes the current state. If the energy of the new state is higher than the energy of the current state, the acceptance probability for s' to become the next state is

$$P_{accept} = Exp(\frac{-\Delta E}{kT}) \tag{18.7}$$

where $\Delta E = E(s') - E(s)$, T is the temperature and k is Boltzmann's constant, which serves to match the units of T and E. A pseudo random number is generated with a uniform distribution on [0,1]. If the random number is less than the acceptance probability, the move is accepted, otherwise it is rejected and the random walk remains at the old state s for another time step [11]—[12].

The formal algorithm is:

```
Metropolis (T)
float T;
{
initialize;

    GetNewState(CurrentState, NewState);
    ΔE = E(NewState) - E (CurrentState) ;
    if (ΔE < 0.0)
    CurrentState = NewState;
    else {
    if (Random < (exp (−ΔE/kT)))
    CurrentState = NewState;
    }

}
```

Simulated annealing uses the Metropolis algorithm with a decreasing temperature. Initially the control parameter T is given a very high value which is lowered in each step until a given time temperature is reached:

```
SimulatedAnnealing()
{
initialize;

    CurrentState = GenerateState();
    T = SomeConstant;
    stop = SomeOtherconstant;
    While ( T ≠ stop) {
    Metropolis(T);
    T= decrease(T);
    }

}
```

Simulated Annealing probabilistically transforms each current state to one of its neighbors. This random walk can be described by a Markov chain. Markov chains are discrete time and state stochastic processes with the property that the next state does not depend on the states that have preceded the current state [13]. For a Markov chain on n states the transition probabilities form an $n * n$ matrix $P = p_{ss'}$

$$p_{ss'} = \begin{cases} \frac{1}{|N(s)|} & \text{if } s' \in N(s) \\ 0 & \text{otherwise} \end{cases}$$

where $|N(s)|$ is the number of neighbors of s. Also

$$\forall s, = 1, \ldots, n \qquad \sum_{s'=1}^{n} p_{ss'} = 1 \tag{18.8}$$

which means that P is a stochastic matrix.

5 RESULTS AND CONCLUSIONS

The results from these experiments show that there exist gaps in the spectrum values of the permanent of an $n * n$ fully irreducible (0,1) matrix. The problem was tested for $d = 2, \ldots, 8$ where d represents the number of handles in a handle basis representation of a digraph G. Applying simulated annealing was very useful in our experiment. It enabled us to escape from local maxima and reach the maximum value of the permanent. The following tables and chart display the values of the permanents which we were able to find. For $d = 8$ we found the permanent value 72 which has not been seen before. Also, for $d = 2, ..., 8$, we present the spectrum of values of the permanent and a plot for observed frequencies with temperatures $T = 3$, 6, and 10 for $d = 7$.

Acknowledgments

I wish to express my sincere gratitude to Prof. P. Salamon and Prof. B. Andersen for their generous guidance as well as my deep appreciation to Prof. R, Hager for all the valuable discussions we had. I also would like to thank the Ørsted Laboratory, NBI, University of Copenhagen for their support and hospitality.

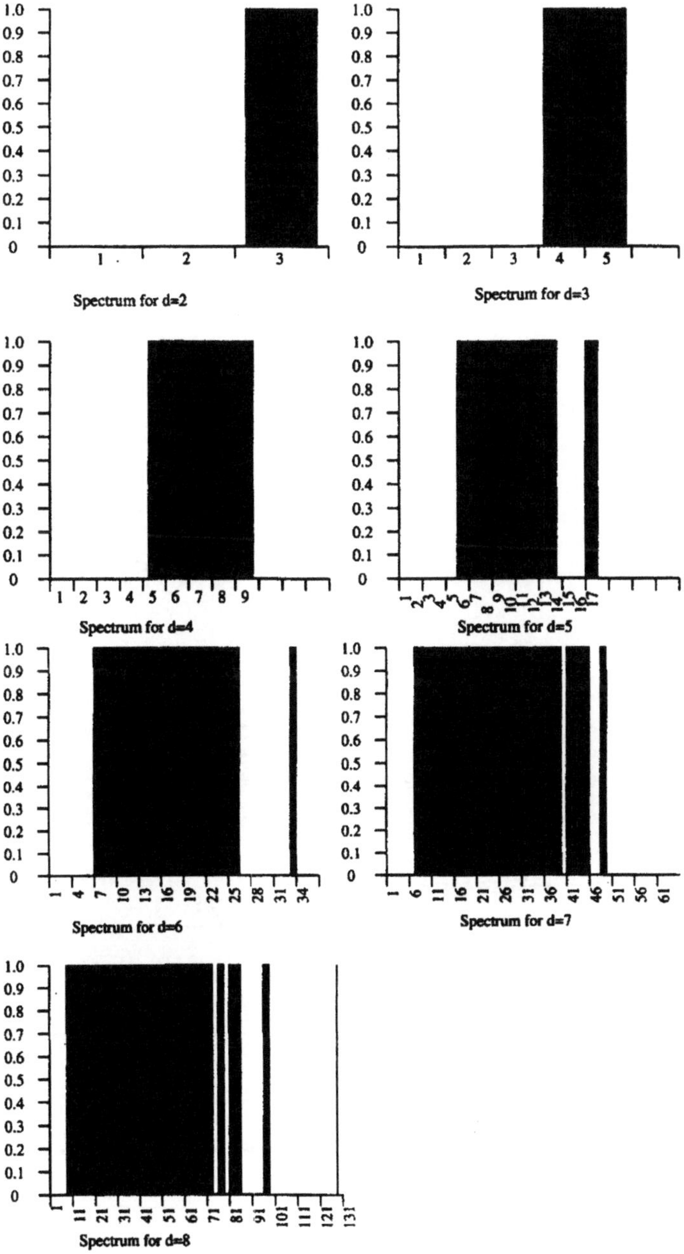

Figure 4 Spectrums of values of the permanent for $d = 2, \ldots, 8$. One presents the permanent values which have been seen and zero presents the gaps.

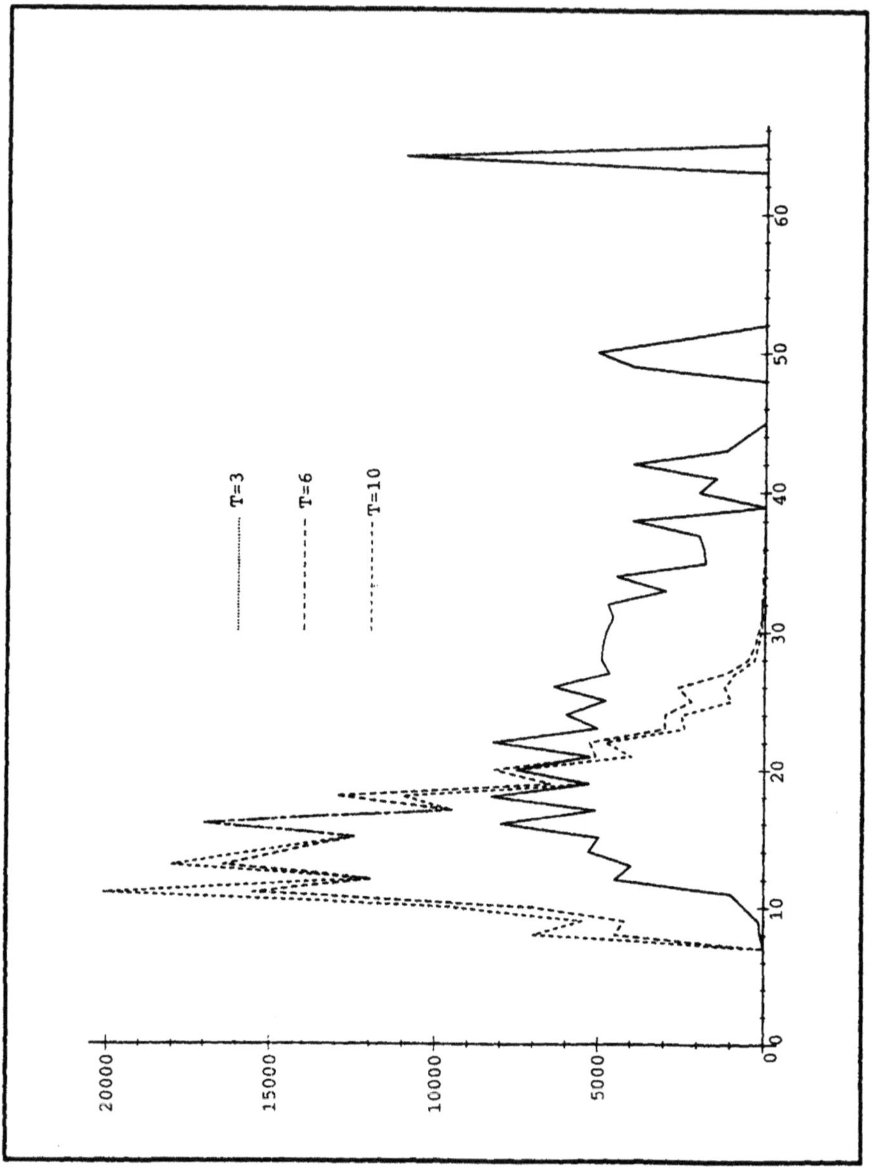

Figure 5 Chart for frequencies of states for $d = 7$ and $T = 3$, 6 and 10. The frequency distribution of the energy, *−permanent*, is different for each temperature

REFERENCES

[1] A. L. Cauchy, "Memoire sur les fonctions qui ne peuvent obtenir que deux valeurs egales et de signs contraires par suite des transpositions operees entre les variables quelles renferment", J. Ec. Polyt. **10** Cah. **17**, 29–112, Oeuvres (2)i.

[2] R. M. Karp, "Reducibility among combinatorial problems", in R. E. Miller and J. W. Thatcher, (Eds.) *Complexity of computer computations* (Plenum Press, New York, 1972), 85-103.

[3] L. G. Valiant, "The complexity of computing the permanent", Theor. Comput. Sci. **8**, 189-201 (1978).

[4] M. R. Garey and D. S. Johnson, *Computers and Intractability, A guide to the Theory of NP-Completeness* (W. H. Freeman and Company, New York, 1979).

[5] H. Minc, *Permanents, Encyclopedia of mathematics and its applications*, **v. 6**, section on Linear algebra (Addison-Wesley, 1978).

[6] J. Donald and J. Elwin, "On The Structure Of The Strong Orientation Of A Graph", SIAM J. Disc. MATH. **9**, No. 1, 30-43, (February 1993).

[7] J. Donald, J. Elwin, R. Hager, and P. Salamon, "Handle bases and bounds on the number of subgraphs", J. Combin. Theory, Ser. B, **42** 1-13 (1987).

[8] J. Donald, J. Elwin, R. Hager, and P. Salamon, "A Graph Theoretic Upper Bound on the Permanent of a Nonnegative Integer Matrix", I, II, Linear Algebra and its Applications **61**, 187-198 (1984).

[9] A. Nijenhuis and H. Wilf, *Combinatorial Algorithms* (Academic Press, Inc., 1975).

[10] N. Metropolis, A. Rosenbluth, M. Rosenbluth, A. Teller, and E. Teller, "Equation of state calculation by fast computing machines", J. Chem. Phys. **21**, 1087-1092 (1953).

[11] P. J Van Laarhoven and E. H. L. Aarts, *Simulated Annealing and Applications* (D. Reidel Publishing Company, 1987).

[12] R. H. J. M Otten and L. P. P. P. Van Ginneken, *The Annealing Algorithm* (Kluwer Academic Publishers, 1989).

[13] R. G. Kemeny and J. L. Snell, *Finite Markov chains* (D. Van Nostrand Company, Inc., 1960).

VECTOR–WEIGHTED MATCHINGS

Dietmar Schweigert

Universität Kaiserslautern
Fachbereich Mathematik
Postfach 3049
D 6750 Kaiserslautern
Germany

1 INTRODUCTION

The theorem of efficiency shows that every efficient solution of a multiple objective linear program (MOLP) is the optimal solution of a parametric optimization problem. For the parametric optimization one uses weighted sums

$$\lambda_1 c_1(x_1, \ldots, x_n) + \cdots + \lambda_k c_k(x_1, \ldots, c_n)$$

for the k objective functions $c_i(x_1, \ldots, x_n)$, $i = 1, \ldots, k$ where the weights have the properties that $\lambda_i > 0$ and $\lambda_1 + \cdots + \lambda_k = 1$. To get every efficient solution one has to organize effectively a search of these vectors $(\lambda_1, \ldots, \lambda_k)$ which delivers different efficient solutions. We decompose the polytope given by the unit vectors $(1, 0, \ldots, 0), \ldots, (0, \ldots, 0, 1)$ into the polytopes which correspond to the efficient solutions. This decomposition can be performed by an algorithm using hyperplanes. Our methods will be examined for vector-weighted matchings but of course apply to every multiobjective linear program. We consider a simple bipartite graph G where the edges have weights $w(e) = [w_1(e), \ldots, w_n(e)]$ of vectors of positive reals. A matching M of G is called efficient if there is no other matching M' with

$$w(M) = [w_1(M), \ldots, w_n(M)] < [w_1(M'), \ldots, w_n(M')] = w(M')$$

where $w_i(M)$ is the sum of the ith components of the weights of all edges contained in the matching M. The problem to find the efficient matchings can be considered as a problem in multicriteria optimization. One can even present this problem as multiple objective linear program (MOLP). The solution of such a problem provides some unexpected difficulties. In linear optimization the assignment problem can be solved without much effort despite the fact

C. J. Colbourn and E. S. Mahmoodian (eds.), Combinatorics Advances, 267–276.

that it is an integer problem. In contrast we can show that the approaches neglecting the restriction to integers fail to find all efficient solutions. To solve this problem we use an algorithm for the K best matchings. In the first part of the paper we present a method to determine all preference solutions. These solutions correspond to the subdivision with polytopes of a set of parameters. In the second part we use an algorithm for the K-best solution to determine all efficient solutions.

2 PRELIMINARIES

We study a multicriteria optimization problem in two forms, as a graph theoretical optimization problem and a multilinear program.

2.1 Efficient Matching

Let $K_{m,m} = (V, E)$ be a simple, complete, bipartite graph with a weight function $w : E \to R^n$ where R is the set of nonnegative real numbers. Every edge $e \in E$ has a weight $w(e) = [w_1(e), \ldots, w_n(e)]$ with n components. If M is a set of edges we define $w(M) := \sum_{e \in M} w(e)$ and denote the ith component by $w_i(M) := \sum_{e \in M} w_i(e)$ for $i \in \{1, \ldots, n\}$. A perfect matching M is called efficient if there exist no perfect matching M' with $w(M) < w(M')$ ($w(M) < w(M')$ is defined by $w_i(M) \leq w_i(M')$ for every $i = 1, \ldots, n$ and $w_j(M) \nleq w_j(M')$ for at least one $j \in \{1 \ldots, n\}$). See Figure 1.

The list of all perfect matchings $M_1, \ldots, M_6$.

$$
\begin{aligned}
M_1, E(M_1) &= \{e_1, e_6, e_9\}, w(M_1) = (9, 3, 5) \\
M_2, E(M_2) &= \{e_3, e_4, e_9\}, w(M_2) = (5, 5, 4) \\
M_3, E(M_3) &= \{e_1, e_7, e_8\}, w(M_3) = (5, 5, 1) \\
M_4, E(M_4) &= \{e_6, e_5, e_2\}, w(M_4) = (6, 4, 6) \\
M_5, E(M_5) &= \{e_5, e_3, e_7\}, w(M_5) = (3, 4, 4) \\
M_6, E(M_6) &= \{e_2, e_4, e_3\}, w(M_6) = (4, 7, 2)
\end{aligned}
$$

One can recognize that M_1, M_2, M_4 and M_6 are efficient matchings whereas M_3 and M_5 are not because of $w(M_3) < w(M_2)$ and $w(M_5) < w(M_2)$ respectively.

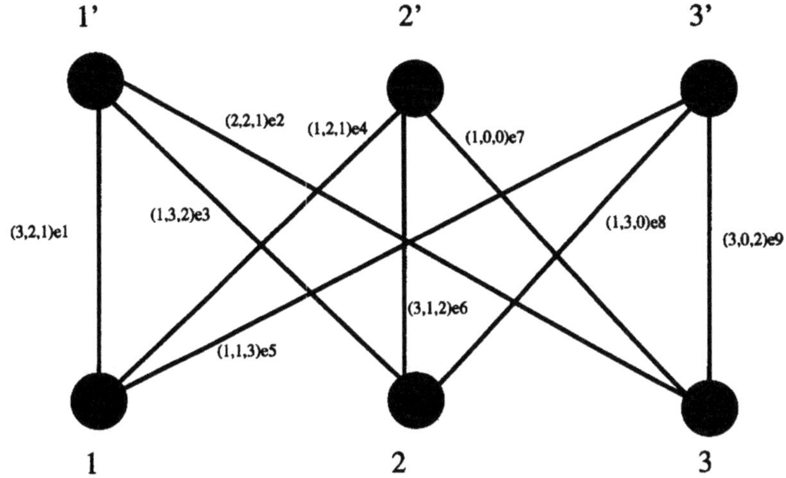

Figure 1 Example for $K_{3,3}$

2.2 Multilinear Assignment Problem

Finding the efficient matchings on the bipartite graph as above $K_{m,m}$ can be considered as an assignment problem. We formulate our example as multiple objective linear program (MOLP). We have three objective functions $\sum c_{ij}^{(k)} x_{ij}$, $k = 1, 2, 3$ which have simultaneously to be maximized under six constraints

$$\max \begin{cases} \sum_{i,j=1}^{3} c_{ij}^{(1)} x_{ij} \\ \sum_{i,j=1}^{3} c_{ij}^{(2)} x_{ij} \\ \sum_{i,j=1}^{3} c_{ij}^{(3)} x_{ij} \end{cases} \qquad x_{ij} \in \{0, 1\}$$

$$\sum_{i,j=1}^{3} x_{ij} = 1 \quad \text{for} \quad j = 1, 2, 3$$

$$\sum_{i,j=1}^{3} x_{ij} = 1 \quad \text{for} \quad i = 1, 2, 3$$

We have that $x_{ij} = 1$ if the edge (i, j) with the vertices i and j is in the matching and $x_{ij} = 0$ else. The values $c_{ij}^{(k)}$ are given as the kth component $w_k((i, j))$ of the weight of the edge (i, j) with the vertices i and j.

3 PREFERENCE MATCHINGS

A common approach to solve a multiple objective linear program (MOLP) is the method of preference functions (or weighted sums). Whenever the MOLP is not restricted to integers then this approach delivers every efficient solution [6]. Although we have a 01-program the efficient solutions found by preference functions play an essential role. They will be called preference solutions or preference matchings in the following.

Notation 3.1 *Let $(R^n; \leq)$ be the poset of nonnegative real vectors ordered component wise. The function $f : R^n \longrightarrow R$ is called a preference function if $f(x_1 \ldots x_n) = \lambda_1 x_1 + \cdots + \lambda_n x_n$ with $\lambda_i > 0$, $i = 1, \ldots, n$,* and *$\lambda_1 + \cdots + \lambda_n = 1$. A preference function is strictly monotone and additive.*

Because a preference function is strictly monotone the poset $(R^n; \leq)$ is mapped into linear order $(R; \leq)$. With the linear order one can reduce the multiple objective linear program (MOLP) to the usual linear program.

Notation 3.2 *If every pair $a, b \in L$ of a poset $(L; \leq)$ is comparable, i.e. either $a \leq b$ or $b < a$, then the poset $(L; \leq)$ is called a linear order. A linear order $(L(E); \leq)$ is called a linear extension of a poset $(E; \leq)$ if $(L(E); \leq)$ has the same underlying set $L(E) = E$ and if we have $a < b$ in $(L(E); \leq)$ whenever $a < b$ in $(E; \leq)$.*

Every preference function $f : R^n \longrightarrow R$ defines a linear extension $(L_f(E); \leq)$ of the poset $(E; \leq)$ of weighted edges by $e_i < e_j$ iff $f(w_1(e_i), \ldots, w_n(e_i)) < f(w_1(e_j), \ldots, w_n(e_j))$ or $f(w_1(e_i), \ldots, w_n(e_i)) = f(w_1(e_j), \ldots, w_n(e_j))$ and $i < j$.

Illustration. Consider the poset $(E; \leq)$ of weighted edges of the above bipartite graph G and apply the preference function $f(x_1, x_2, x_3) = \frac{1}{3}x_1 + \frac{1}{3}x_2 + \frac{1}{3}x_3$. One gets the following linear extensions $e_7 < e_4 < e_8 < e_2 < e_5 < e_9 < e_3 < e_6 < e_1$ with the new weights: $w(e_7) = 1$, $w(e_4) = w(e_8) = 4$, $w(e_2) = w(e_5) = w(e_9) = 5$, $w(e_3) = w(e_6) = 6, w(e_1) = 7$.
The linear extension $(L_f(E); \leq)$ is called a preference extension. We consider the following

Algorithm 2.1

a) Input: Vector-weighted graph $G = (V, E)$

b) Choose a linear extension $(L(E); \leq)$ with weights $w(e) \in R$ for all $e \in E$.

c) Apply the algorithm of Kuhn Munkres (or simplex-algorithm or some other algorithm for a maximal matching)

d) Output: Perfect Matching.

Theorem 3.3 *If the linear extension in Algorithm 2.1 is a preference extension $(L_f(E); \leq)$ then the perfect matching is an efficient matching.*

This theorem can be slightly generalized as we use in the proof essentially the strict monotonicity of the preference function. One can weaken the additivity to get the same result. On the other hand a function $f : R^n \longrightarrow R$ is strictly monotone and additive if and only if f is a weighted sum.

4 PREFERENCE POLYTOPES

If we want to find all preference solutions for k objective functions we have to organize the selections of $(\lambda_1, \ldots, \lambda_k)' \in (0,1)^k$. The vectors $(\lambda_1, \ldots, \lambda_k)'$ will be collected in a polytope whenever they produce the same preference solution. We present a method to find all these preference polytopes.

Notation 4.1 *We say that the preference function*

$$f_{\lambda_1 \ldots \lambda_k}(x_1, \ldots, x_k) = \lambda_1 x_1 + \cdots + \lambda_k x_k$$

with $\lambda_i \in (0,1), \quad i = 1, \ldots, k, \quad \sum_{i=1}^{k} \lambda_i = 1$ *belongs to a preference solution* M *if* $f_{\lambda_1 \ldots \lambda_k}(w(M')) \leq f_{\lambda_1 \ldots \lambda_k}(w(M))$ *for every admissible solution* M'

Remark 4.2 *In the case $\lambda_i = 1, \lambda_j = 0$ for $j \neq i$, we know that the function $f_{\lambda_1 \ldots \lambda_k}$ is not strictly monotone anymore and we can get a solution which may not be efficient. There are several possibilities to correct this situation. One is to put $\lambda_j = \epsilon$ for $j \neq i$ and $\lambda_i = 1 - \epsilon$ for some small $\epsilon > 0$. Another is to calculate all linear extension for these cases. To simplify the following presentation we assume $\lambda_i \in [0,1], \quad i = 1, \ldots, k$.*

Definition 4.3 *Let M be a preference solution. The set*

$$K(M) := \left\{(\lambda_1, \ldots, \lambda_k)' \in [0,1]^k \quad | \quad f_{\lambda_1 \ldots \lambda_k} \text{ belongs to } M\right\}$$

is called the preference set of M.

Lemma 4.4 *The preference set $K(M)$ of a preference solution M is convex.*

Proof. If $f_{\alpha_1 \ldots \alpha_k}$ and $f_{\beta_1 \ldots \beta_k}$ belongs to M then we have to show that $f_{\lambda_1 \ldots \lambda_k}$ also belongs to M where

$$\begin{bmatrix} \lambda_1 \\ \vdots \\ \lambda_k \end{bmatrix} = t \begin{bmatrix} \alpha_1 \\ \vdots \\ \alpha_k \end{bmatrix} + (t-1) \begin{bmatrix} \beta_1 \\ \vdots \\ \beta_k \end{bmatrix} \quad t \in [0,1]$$

$f_{\lambda_1 \ldots \lambda_k}$ belongs to M whenever we have $f_{\lambda_1 \ldots \lambda_k}(w(M')) \leq f_{\lambda_1 \ldots \lambda_k}(w(M))$ for every admissible solution M'. This inequality can be shown by direct calculation.

Definition 4.5 *The efficient solutions M_a and M_b are called equivalent if $w(M_a) = w(M_b)$.*

Theorem 4.6 *If M_a and M_b are non-equivalent preference solutions, then $K(M_a) \cap K(M_b)$ is either empty or a convex set on a hyperplane.*

Proof. If $K(M_a) \cap K(M_b)$ is not empty then there exist $\lambda_1, \ldots, \lambda_k$ such that $f_{\lambda_1 \ldots \lambda_k}$ belongs to M_a as well as to M_b. Hence we have $f_{\lambda_1 \ldots \lambda_k}(w(M_a)) = f_{\lambda_1 \ldots \lambda_k}(w(M_b))$ and therefore

$$\lambda_1 w_1(M_a) + \cdots + \lambda_k w_k(M_a) = \lambda_1 w_1(M_b) + \cdots + \lambda_k w_k(M_b)).$$

We put $\lambda_k = 1 - (\lambda_1 + \cdots + \lambda_{k-1})$ and have

$$\begin{aligned} \lambda_1(w_1(M_a) - w_1(M_b) + w_k(M_b) - w_k(M_a)) + \cdots \\ + \lambda_{k-1}(w_{k-1}(M_a) - w_{k-1}(M_b)) + w_k(M_b) - w_k(M_a) = w_k(M_b) - w_k(M_a) \end{aligned}$$

We put $x_i := \kappa_i, \quad i = 1, \ldots, k$

$$\begin{aligned} w_i &:= w_i(M_a) - w_i(M_b) + w_k(M_b) - w_k(M_a) \\ w &:= w_k(M_b) - w_k(M_a) \end{aligned}$$

All these $(\lambda_1, \ldots, \lambda_{k-1})$ fulfill the following hyperplane equation:

$$(*) \qquad x_1 w_1 + \cdots + x_{k-1} w_{k-1} = w.$$

Remark 4.7 *The preference set $K(M)$ is a convex polytope and hence is called preference polytope. The union of these polytopes is*

$$L = \left\{ (\lambda_1, \ldots, \lambda_k)' \quad | \quad \sum_{i=1}^{n} \lambda_i = 1, \quad 1 \geq \lambda_i \geq 0, \quad i = 1, \ldots, n \right\}.$$

With the following theorem we can decompose L into its preference polytopes and get an overview about all preference solutions.

Theorem 4.8 *Let M_a and M_b be non equivalent preference solutions such that $K(M_a) \cap K(M_b) = \emptyset$. Then the hyperplane*

$$(*) \qquad x_1 w_1 + \cdots + x_{k-1} w_{k-1} = w$$

separates the preference polytopes $K(M_a)$ from $K(M_b)$ without having a point in common with $K(M_a)$ or $K(M_b)$.

Proof. Let $f_{\alpha_1 \ldots \alpha_k}$ be a preference function which belongs to the efficient solution M_a. By hypothesis we have $f_{\alpha_1 \ldots \alpha_k}(w(M_b)) < f_{\alpha_1 \ldots \alpha_k}(w(M_a))$ and hence

$$\alpha_1 w_1(M_b) + \cdots + \alpha_k w_k(M_b) < \alpha_1 w_1(M_a) + \cdots + \alpha_k w_k(M_a).$$

For $\alpha_k = 1 - (\alpha_1 + \cdots + \alpha_{k-1})$ we have

$$\alpha_1 w_1 + \cdots + \alpha_{k-1} w_{k-1} > w$$

where $w_i = w_i(M_a) - w_i(M_b) + w_k(M_b) - w_k(M_a)$ and $w = w_k(M_b) - w_k(M_a)$. Let $f_{\beta_1 \ldots \beta_k}$ belong to M_b. We proceed as above and have

$$\beta_1 \bar{w}_1 + \cdots + \beta_{k-1} \bar{w}_{k-1} > \bar{w}$$

where $\bar{w}_i = w_i(M_b) - w_i(M_a) + w_k(M_a) - w_k(M_b)$ and $\bar{w} = w_k(M_a) - w_k(M_b)$. We conclude that

$$\beta_1 w_1 + \cdots + \beta_{k-1} w_{k-1} < w.$$

From the above results we get the following method which determines the set of all preference polytopes. At first we search the preference solutions with $\lambda_i = 1$ $(i = 1, \ldots, n)$. In this case the preference function is not strictly monotone anymore and we have to verify that we find efficient solutions. This preference functions corresponds to vertices of preference polytopes. The vertices are given as the intersection of k hyperplanes. The vertices which are obtained, the k

hyperplanes and the preference solution will be put into a list. We consider a vertex of this list whenever it is not saturated. From such a vertex we traverse every possible edge till we meet the next vertex. The next vertex is determined as the intersection of the edge with the surrounding polytope. If this vertex does not belong any more to the preference polytope then we have to calculate new hyperplanes according to Theorem 4.8. A vertex is saturated if every edge has been traversed. If we apply this method to our example we get the subdivision of preference polytopes depicted in Figure 2.

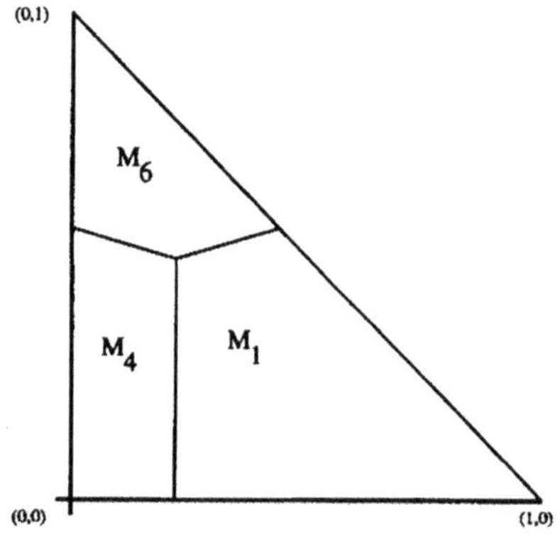

Figure 2

We recognize that M_2 is not contained in this subdivision.

5 THE SET OF EFFICIENT SOLUTIONS

If vectors $x \in R^m$ are admissible then we employ the following result for a multiple objective linear program (MOLP) ([3] p. 713): $x \in R^m$ is an efficient solution of $\{(c_1(x), \ldots, c_n(x) \mid x \in S\}$ if and only if x is an optimal solution of $\{\lambda_1 c_1(x) + \cdots + \lambda_n c_n(x) \mid x \in S\}$ for some $(\lambda_1, \ldots, \lambda_n)' \in R^n$. If only vectors $x \in N_0^m$ are admissible then the above result does not hold. We could not find the efficient solution M_2 by the method of preference polytopes. This can also be shown in an elementary way (compare [4] p.226). As an integer problem the assignment or matching problem does not provide any difficulties and can be solved for instance by the simplex algorithm. But as a multicriteria optimization problem it behaves like a typical integer problem and we have to search for a method to find every efficient solution. Again we consider the simple, complete, bipartite graph $K_{m,m}$ with a weight function $w : E \longrightarrow R^n$

to determine the set of all efficient matchings. We restrict us to the *ith* component and search for the K-best perfect matchings in this component. There are several algorithms [1], [2] available. We select one and denote it as algorithm X-BEST(i) whenever it is applied to the *ith* component.

An algorithm to find the K-best solutions will stop according to the agreement on K. We want X-BEST(i) to stop whenever a given set X is contained the K-best solutions. For instance if we want to find the set C_1 of all first corner we execute X-BEST(1) as long for K as the first component of the K best solution does not change. X-Best plays an essential role in the following.

5.1 Algorithm "Descent on a chain"

1) Apply X-Best(1) to determine the set C_i of all *ith* corner, $X = C_i$.
2) Begin with the first corner in list C_1 and collect it in the list L and K.
3) "Descent on a chain". Let $K = \{a_1, \ldots, a_l\}$. Apply Y-Best(1) to determine the next solution a_{l+1} according to the first component.
4) Let $L = \{e_1, \ldots, e_r\}$. If $w(a_{l+1}) < w(e_i)$ for $i = 1, \ldots, r$ then we will not take a_{l+1} into the list. Otherwise a_{l+1} will be put into L. We will drop every e_j from L if $w(e_j) < w(e_{l+1})$ and enumerate the solution to get the list $L' = \{e_1, \ldots, e_s\}$.
5) If $Y \not\subseteq K$ then we go to 3) putting $L := L'$. Otherwise stop.

Proposition 5.1 *Let a bicriterial assignment problem be given and let (a, s_2) be the second corner for this problem. If $Y = \{(a, s_2)\}$ then the algorithm "Descent on a chain" finds every efficient solution.*

Proof. Let (c, d) be an efficient solution. We have that $a \leq c \leq s_1$. The algorithm descends the chain from s_1 to a and will find the solution (c, d).

If there are more than two criteria the determination of Y may be difficult. For our example we take Y as the set of all corners.

Example. $Y := \{M_1, M_4, M_6\}$
We begin with M_1 and descend the chain according to the first component. We get the following sequence M_4, M_2, M_6 and are ready.

While algorithms for the K best solutions are usually of good complexity we have here that the number of efficient solutions may grow exponentially with number of vertices. This is demonstrated with the example in Figure 3.

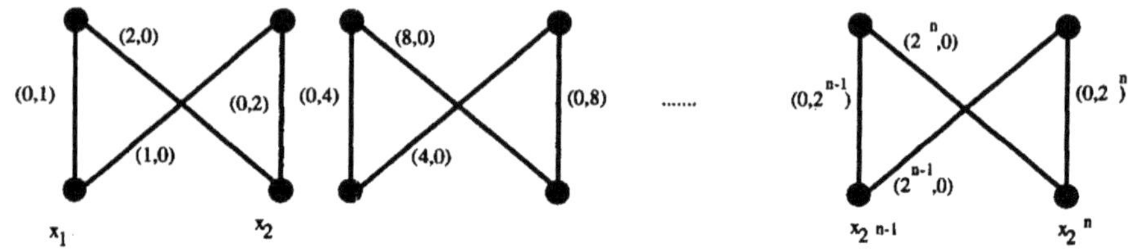

Figure 3

REFERENCES

[1] Ch. R. Chegireddy, H. W. Hamacher, Algorithms for finding the k-best perfect matchings. Discrete Appl. Math. 18(1987) 155-165

[2] H. W. Hamacher, M. Queyranne, K-best solutions to combinatorial optimization problems. Ann. Oper. Res. 4(1985/6) 12233-1143

[3] H. Isermann, The enumeration of the set of all efficient solutions for a linear multiple objective program. Opl. Res. Q. 28(1977) 711-725

[4] D. Schweigert, Linear extensions and vector-valued spanning trees. Methods of operations research 60, Anton Hain Frankfurt 1990.

[5] D. Schweigert, Linear extensions and efficient trees. Uni. Kaiserslautern, preprint 172 (1990)

[6] R. E. Steuer, Multiple criteria optimization: Theory, computation and application. Wiley, New York, 1986

DIRECTED QUADRUPLE DESIGNS

Nasrin Soltankhah

Department of Mathematical Sciences
Sharif University of Technology
P.O. Box 11365-9415
Tehran, Iran
soltan@irearn.bitnet

ABSTRACT

A t-(v,k,λ) directed design (or simply a t-(v,k,λ)DD) is a pair $(V,\mathcal{B})$, where V is a v-set and $\mathcal{B}$ is a collection of ordered k-tuples of distinct elements of V, such that every ordered t-tuple of distinct elements of V belongs to exactly λ elements of $\mathcal{B}$. (We say that a t-tuple belongs to a k-tuple, if its components appear in that k-tuple as a set, and they appear with the same order.) In this paper we show that the necessary conditions for the existence of a 3-$(v,4,1)$DD are sufficient. Also for each even v, and $v \equiv 1, 5, 7$ or $9 \pmod{12}$ there exists a simple 3-$(v,4,2)$DD. Also we exhibit a construction in the case when $\lambda = 2$.

1 INTRODUCTION

Let $0 < t \leq k \leq v$ and $\lambda > 0$ be integers, and V be a set of v elements. In this note, by an n-tuple of V, we mean an ordered n-subset of V. Each k-tuple of distinct elements of V is called a *block*. Note that a t-tuple is said to appear in a block, if its components are contained in that block as a set, and they appear with the same order.

Definition. *A* collection $\mathcal{B}$ of blocks is called a t-(v,k,λ) *directed design* (or simply a t-(v,k,λ)DD) if each t-tuple of V appears in precisely λ blocks. A directed design is said to be *simple* if no two of its blocks are identical, otherwise it is said to have repeated blocks. In this paper unless otherwise stated we allow our designs to have repeated blocks.

C. J. Colbourn and E. S. Mahmoodian (eds.), Combinatorics Advances, 277–291.

It has been shown [2,7,8,9] that when $t = 2$ and $k = 3, 4, 5$ or 6 the necessary conditions for the existence of a 2-(v, k, λ)DD (given in Theorem 1.1 below) are also sufficient, with two exceptions, namely, the nonexistence of a 2-$(15, 5, 1)$DD and a 2-$(21, 6, 1)$DD.

In this paper we show that for $t = 3$, $k = 4$, the necessary conditions for the existence of a 3-$(v, 4, 1)$DD are sufficient, besides which there exist at least four disjoint 3-$(v, 4, 1)$DDs on a given set, for all even v. So for each even v, and $v \equiv 1, 5, 7$ or $9 \pmod{12}$ there exists a simple 3-$(v, 4, 2)$DD. Also we introduce a construction which enable us to prove:

- If there exists a simple 3-$(v, 4, 2)$DD and a simple 3-$(u, 4, 2)$DD, where v or u is odd, then there exists a simple 3-$(uv, 4, 2)$DD.

Finally in the last section we introduce 3-$(v, 4, 2)$DDs for some small values of v.

First we prove the following theorem which gives some necessary conditions for the existence of a t-(v, k, λ)DD.

Theorem 1.1 *Each t-(v, k, λ)DD is a $(t-1)$-(v, k, λ')DD, where*

$$\lambda' = \frac{\lambda t(v - t + 1)}{(k - t + 1)}.$$

Proof. Let $V = \{x_1, \ldots, x_v\}$ be a v-set and A be a t-(v, k, λ)DD on V. Suppose that a $(t-1)$-tuple of V, $X = x_{\alpha_1} \cdots x_{\alpha_{t-1}}$, appears in λ_X blocks of A. In each block which contains X, the number of t-tuples containing X is $(k - t + 1)$. Thus the number of t-tuples in A which contain X is $(k - t + 1)\lambda_X$. On the other hand the number of t-tuples on V that contain X is $(v - t + 1)t$. Since each t-tuple appears λ times in A, the number of t-tuples in A which contain X is $(v - t + 1)\lambda t$. Therefore

$$(v - t + 1)\lambda t = (k - t + 1)\lambda_X$$

so that

$$\lambda_X = \tfrac{(v-t+1)\lambda t}{(k-t+1)}.$$

So λ_X does not depend on the choice of $(t-1)$-tuple X. This constant number is denoted by λ_{t-1}. Thus A is a $(t-1)$-(v,k,λ_{t-1})DD. ■

The following corollary is immediate.

Corollary 1.2. Each t-(v,k,λ)DD is also an i-(v,k,λ_i)DD, for $0 \le i \le t-1$, where λ_i, the number of appearances of each i-tuple, is:

$$\lambda_i = \frac{\lambda\binom{k}{i}\,P_{v-i}^{k-i}}{\binom{k}{t}\,P_{v-t}^{k-t}}, \quad P_v^k = \frac{v!}{(v-k)!},$$

where λ_0 is the number of blocks and usually is denoted by b. By the corollary above, if there exists a 3-$(v,4,\lambda)$DD, then,

$$b = \frac{v(v-1)(v-2)\lambda}{4}, \; \lambda_1 = (v-1)(v-2)\lambda, \; \lambda_2 = \frac{3(v-2)\lambda}{2}.$$

Therefore when λ is odd, v must be even, but when λ is even, there is no restriction on v. Thus it is sufficient to show that there exists a 3-$(v,4,1)$DD for each even v, and there exists a 3-$(v,4,2)$DD for every v ($v \neq 3$).

2 EXISTENCE OF 3-$(v,4,1)$DDS

In this section we concentrate on the existence of a 3-$(v,4,\lambda)$DD when $\lambda = 1$. First we state the following definition:

Let $K = \{k_1,\ldots,k_l\}$ be a set of numbers. A t-(v,K,λ) *design* is a v-set V and a collection of k_i-subsets, also called blocks, such that every t-subset of V appears exactly λ times in the blocks. When $t=2$, these designs are also called pairwise balanced designs . The following lemma may be accepted without proof.

Lemma 2.1 *If there exists a t-(v,K,λ_1) design, and a t-(k',k,λ_2)DD for each $k' \in K$, then there exists a t-$(v,k,\lambda_1\lambda_2)$DD.*

This lemma is used frequently in this paper.

In this case v must be even. The smallest possible numbers are $v = 2,4,6$.

There are four disjoint 3-(4, 4, 1)DDs on $V = \{1, 2, 3, 4\}$, as given below.

D_1 :	1234	1432	2413	3412	3214	4231
D_2 :	1243	1342	2314	3241	4213	4312
D_3 :	1423	2431	2134	3421	3124	4132
D_4 :	1324	2341	2143	3142	4123	4321

Further there are at least four disjoint 3-(6, 4, 1)DDs on $V = \{1, 2, 3, 4, 5, 6\}$.

D'_1 :	1234	1256	1352	1436	1642	1653	2145	2365	2416	2513
	2631	3154	3264	3412	3516	3621	4325	4526	4531	4615
	4623	5243	5346	5421	5614	5632	6254	6345	6413	6512

D'_2 :	1235	1246	1342	1536	1643	1652	2316	2413	2514	2563
	2645	3145	3461	3512	3564	3625	4215	4326	4516	4523
	4635	5261	5324	5431	5462	6154	6234	6321	6412	6513

D'_3 :	1236	1245	1354	1432	1562	2165	2351	2543	2634	3146
	3241	3256	3612	3645	4163	4261	4315	4512	4536	4625
	5164	5213	5246	5342	5631	6153	6214	6352	6423	6541

D'_4 :	1243	1265	1345	1362	1546	2164	2315	2461	2536	3254
	3416	3526	3614	4153	4251	4263	4312	4365	4562	5142
	5163	5234	5321	5641	6152	6213	6245	6342	6351	6543

A cyclic 3-(6, 4, 1)DD with base blocks (mod 6) on the set $Z_6 = \{0, 1, 2, 3, 4, 5\}$ is given below:

0135 0421 0312 0514 0523.

Theorem 2.2 *There exists a* 3-$(v, 4, 1)$*DD for each even* v.

Proof. Hanani [3] has shown that for $v \equiv 2$ or 4 (mod 6), there exists a 3-$(v, 4, 1)$ design (Steiner Quadruple System). Thus by Lemma 2.1 with the

3-(4, 4, 1)DD constructed above there is a 3-$(v, 4, 1)$DD for all $v \equiv 2$ or 4 (mod 6). Hanani [4] has shown that for $v \equiv 0$ (mod 6), there exists a 3-$(v, \{4, 6\}, 1)$ design. So applying Lemma 2.1 with the 3-(4, 4, 1)DD and 3-(6, 4, 1)DD constructed above we see that there is a 3-$(v, 4, 1)$DD for all $v \equiv 0$ (mod 6). ∎

This solves the existence problem of 3-$(v, 4, 1)$DD. Moreover, since there are at least four disjoint 3-(4, 4, 1)DDs and at least four disjoint 3-(6, 4, 1)DDs, the following theorem is immediate.

Theorem 2.3 *There exist at least four disjoint* 3-$(v, 4, 1)$*DDs on a given set, for all even* v.

3 ON THE EXISTENCE OF 3-$(v, 4, 2)$DDS

In this section we show that there exists a simple 3-$(v, 4, 2)$DD for each even v and for $v \equiv 1, 5, 7$, or 9 (mod 12). Also we exhibit a recursive construction.

When $\lambda = 2$, there is no restriction on v. There is a simple 3-$(v, 4, 2)$DD for all even v since it may be obtained from the union of two disjoint 3-$(v, 4, 1)$DDs.

In [5] it has been shown that there exists a simple 3-$(v, 4, 2)$ design for $v \equiv 1$ or 5 (mod 12). Applying Lemma 2.1 with a 3-(4, 4, 1)DD implies the existence of simple 3-$(v, 4, 2)$DDs for all $v \equiv 1$ or 5 (mod 12).

Now we state three definitions and two lemmas which are generalizations of Hanani [4].

Definition 1(cf. [4, Definition 2]). Let a set E of $v = mt + r$ elements be given. We split it in to $t + 1$ disjoint subsets, viz. t sets E_i $(i = 1, \ldots, t)$ of m elements each, and a set E_{t+1} of r elements, and denote $E_i' = E_i \cup E_{t+1}$ $(i = 1, \ldots, t)$. If there exist a system 3-(v, K, λ)DD, D, on the set E, 3-$(m + r, K, \lambda)$DDs, D_i, on the sets E_i' and if for $r > 2$ also a 3-(r, K, λ)DD, D_r, on the set E_{t+1} such that $D_i \subset D$ and $D_r \subset D_i$ $(i = 1, \ldots, t)$, then the system D will be denoted by $D_m[K, \lambda, v]$.

It is easily seen that

$$D_m[K, \lambda, v] = \cup_{i=1}^{t} D_i \cup D_m{}'[K, \lambda, v]$$

where $D_m'[K, \lambda, v]$ is a subsystem of $D_m[K, \lambda, v]$ consisting of precisely those blocks which have no more than two elements in common with each of the E_i' $(i = 1, 2, \ldots, t)$.

Definition 2(cf. [4, Definition 3]). Let a set E of $v = mt$ elements be a union of its t disjoint subsets E_i $(i = 1, 2, \ldots, t)$ each having m elements, let λ be a positive integer, and let $K = \{k_i\}_{i=1}^{n}$ be a finite set of integers $k_i \geq 4$ $(i = 1, 2, \ldots, n)$. If it is possible to form a system of blocks in such a way that

(*i*) the number of elements in each block is some $k_i \in K$,

(*ii*) each block has at most one element in common with each of the sets E_i $(i = 1, 2, \ldots, t)$, and

(*iii*) every 3-tuple of distinct elements of E having at most one element in common with each of the sets E_i is contained in exactly λ blocks,

then we shall denote such a system by $D_m''[K, \lambda, v]$.

Definition 3(cf. [4, Definition 5]). Let a set E of $v = mt + r$ elements be subdivided into subsets E_i and E_i' as in Definition 1. Denote $E_i' = E_i \cup E_{t+1}$ $(i = 1, \ldots, t)$. Denote by T_1 the system of all those 3-tuples which have at most one element in common with each of the sets E_i, and by T_2 the system of all those 3-tuples which have exactly two elements in common with at least one of the sets E_i', each 3-tuple taken λ times (where λ is a positive integer). Subdivide the system T_2 into two disjoint classes T_2' and T_2'' in such a way that all the 3-tuples which have an element in E_{t+1} belong to T_2' and that the classes T_2' and T_2'' are invariant with respect to any permutation of the sets E_i; otherwise the subdivision is arbitrary.

Let $K = \{k_i\}_{i=1}^{n}$ be a finite set of integers $k_i \geq 4$. If it is possible to form on E two systems of blocks in such a way that

(*i*) the number of elements in each block is some $k_i \in K$, and

(*ii*) the 3-tuples occurring in the blocks of the first system are exactly those of $T_1 \cup T_2'$ and in the blocks of the second system those of $T_1 \cup T_2''$,

then we denote such systems by $DQ_m'[K, \lambda, v]$ and $DQ_m''[K, \lambda, mt]$ respectively.

The class of integers v for which systems $D_m[K,\lambda,v]$ exist will be denoted by $D_m(K,\lambda)$. The classes $D_m''[K,\lambda,v]$, $DQ_m'[K,\lambda,v]$ and $DQ_m''[K,\lambda,mt]$ are defined in a similar way.

Lemma 3.1 *(cf. [4, Proposition 8]). Suppose $v = mt+r$ where there exists a 3-$(t+1, K', \lambda')$ design, and that $m(k'-1)+r \in D_m(K,\lambda'')$ and $mk' \in D_m''(K,\lambda'')$ for every $k' \in K'$, then there exists a 3-$(v,K,\lambda)DD$, where $\lambda = \lambda'\lambda''$.*

Proof. Denote the set of v elements by $\{(i,j) \mid i = 0,1,\ldots,m-1; j = 0,1,\ldots,t-1\} \cup \{\infty_l \mid l = 0,1,\ldots,r-1\}$. The blocks are constructed as follows,

(*i*) on each of the sets

$$E_j' = \{\infty_l, (i,j) \mid i = 0,1,\ldots,m-1; l = 0,1,\ldots,r-1\} \quad (j = 0,1,\ldots,t-1)$$

construct a 3-$(m+r, K, \lambda'')$DD, D_j, for every j ($j = 0,1,\ldots,t-1$), and take these directed designs λ' times each. If, however, $r > 2$ and as required by Definition 1 there exists a 3-(r,K,λ'')DD, D_r, on the set $\{\infty_l \mid l = 0,1,\ldots,r-1\}$ which is contained in D_j for every j, then take D_r and $D_j - D_r$ for every j ($j = 0,1,\ldots,t-1$), λ' times each;

(*ii*) in addition, construct a 3-$(t+1, K', \lambda')$ design, B, on the set $\{j \mid j = 0,1,\ldots,t\}$; for each block b' of B which contains the element t, omit t, and construct a $D_m'[K,\lambda'', m(|b'|-1)+r]$ on the set $\{(i,j) \mid i = 0,1,\ldots,m-1; j \in b' - \{t\}\} \cup \{\infty_l \mid l = 0,1,\ldots,r-1\}$, and for each block b of B which does not contain the element t, construct a $D_m''[K,\lambda'', m|b|]$ on the set $\{(i,j) \mid i = 0,1,\ldots,m-1; j \in b\}$. ∎

Lemma 3.2 *(cf. [4, Proposition 9]). Suppose $v = mt+r$ where there exist a 3-$(t+2, K', \lambda')$ design and 3-$(m(k'-2)+r, K, \lambda'')DD$ for every $k' \in K'$. Let there exist a 3-$(r,K,\lambda'')DD$ such that 3-$(r,K,\lambda'')DD \subset$ 3-$(m(k'-2)+r, K, \lambda'')DD$, if $r > 2$. Finally, let $mk' \in D_m''(K,\lambda'')$ and $m(k'-1)+r \in DQ_m'(K,\lambda'')$ and $m(k'-1) \in DQ_m''(K,\lambda'')$ for suitably chosen classes T_2', T_2''. Then there exists a 3-$(v,K,\lambda)DD$, where $\lambda = \lambda'\lambda''$.*

Proof. Denote the set of v elements by $\{(i,j) \mid i = 0,1,\ldots,m-1; j = 0,1,\ldots,t-1\} \cup \{\infty_l \mid l = 0,1,\ldots,r-1\}$. The blocks are constructed as follows. First form a 3-$(t+2, K', \lambda')$ design, B, on the set $\{j, x', x'' \mid j = 0,1,\ldots,t-1\}$. Next, (*i*) if b^* is a block in B which contains both of the elements x' and x'', omit these elements and construct a 3-$(m(|b^*|-2)+r, K, \lambda'')$DD, D^*, on the set $\{(i,j) \mid i = 0,1,\ldots,m-1; j \in b^* - \{x', x''\}\} \cup \{\infty_l \mid l = 0,1,\ldots,r-1\}$. However, if $r > 2$, take the 3-(r, K, λ'')DD, D_r, on the set $\{\infty_l \mid l = 0,1,\ldots,l-1\}$, λ' times and instead of D^* take $D^* - D_r$;

(*ii*) if b' is a block in B which contains the element x' but not x'', construct $DQ_m{}'[K, \lambda'', (m(|b'|-1)+r]$ on the set $\{(i,j) \mid i = 0,1,\ldots,m-1; j \in b' - \{x'\}\} \cup \{\infty_l \mid l = 0,1,\ldots,r-1\}$;

(*iii*) if b'' is a block in B which contains the element x'' but not x', construct $DQ_m{}''[K, \lambda'', (m(|b''|-1)]$ on the set $\{(i,j) \mid i = 0,1,\ldots,m-1; j \in b'' - \{x''\}\}$;

(*vi*) if b is a block in B which contains neither x' nor x'', construct a $D_m{}''[K, \lambda'', m(|b|]$ on the set $\{(i,j) \mid i = 0,1,\ldots,m-1; j \in b\}$. ∎

Theorem 3.3 *There exists a simple* 3-$(v,4,2)$*DD for all* $v \equiv 7 \pmod{12}$.

Proof. Here $v = 6t+1$ and by [4] there exists a 3-$(t+1, \{4,6\}, 1)$ design. By Lemma 3.1 this theorem will follow from the fact that $19 \in D_6(4,2)$, $31 \in D_6(4,2)$, $24 \in D_6{}''(4,2)$ and $36 \in D_6{}''(4,2)$. For the two first cases, see section 4. The existence of the last two systems can be obtained as a consequence of the results in Hanani [4, Lemma 9]. Since the above systems are simple, the desired directed design is also simple. ∎

Theorem 3.4 *There exists a simple* 3-$(v,4,2)$*DD for all* $v \equiv 9 \pmod{12}$.

Proof. Here $v = 2t+1$ and by [6] there exists a 3-$(t+2, \{4,6\}, 1)$ design, such that the blocks of size six, as a set, form a parallel class. We choose x', x'' such that they belong to the same block of order six. Considering Lemma 3.2 we have to prove that there exist a simple 3-$(5,4,2)$DD, 3-$(9,4,2)$DD, $D_2{}''[4,2,8]$, $D_2{}''[4,2,12]$. Further, for suitably chosen classes T_2' and T_2'', there exist $DQ_2{}'[4,2,7]$ and $DQ_2{}''[4,2,6]$. For the first and the second see, Section 4. The others are proved as follows.

$8 \in D_2{}''(4,2)$:

Elements: $\{(i,j) \mid i = 0,1\ ;\ j = 0,1,2,3\}$.

Blocks: Form a 3-$(4,4,1)$DD on quadruples

$\{(i_0,0),(i_1,1),(i_2,2),(i_3,3)\}$ $\quad\left(\Sigma_{j=0}^{3} i_j \equiv h \pmod 2 \quad h=0,1\right).$

$12 \in D_2''(4,2)$:

Elements: $\{(i,j) \mid i=0,1\ ;\ j=0,1,2,3,4,5\}$.

Blocks: Form a 3-(6,4,1)DD on sets

$$\{(i,0),(i,1),(i,2),(i,3),(i,4),(i,5)\} \quad (i=0,1);$$

further construct a 3-(4,4,1)DD on sets
$\{(i,j),(i,j+2),(i+1,j+1),(i+1,j+4)\}$ $\quad (i=0,1; j=0,1,\ldots,5)$;
$\{(0,j),(0,j+2),(1,j+3),(1,j+5)\}$;
$\{(0,2\gamma),(0,2\gamma+\epsilon),(1,2\gamma'),(1,2\gamma'+\epsilon)\}$ $\quad (\gamma,\gamma'=0,1,2;\gamma'\neq\gamma;\epsilon=\pm 1)$.

In order to proceed, we choose the class T_2' to consist of all the 3-tuples of T_2 which contain the element ∞, together with all those 3-tuples which contain pairs of the form $(a,0)(a+1,0)$ or $(a+1,0)(a,0)$. Then $T_2''=T_2-T_2'$.

$7 \in DQ_2'(4,2)$:

Elements: $\{(i,j) \mid i=0,1\ ;\ j=0,1,2\} \cup \{\infty\}$.

Blocks: Form a 3-(4,4,1)DD on quadruples
$\{\infty,(i_0,0),(i_1,1),(i_2,2)\}$ $\quad (i_0+i_1+i_2 \equiv h \pmod 2, h=0,1)$;
$\{(i_0,0),(i_0+1,0),(i_1,1),(i_2,2)\}$ $\quad (i_0+i_1+i_2 \equiv 0 \pmod 2)$).

$6 \in DQ_2''(4,2)$:

Elements: $\{(i,j) \mid i=0,1; j=0,1,2\}$.

Blocks: Form a 3-(4,4,1)DD on quadruples
$\{(i_0,0),(i_1,1),(i_1+1,1),(i_2,2)\}$ $\quad (i_0+i_1+i_2 \equiv 0 \pmod 2)$);
$\{i_0,0),(i_1,1),(i_2,2),(i_2+1,2)\}$ $\quad (i_0+i_1+i_2 \equiv 1 \pmod 2)$).

Since the above systems are simple, the desired directed design is also simple. ■

Now we introduce a recursive construction. Although it does not fill any congruence classes, it settles the existence of 3-$(v,4,2)$DDs for infinitely many values of v. We use the following theorem from graph theory for the recursive construction.

Theorem 3.5 *[1, page 167]. For every positive integer n, the complete graph K_{2n+1} can be decomposed into n Hamiltonian cycles.*

Note. Assume that $F = \{C_1, \ldots, C_n\}$ and $F' = \{C'_1, \ldots, C'_n\}$ are two decompositions into Hamiltonian cycles of two copies of the graph K_{2n+1} on vertex sets $V_1 = \{x_1, \ldots, x_{2n+1}\}$ and $V_2 = \{x'_1, \ldots, x'_{2n+1}\}$ respectively. We consider a correspondence between C_is and C'_is and take the quadruples $\{x, y, z, w\}$ such that $\{x, y\} \in C_i$ and $\{z, w\} \in C'_i$.

The number of such quadruples is $n(2n+1)^2$. We claim that, if a triple appears in these quadruples then, it appears in precisely two of them. For let $\{x, y, z\}$ be a triple which is contained in some of these quadruples. At most two elements from $\{x, y, z\}$ are in V_1 or V_2. Assume $x, y \in V_1$, $z \in V_2$. Thus there exists $C_i \in F$ and there exists $C'_i \in F'$ such that $\{x, y\} \in C_i$, $\{w, z\}$, $\{w', z\} \in C'_i$. Thus we have quadruples $\{x, y, z, w\}$, $\{x, y, z, w'\}$. If there is a third quadruple which contains $\{x, y, z\}$, say $\{x, y, z, w''\}$, then we have taken $\{x, y\} \in C_i$ and $\{z, w''\} \in C'_i$. This is a contradiction, because C'_i is a cycle.
Also we need some orthogonal arrays in our construction.

Definition 4. Let $1 \leq t \leq k$ and $2 \leq v$. An *orthogonal array of strength t, depth k and index* λ based on v symbols, say, $0, 1, \ldots, v-1$, is an $N \times k$ array such that for any $N \times t$ sub-array each ordered t-tuple (not necessarily of distinct elements) from $V = \{0, 1, \ldots, v-1\}$ occurs precisely λ times as a row of the sub-array. It is denoted by t-(v, k, λ) OA.
Note that a 3-$(u, 4, 1)$ OA always exists for every u. For we may consider an array on u symbols, say $\{0, 1, 2, \ldots, u-1\}$, whose rows are the solutions of $a_1 + a_2 + a_3 + a_4 \equiv 0 \pmod{u}$.

Theorem 3.6 *Let A be a* 3-$(u, 4, 2)DD$ *and B a* 3-$(v, 4, 2)DD$*, such that at least one of the parameters u or v is odd. Then there exists a* 3-$(uv, 4, 2)DD$.

Proof. Assume u is odd. Let $X = \{1, \ldots, u\}$ and $Y = \{1, \ldots, v\}$, and let

$$X_i = X \times \{i\}, \quad i = 1, \ldots, v.$$

We construct a 3-$(u, 4, 2)$DD on X_i, and denote it by A_i. Let H be a 3-$(u, 4, 1)$ OA on the set X, and $F_i = \{C_{1i}, \ldots, C_{\frac{u-1}{2}i}\}$ be a decomposition into Hamiltonian cycles of the complete graph K_u on vertex set X_i. The blocks for the desired 3-$(uv, 4, 2)$DD are:

(*i*) the blocks of A_i for every $i \in Y$, $i = 1, \ldots, v$. The number of these blocks is $uv(u-1)(u-2)/2$;

(*ii*) the blocks which are obtained from constructing a 3-(4, 4, 1)DD on the quadruples $\{\alpha, \beta, \gamma, \xi\}$ such that $\{\alpha, \beta\} \in C_{ij}$ and $\{\gamma, \xi\} \in C_{ik}$ for $j < k$, $j, k \in Y$. The number of these blocks is $3u^2v(u-1)(v-1)/2$;

(*iii*) the blocks obtained from the Kronecker product $H \times B$. The number of these blocks is $u^3v(v-1)(v-2)/2$.

Every 3-tuple is contained in at least two of the blocks. Consider, for example, $(x, i)\ (y, j)\ (z, k)$. This is contained in (*i*) if $i = j = k$, in (*ii*) if $i = j \neq k$, $i = k \neq j$, or $j = k \neq i$ and in (*iii*) if $i \neq j \neq k \neq i$. Also the total number of blocks obtained is $\frac{uv(uv-1)(uv-2)}{2}$. This completes the proof. ∎

Corollary 3.7. If in Theorem 3.6, A and B are simple directed designs, then the resulting directed design is also simple.

Corollary 3.8. If u is odd and there exists a 3-$(u, 4, 2)$DD then there exists a 3-$(2u, 4, 2)$DD.

Proof. This is the special case of Theorem 3.6 obtained by taking B to be the empty directed design.

4 SOME SMALL CASES (FOR $\lambda = 2$)

In this section we give some constructions of 3-$(v, 4, 2)$DDs for some small values of v.

4.1 $v = 5$

The element set is Z_5. Developing the following base blocks (mod 5) into 30 blocks gives a cyclic 3-(5, 4, 2)DD.

0123 0134 0214 0321 0324 0431.

4.2 $v = 7$

The element set is Z_7. Developing the following base blocks (mod 7) into 105 blocks gives a cyclic 3-$(7, 4, 2)$DD.

0123 0134 0145 0156 0216 0241 0351 0361
0432 0461 0462 0532 0536 0643 0652

4.3 $v = 9$

First we show that there exists a simple 3-$(9, \{4, 6\}, 2)$ design. Then by applying Lemma 2.1 with a 3-$(4, 4, 1)$DD and a 3-$(6, 4, 1)$DD, we will have a simple 3-$(9, 4, 2)$DD.

Elements: $\{(i, j) \mid i = 0, 1, 2\ ;\ j = 0, 1\} \cup \{\infty_l \mid l = 0, 1, 2\}$.

Blocks: $\{\infty_l, \infty_{l+1}, (i, 0), (j, 1)\}$ $\quad i + j \equiv l + 1 \pmod 3$;
$\{\infty_l, (i, j), (i + 1, j), (k, j + 1)\}$ $\quad i + k \equiv l + j + 1 \pmod 3$;
$\{\infty_l, (i, j) \mid l = 0, 1, 2,\ i = 0, 1, 2\}$ $\quad j = 0, 1$;
$\{(i, j) \mid i = 0, 1, 2\ ;\ j = 0, 1\}$.

Also the following base blocks (mod 9) give a cyclic 3-$(9, 4, 2)$DD.

0123 0134 0145 0156 0167 0178 0218 0241 0247 0257
0326 0461 0361 0371 0382 0581 0638 0742 0842 0483
0543 0846 0751 0652 0752 0654 0763 0768

4.4 $v = 19$

We show that there exists a simple 3-$(19, 4, 2)$DD, which is also a system $D_6[4, 2, 19]$.

Elements: $\{(x, y, j) \mid x = 0, 1\ ;\ y = 0, 1, 2\ ;\ j = 0, 1, 2\} \cup \{\infty\}$.

Blocks: Form a 3-$(7, 4, 2)$DD on each of the sets

$$\{(x, y, j) \mid x = 0, 1\ ;\ y = 0, 1, 2\} \cup \{\infty\} \quad (j = 0, 1, 2).$$

Further, construct a 3-$(4, 4, 1)$DD on each of the quadruples obtained as follows.

Let $F^j = \{F_i^j \ ; \ i = 1,2,3,4,5\}$ $(j = 0,1,2)$ be a 1-factorization of K_6 on vertex sets $A^j = \{(x,y,j) \mid x = 0,1 \ ; \ y = 0,1,2\}$ $(j = 0,1,2)$, such that

$$\begin{aligned}
F_1^j &= \{\{(0,0,j),(0,1,j)\},\{(0,2,j),(1,1,j)\},\{(1,0,j),(1,2,j)\}\};\\
F_2^j &= \{\{(0,0,j),(0,2,j)\},\{(0,1,j),(1,0,j)\},\{(1,1,j),(1,2,j)\}\};\\
F_3^j &= \{\{(0,0,j),(1,2,j)\},\{(0,1,j),(0,2,j)\},\{(1,0,j),(1,1,j)\}\};\\
F_4^j &= \{\{(0,y,j),(1,y,j)\} \ \ y = 0,1,2\};\\
F_5^j &= \{\{(0,y,j),(1,y+1,j)\} \ ; \ y = 0,1,2\};
\end{aligned}$$

for $j = 0,1,2$.

Now take quadruples $\{x,y,z,w\}$ such that $\{x,y\} \in F_i^l$ and $\{z,w\} \in F_{\sigma(i)}^k$ where $0 \le l < k \le 2$ and $\sigma = (1\ 3\ 2)(4\ 5)$ is a permutation on the set $\{1,2,3,4,5\}$. Add to these the following quadruples:
$\{\infty,(x_0,b_0,0),(x_1,b_1,1),(x_2,b_2,2)\}$

$$\left(x_0,x_1,x_2 = 0,1 \text{ independently}; \ \textstyle\sum_{h=0}^{2} b_h \equiv 0 \pmod 3\right);$$

$\{(a_0,b_0,j),(a_0+1,b_0+1,j),(a_1,b_1,j+1),(a_2,b_2,j+2)\}$

$$\left(\textstyle\sum_{h=0}^{2} a_h \equiv 0 \pmod 2 \ ; \textstyle\sum_{h=0}^{2} b_h \equiv j \pmod 3\right);$$

$$\begin{aligned}
&\{(x,y+1,j),(x,y-1,j),(x+\beta,y,j-1),(x+\beta,y,j+1)\} && (\beta = 0,1);\\
&\{(x,y,j),(x,y+1,j),(x+1,y,j-\epsilon),(x+1,y+1,j+\epsilon)\} && (\epsilon = \pm 1);\\
&\{(x,y,j),(x,y+1,j),(x,y,j+1),(x,y+1,j+1)\};\\
&\{(0,y,j),(1,y,j),(0,y+\gamma,j+1),(1,y+\gamma,j+1)\} && (\gamma = 0,1,2).
\end{aligned}$$

4.5 $v = 31$

We show that there exists a simple 3-(31, 4, 2)DD, which is also a system $D_6[4,2,31]$.

Elements: $\{(x,y,j) \mid x = 0,1 \ ; \ y = 0,1,2 \ ; \ j = 0,1,2,3,4\} \cup \{\infty\}$.

Blocks: Form a 3-(7, 4, 2)DD on each of the sets

$$\{(x,y,j) \mid x = 0,1 \ ; \ y = 0,1,2\} \cup \{\infty\} \quad (j = 0,1,2,3,4).$$

Further, construct a 3-(4, 4, 1)DD on each of the following quadruples.

Let $F^j = \{F_i^j \ ; \ i = 1,2,3,4,5\}$ $(j = 0,1,2,3,4)$ be a 1-factorization of K_6 on vertex sets $A^j = \{(x, y, j) \mid x = 0,1 \ ; \ y = 0,1,2\}$ $(j = 0,1,2,3,4)$, such that

$$F_1^j = \{\{(0,0,j),(0,1,j)\},\{(1,0,j),(1,1,j)\},\{(0,2,j),(1,2,j)\}\},$$

$$F_2^j = \{\{(0,0,j),(0,2,j)\},\{(1,0,j),(1,2,j)\},\{(0,1,j),(1,1,j)\}\},$$

$$F_3^j = \{\{(0,0,j),(1,1,j)\},\{(0,1,j),(1,2,j)\},\{(0,2,j),(1,0,j)\}\},$$

$$F_4^j = \{\{(0,0,j),(1,2,j)\},\{(0,1,j),(1,0,j)\},\{(0,2,j),(1,1,j)\}\},$$

$$F_5^j = \{\{(0,0,j),(1,0,j)\},\{(0,1,j),(0,2,j)\},\{(1,1,j),(1,2,j)\}\},$$

for $j = 0,1,2,3,4$.
Now take quadruples $\{x, y, z, w\}$ such that $\{x, y\} \in F_i^l$ and $\{z, w\} \in F_{\sigma(i)}^k$ where $0 \leq l < k \leq 4$ and $\sigma = (132)(45)$ is a permutation on the set $\{1,2,3,4,5\}$. Add to these the following quadruples:

$$\{\infty, (x + a_0, y + 1 + a_0, j), (x + a_1, y, j - \alpha), (x + a_2, y, j + \alpha)\}$$

$$\left(a_h = 0,1; \textstyle\sum_{h=0}^{2} a_h \equiv 1 \pmod 2; \alpha = 1,2\right);$$

$$\{(a_0, y, j - \alpha), (a_1, y, j + \alpha), (a_2, y - \varepsilon, j - 2\alpha), (a_2, y + \varepsilon, j + 2\alpha)\}$$

$$\left((\alpha = 1,2; \varepsilon = \pm 1; \textstyle\sum_{h=0}^{3} a_h \equiv 0 \pmod 2)\right);$$

$$\{(a_0, y, j - 1), (a_1, y, j + 1), (a_2, y + \gamma, j - 2), (a_3, y + \gamma, j + 2)\}$$

$$\left(\gamma = 0,1,2; \textstyle\sum_{h=0}^{3} a_h \equiv 0 \pmod 2\right);$$

$$\{(0, y + 1, j), (1, y + 1, j), (\beta, y - \eta, j - \alpha), (\beta + \eta, y - \eta, j + \alpha)\}$$

$$(\alpha = 1,2; \beta = 0,1; \eta = 0,1);$$

$$\{(x, y + 1, j), (x + \eta, y - 1, j), (\beta, y, j - \alpha), (\beta + \eta, y, j + \alpha)\}$$

$$(\alpha = 1,2; \beta = 0,1; \eta = 0,1);$$

$$\{(x,y,j),(x+\eta,y+1,j),(\beta,y,j+\alpha),(\beta+\eta,y+1,j+\alpha)\}$$

$$(\alpha = 1,2; \beta = 0,1; \eta = 0,1);$$

$$\{(0,y,j),(1,y,j),(0,y,j+\alpha),(1,y,j+\alpha)\} \qquad (\alpha = 1,2).$$

Acknowledgment. I am very grateful to my supervisor Professor Ebadollah S. Mahmoodian for his important and most helpful comments and encouragement. I also thank to Professor Anne Penfold Street for reading the manuscript and for her helpful suggestions.

REFERENCES

[1] M. Behzad, G. Chartrand and L. Lesniak, "Graphs and Digraphs", *Prindle, Boston*, (1979).

[2] F.E. Bennett, A. Mahmoodi, R. Wei and J. Yin, "Existence of DBIBDs with Block Size Six", *Utilitas Math.* **43** (1993), 205–217.

[3] H. Hanani, "On Quadruple Systems", *Canad. J. Math.* **12** (1960), 145–157.

[4] H. Hanani, "On Some Tactical Configurations", *Canad. J. Math.* **15** (1963), 702–722.

[5] G.B. KHosrovshahi and S. Ajoodani-Namini, "Combining t-Designs", *J. Combin. Theory* **58** (1991), 26–34

[6] W.H. Mills, "On The Covering of Triples by Quadruples", *Proc. Fifth Southeastern Conf. Combinatorics, Graph Th. and Computing* (1974), 563–581.

[7] J. Seberry and D. Skillicorn, "All Directed BIBDs with $k = 3$ Exist", *J. Combin. Theory* A **29** (1980), 244–248.

[8] D.J. Street and J. Seberry, "All DBIBDs with Block Size Four Exist", *Utilitas Math.* **18** (1980), 27–34.

[9] D.J. Street and W.H. Wilson, "On Directed Balanced Incomplete Block Designs with Block Size Five", *Utilitas Math.* **18** (1980), 161–174.

BOUNDING TWO-TERMINAL NETWORK RELIABILITY VIA SURFACE DUALITY

Heidi J. Strayer

Department of Computer Science
University of Waterloo
Waterloo, Ontario, CANADA N2L 3G1

ABSTRACT

Given a two-terminal (possibly directed) network with assigned edge probabilities, we establish that upper bounds on the reliability of any surface dual can be used to obtain lower bounds on the reliability of the primal. This observation leads to valuable new techniques for obtaining efficiently computable lower bounds on two-terminal reliability.

1 INTRODUCTION

Computing two-terminal reliability is #P-complete even for planar graphs [12]. Thus many polynomial time bounding techniques have been developed. Computational results show that the upper bound strategies for two-terminal reliability tend to be much more accurate than the available lower bounds. In fact there exist upper bounds for which a lower bound counterpart is not known. Thus improvements in lower bound techniques are desirable.

Shogan [16] observed that the two-terminal reliability of an $s-t$ planar graph (a graph in which the two terminals both lie on the outer face boundary) is linearly related to the two-terminal reliability of its $s-t$ dual (the dual of the planar graph with an (s,t) edge added, edge directions and operational probabilities defined relative to their primal counterparts). Shogan's result permits the calculation of two-terminal reliability lower bounds of an $s-t$ planar graph from two-terminal reliability upper bounds of its $s-t$ planar dual and vice

C. J. Colbourn and E. S. Mahmoodian (eds.), Combinatorics Advances, 293–306.

versa. This motivates us to consider topological dualization to obtain graphical duals for any graph.

We first define a generalization of Shogan's $s-t$ dual and then use results of Richter and Shank [13] to show that there is a linear inequality between the two-terminal reliability of any general graph and the two-terminal reliability of any of its $s-t$ duals. A direct consequence of this inequality is that two-terminal reliability upper bound techniques can be applied to any $s-t$ dual of a graph to obtain two-terminal reliability lower bounds for the graph.

2 DEFINITIONS

A *two-terminal network* $G = (V, E, \rho, s, t)$ consists of a set of nodes V, a set of edges E, a set of edge operational probabilities ρ, a source terminal s and a destination terminal t. The value $p_{i,j} \in \rho$ represents the probability that edge (i, j) is operational at any instant of time. It is assumed that the edge operational probabilities are independent. Then $q_{i,j} = 1 - p_{i,j}$ represents the probability that edge (i, j) is non-operational at any instant of time, and $\overline{\rho} = \{q_{i,j} : (i,j) \in E\}$. In this paper the number of nodes in a network is denoted n and the number of edges as m. If the edge set of G contains directed edges, then G is a *directed two-terminal network*. We assume that the network is simple and two-connected, since two-terminal reliability of a 1-connected multigraph can be linearly reduced to the calculation of the two-terminal reliability of at most n simple two-connected subgraphs.

An $s-t$ *pathset* is a subgraph in which there is a path from s to t. An $s-t$ *minimal pathset* is just a path from s to t. A *cutset* is a set of edges whose removal makes at least one node unreachable from another; in particular, an $s-t$ cutset is a cutset whose removal makes t unreachable from s. C is a *minimal cutset* if there does not exist a cutset C' such that $C' \subset C$. The set of all $s-t$ pathsets of size i is ν_i, and $\nu = \cup \nu_i$. Similarly the set of all $s-t$ cutsets of size i is κ_i, and $\kappa = \cup \kappa_i$. Define a *cycle* as a loop or closed path and a *circuit* to be a connected set of one or more edge disjoint cycles.

Given a two-terminal (possibly directed) network $G = (V, E, \rho, s, t)$, the *two-terminal reliability of* G, $Rel_2[G]$, is the probability that s and t are connected at any instant of time. The two-terminal reliability can be expressed in terms of pathsets as $Rel_2[G] = \sum_{P \in \nu} \prod_{e \in P} p_e \prod_{e \in E \setminus P} q_e$, or cutsets as $Rel_2[G] = 1 - \sum_{C \in \chi} \prod_{e \in C} q_e \prod_{e \in E \setminus C} p_e$.

An *orientable imbedding* of a network on an orientable surface is the drawing of the network on the surface such that no edges of the network cross. A *two-cell orientable imbedding* is an orientable imbedding in which all the regions are open disks or cells. Spheres and planes (which are just spheres with a point removed) are genus 0 surfaces. For $i \geq 1$, a surface of genus i can be represented by a string of i adjacent tori, or equivalently as a sphere with i loops called *handles* added to it. A planar imbedding is just a special case of a two-cell orientable imbedding in which the orientable surface being imbedded on is the plane. All imbeddings in this paper are assumed to be orientable.

A topological dual G_I^D with respect to an imbedding I is defined in the same way as a planar dual. Each region of I becomes a vertex in G_I^D and edge e of G is added between nodes i and j in G_I^D if e is common to the boundaries of regions i and j in G.

A generalization of an $s-t$ dual $G_{I_{st}}^D$ can be defined for any two-connected simple network $G = (V, E, \rho, s, t)$ as follows. Find some circuit containing both s and t, $C = (s, v_2, v_3, \ldots v_{i-1}, t, v_{i+1} \ldots v_k)$ (such a circuit must exist since G is two-connected). Add edge $e_{st} = (s, t)$ to G. Adding e_{st} to C creates two circuits $C_1 = (s, v_2, v_3, \ldots, v_{i-1}, t)$ and $C_2 = (t, v_{i+1}, \ldots, v_k, s)$. One can then find an imbedding of $G \cup e_{st}$ such that these two circuits form region boundaries [17]. The edge probabilities in $G_{I_{st}}^D$ are $\overline{\rho}$ and the terminals s' and t' of $G_{I_{st}}^D$ are the dual nodes corresponding to the regions whose boundaries C_1 and C_2 respectively. Thus $G_{I_{st}}^D = (V^D, E^D, \overline{\rho}, s', t')$.

If G is directed, we direct e_{st} from s to t and ensure that e_{st} is directed counterclockwise around C_1 and clockwise around C_2. We then assign directions to the dual edges as follows: if edge e is directed clockwise around region $R1$ and thus counterclockwise around region $R2$ then in the dual edge e^D is directed from node $R1$ to node $R2$. Furthermore for all edges $e^D = (i, j) \in E^D, i \neq j$, add edge $e'^D = (j, i)$ to E^D with operational reliability 1. Lastly delete all edges directed into s' and all edges out of t'. The $s-t$ dual of a directed network G with respect to some I_{st} imbedding is denoted $G_{I_{st}}^D = (V^D, E^{D+}, \overline{\rho}^+, s', t')$.

One algebraic method of representing a two-cell orientable imbedding is a *pure rotation system.* If G is directed, the positive edge direction is taken to be from tail to head. If G is undirected, assign an arbitrary positive direction to each edge $+e_i$. Given a two-cell orientable imbedding I, the pure rotation system consists of a *rotational vector* r_i for each vertex i containing entries for edges incident to i in the order they are encountered when making a clockwise circulation around i. Let $e = (i, j)$ be an edge incident to i. If the assigned positive direction is from i to j then the entry for e in r_i is $-e$ and in r_j it is $+e$. It is easily seen from this algebraic representation that the number of two-cell orientable $s-t$ imbeddings for a network G is $\prod(d(v_i)-1)!$. For more details see [6].

3 RESULTS

This first proposition is due to Richter and Shank [13].

Proposition 3.1 *Let $G=(V,E)$ be a simple undirected graph and let I denote any imbedding of G on some genus g orientable surface S_g. Let $G_I^D=(R,E)$ be the dual of G with respect to I. Let C be a minimal cutset of G and C^D its corresponding subgraph in G_I^D.*

Then the degree of every vertex in C^D is even, i.e. C contains an even number of boundary edges from every region in G.

Corollary 3.2 *C^D can be expressed as a set of circuits.*

This follows from the fact that all vertices in C^D have even degree. Thus the number of cyclic subgraphs of size i in G_I^D is an upper bound for the number of cutsets of size i in G.

The following proposition yields some useful relationships between a network and its $s-t$ dual.

Proposition 3.3 *An s–t cutset C in an undirected network $G=(V,E,\rho,s,t)\cup e_{st}$ is a cyclic subgraph C^D in any $s-t$ dual $G_{I_{st}}^D=(V^D,E^D,\overline{\rho},s',t')$ of G. Furthermore s' and t' both lie on some common circuit of C^D.*

Proof:

Consider cutset C. It must contain a minimal $s-t$ cutset $\tilde{C}$. Since $\tilde{C}$ is an $s-t$ cutset, it must contain e_{st}. Therefore s' and t' are connected in $G_{I_{st}}^D$ since regions s' and t' share edge e_{st} by construction.

Now $\tilde{C}$ is also a minimal network cutset. Thus by proposition 3.1, $\tilde{C}^D$ is a set of circuits. Since s' and t' are connected in $\tilde{C}^D$, they must lie on a common circuit in $\tilde{C}^D$ and therefore in C^D. □

Corollary 3.4 *C^D is an $s' - t'$ pathset in $G^D_{I_{st}} - e_{st}$.*

By the above proposition any $s - t$ cutset $\hat{C}$ in $G \cup e_{st}$ forms a cyclic subgraph $\hat{C}^D$ in $G^D_{I_{st}}$ where s' and t' are contained on some circuit. $C = \hat{C} - e_{st}$ is a cutset in G and C^D is a subgraph of $G^D_{I_{st}} - e^D_{st}$ in which s' and t' are connected.

From this point on, $G^D_{I_{st}}$ refers to the topological $s - t$ dual with edge e^D_{st} removed.

Corollary 3.5 *Let $G = (V, E, \rho, s, t)$ be an undirected two-terminal network and let $G^D_{I_{st}} = (V^D, E^D, \overline{\rho}, s', t')$ be any $s - t$ dual of G. Then $Rel_2[G] \geq 1 - Rel_2[G^D_{I_{st}}]$.*

Let $C = \{C_1, \ldots, C_k\}$ be the set of s–t cutsets of G and $P = \{P_1, P_2, \ldots, P_r\}$ be the set of $s' - t'$ pathsets of $G^D_{I_{st}}$. By corollary 1, $C \subseteq P$. Thus $\sum_{i=1}^{k} \prod_{e \in C_i} q_e \leq \sum_{i=1}^{r} \prod_{e \in P_i} q_e$ so $1 - \sum_{i=1}^{k} \prod_{e \in C_i} q_e \geq 1 - \sum_{i=1}^{k} \prod_{e \in P_i} q_e$ since the sums are between 0 and 1.

Thus $Rel_2[G] \geq 1 - Rel_2[G^D_{I_{st}}]$. This implies that a lower bound for $Rel_2[G]$ can be obtained from an upper bound for $Rel_2[G^D_{I_{st}}]$.

Proposition 3.6 *Let $G = (V, E, \rho, s, t)$ be a directed two-terminal network and $G^D_{I_{st}} = (V^D, E^{D+}, \overline{\rho}^+, s', t')$ any directed $s - t$ topological dual of G. Then the $Rel_2[G] \geq 1 - Rel_2[G^D]$.*

Proof: Let $H^D_{I_{st}} = (V^D, E^D, \overline{\rho}, s', t')$ be a $s - t$ directed dual of $G \cup e_{st} = (s, t)$ without the edges directed into s' and the edges directed out of t' removed. Let $H = G \cup \{e_{st} = (s, t)\} \cup \{e' = (i, j)$, for all $e = (j, i) \in E\}$. If in I_{st} e borders on regions R_j and R_k (j may equal k), we treat e and e' as both bordering R_j and R_k. We do not obtain a region between e and e' unlike standard imbedding schemes. All edges in H but not in G are assigned operational probability 0. Denote the set of operational probabilities for H as ρ^+ and the edge set of H as E^+. Thus $H = (V, E^+, \rho^+, s, t)$.

Note that $Rel_2[H] = Rel_2[G]$ since H is just G with edges having 0 operational probability added. Moreover $Rel_2[H^D_{I_{st}}] = Rel_2[G^D_{I_{st}}]$ since $H^D_{I_{st}}$ is just $G^D_{I_{st}}$ with edges directed into s' and out of t' (edge $(e^D_{st} = (t', s')$ is such an edge). Such edges are contained in no minimal $s' - t'$ cutset and so have no affect on the two-terminal reliability of $H^D_{I_{st}}$. Furthermore there is a 1-1 correspondence between the edges in $H^D_{I_{st}}$ and H. Every minimal $s - t$ cutset C in H contains at most one of e, e'. Thus by the above point and proposition 3.1 C^D has all even-degreed vertices in $H^D_{I_{st}}$ with s' and t' in some common component and contains edge (t', s').

It remains to be shown that t' is reachable from s' in C^D.

Arc (t', s') is directed counterclockwise around region s' so there must be some edge e_1 shared between regions s' and say $R1$ in C which is directed clockwise around region s'. Otherwise, C would contain a backward directed edge, contradicting the definition of a minimal cutset. Thus $e_1^D = (s', R1)$. By the same argument there must be some edge e_2 shared between region $R1$ and say $R2$ in C directed counterclockwise around $R1$ and clockwise around $R2$ implying $e_2^D = (R1, R2)$. Continuing in this fashion we eventually construct a path from s' to t' which contains all edges minus (t', s') in the component of C^D containing both s' and t'. Thus we have every $s - t$ directed cutset in H mapping to an $s' - t'$ pathset in $H^D_{I_{st}}$ so the proposition holds. □

4 IMPLEMENTATION

A major stumbling block in implementing this bounding technique is to find a "good" $s - t$ imbedding. To do this we first must establish what we mean by "good". The goal of our imbedding scheme is to maximize the lower bounds of the primal, which implies that we wish to minimize the upper bounds of the dual. In other words, we wish to find the least reliable $s - t$ dual possible. We attempt to find a dual with long $s' - t'$ paths, as such a dual should be fairly unreliable. The implementation of our heuristic is as follows. Let $G = (V, E, \rho, s, t)$.

If G is directed, we disregard the edge directions when applying the following algorithm, until we reach the last step (step 6). Given a subgraph S, let *edges*(S) denote the set of edges in S.

1. Find the shortest $s-t$ circuit C in G, add edge $e_{st} = (s,t)$ to C to get the two circuits C_1 and C_2. C_1 and C_2 become s' and t' in $G^D_{I_{st}}$. Set $E_u = edges(C_1) \cup edges(C_2)$. Set $E_a = E - E_u - \{e_{st}\}$. Set $\Theta = \{C_1, C_2\}$.

2. For each $e = (v_1, v_2) \in E_u$ find if possible a shortest path P from v_1 to v_2 in E_a. If such a path is found, set $E_a = E_a - edges(P)$, $E_u = E_u + edges(P)$, and $\Theta = \Theta \cup \{P + e\}$. In any case, set $E_u = E_u - \{e\}$. Repeat this step until either $E_u = \emptyset$ or $E_a = \emptyset$.

3. If $E_u = \emptyset$ but $E_a \neq \emptyset$ then find a cycle C in E_a. If such a cycle is found then set $E_a = E_a - edges(C)$, $E_u = edges(C)$, $\Theta = \Theta \cup \{C\}$ and repeat step 2.

4. Find a rotation system for the subgraph $S = (V, E_u)$ that ensures the circuits in Θ form face boundaries. This is always possible (see [18]).

5. If $E_a \neq \emptyset$ then add the remaining edges in E_a at the end of their respective incident vertices' rotation vectors.

6. Assign the "signs" to the edge entries in the rotation system so that they are consistent with the forward direction of the edge, as described in Section 2.

The lower bounds eventually obtained from an imbedding constructed in this manner are quite competitive with other exact techniques as shown in the next section.

5 EXAMPLES

We use the 1979 Arpanet with equal edge probabilities as our sample network because there are already numerous published results for comparison [3, 9, 4]. The terminal pairs we analyze are $(1,49)$, $(1,59)$, $(5,56)$, and $(20,41)$.

The best topological bounds outperformed the direct lower bound methods over the entire probability range tested for terminal pairs $(1,59)$ and $(5,56)$. Moreover for all but extremely high edge probabilities the topological bounds were better for node pair $(20,41)$. However, for node pair $(1,49)$ the direct lower bound methods were decidedly better.

It is apparent from these results that for networks with high edge probabilities, using renormalization on a "good" dual yields very competitive lower bounds.

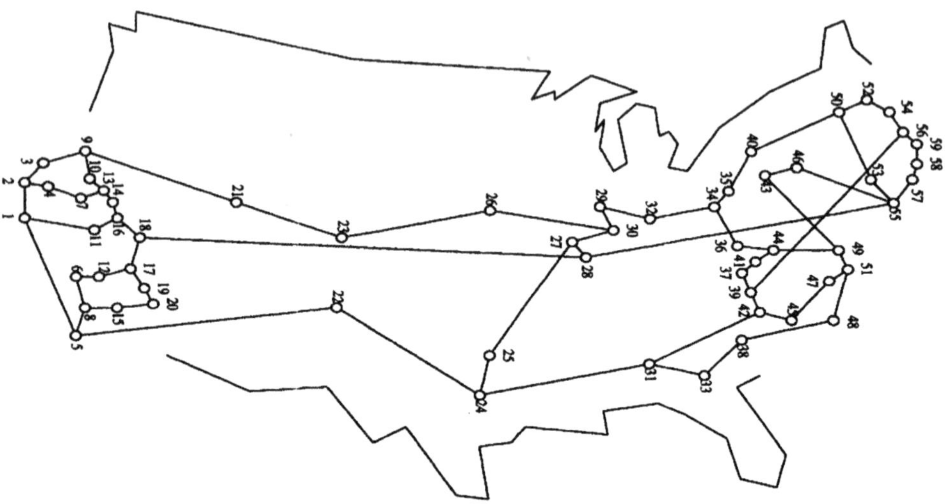

Figure 1 The 1979 Arpanet

This is because the error term in renormalization is small when the edge probabilities are small [7]. Across all edge probabilities, Shanthikumar, Non-crossing k-cuts, and Series-Parallel techniques yield reasonable results when applied to the dual.

A point of interest is that bounds often cross. Some bounds tend to be better at low edge probabilities while others tend to be better at high edge probabilities. In particular, Series-Parallel lower bounds sometimes cross the lower bounds determined from Series-Parallel upper bounds on the dual. This is due to the different series-parallel subgraphs extracted. At low edge probabilities a highly connected series-parallel subgraph is determined while for high edge probabilities a series-parallel graph which attempts to minimize the $s-t$ path lengths is used as a result of the heuristics employed [8].

We give both a graphical and tabular presentation of the computational results for each of the four terminal pairs. The graph drawn shows the ordering of the bounds with respect to the accuracy for fourteen values of p, with the most accurate at the top. In order to reflect how close the bounds are, we report for each pair of bounds that are adjacent in the ordering an integer α, which is larger when the bounds are "close" than when they are "far apart". More specifically, if two bounds A, B with $A > B$ are compared, the number α is chosen so that $e^{-(\alpha-1)}A \leq A - B \leq e^{-\alpha}A$.

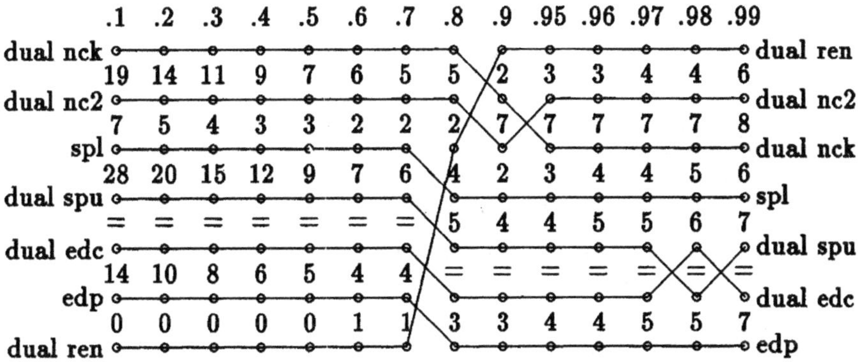

Figure 2 Lower two-terminal bounds for the 1979 Arpanet, $s = 20$, $t = 41$

p	spl	edp	dual spu	dual nc2	dual nck	dual ren	dual edc
.3	.0000118	.0000118	.0000118	.0000119	.0000119	.0000060	.0000118
.4	.0002099	.0002097	.0002099	.0002156	.0002156	.0001118	.0002099
.5	.0019599	.0019521	.0019597	.0020526	.0020537	.0011178	.0019597
.6	.0121741	.0120566	.0121688	.0130317	.0130520	.0078609	.0121688
.7	.0567839	.0556971	.0566659	.0618679	.0620687	.0450263	.0566659
.8	.2096305	.2032191	.2082441	.2288572	.2297486	.2130434	.2082441
.9	.5946042	.5757802	.5878493	.6287570	.6291905	.6685376	.5878493
.95	.8545479	.8389879	.8478182	.8752900	.8745993	.9070316	.8478182
.96	.9002719	.8876628	.8946397	.9159448	.9152270	.9401368	.8946397
.97	.9400237	.9310539	.9358925	.9503814	.9497637	.9663271	.9358925
.98	.9715729	.9665376	.9691856	.9769516	.9765483	.9851278	.9691856
.99	.9924450	.9908572	.9916713	.9940064	.9938626	.9963290	.9916713

Table 1 Lower two-terminal bounds for the 1979 Arpanet, $s = 20$, $t = 41$

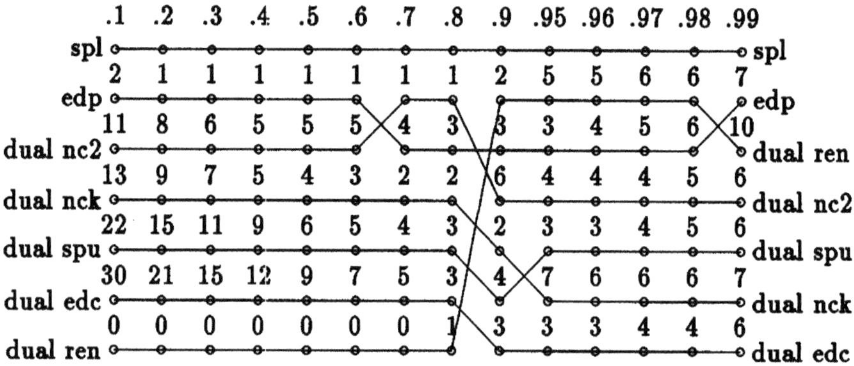

Figure 3 Lower two-terminal bounds for the 1979 Arpanet, $s = 1$, $t = 49$

p	spl	edp	dual spu	dual nc2	dual nck	dual ren	dual edc
.3	.0001049	.0000854	.0000852	.0000853	.0000852	.0000196	.0000852
.4	.0011796	.0009240	.0009173	.0009210	.0009174	.0002627	.0009173
.5	.0078085	.0059731	.0058522	.0059420	.0058577	.0019764	.0058517
.6	.0367784	.0279757	.0267265	.0278952	.0268463	.0105311	.0267045
.7	.1348309	.1044372	.0961654	.1053364	.0974188	.0462366	.0956753
.8	.3906123	.3190833	.2851360	.3253376	.2909274	.2036759	.2794718
.9	.8124641	.7397975	.6767808	.7383872	.6822952	.7635139	.6511158
.95	.9652984	.9394351	.8994119	.9269133	.8988280	.9615655	.8755495
.96	.9810166	.9647239	.9339633	.9527999	.9327338	.9782835	.9143367
.97	.9914938	.9830459	.9622760	.9733428	.9608924	.9893587	.9481483
.98	.9973417	.9942692	.9831884	.9881673	.9821449	.9959437	.9751890
.99	.9996524	.9991816	.9958547	.9970621	.9954482	.9991468	.9933187

Table 2 Lower two-terminal bounds for the 1979 Arpanet, $s = 1$, $t = 49$

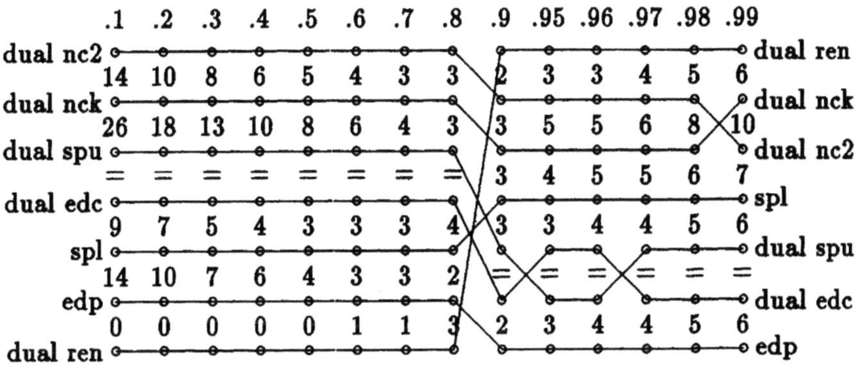

Figure 4 Lower two-terminal bounds for the 1979 Arpanet, $s = 1$, $t = 59$

p	spl	edp	dual spu	dual nc2	dual nck	dual ren	dual edc
.3	.0001312	.0001312	.0001318	.0001318	.0001318	.0000679	.0001318
.4	.0013129	.0013102	.0013296	.0013314	.0013296	.0007154	.0013296
.5	.0078563	.0077972	.0080594	.0080982	.0080613	.0046383	.0080594
.6	.0340291	.0333102	.0353242	.0358198	.0353850	.0222991	.0353242
.7	.1174355	.1119727	.1216841	.1259199	.1227107	.0870798	.1216841
.8	.3332342	.3073968	.3360230	.3586441	.3453954	.3015846	.3360230
.9	.7322978	.6756324	.7136485	.7635615	.7466581	.8036517	.7136485
.95	.9255995	.8867141	.9068495	.9381943	.9330353	.9648106	.9068495
.96	.9526123	.9223762	.9372406	.9609452	.9580246	.9793603	.9372406
.97	.9738407	.9532330	.9628520	.9785067	.9772697	.9893877	.9628520
.98	.9887938	.9777283	.9826330	.9907468	.9904898	.9956929	.9826330
.99	.9973635	.9940316	.9954343	.9977832	.9978283	.9990163	.9954343

Table 3 Lower two-terminal bounds for the 1979 Arpanet, $s = 1$, $t = 59$

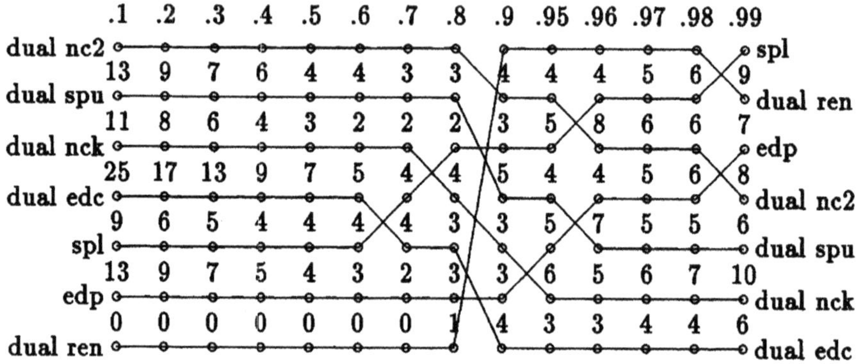

Figure 5 Lower two-terminal bounds for the 1979 Arpanet, $s = 5$, $t = 56$

p	spl	edp	dual spu	dual nc2	dual nck	dual ren	dual edc
.3	.0007313	.0007307	.0007367	.0007371	.0007349	.0000077	.0007349
.4	.0041576	.0041379	.0042454	.0042556	.0042037	.0001502	.0042035
.5	.0163851	.0161131	.0171320	.0172603	.0166603	.0015400	.0166510
.6	.0524439	.0502754	.0565011	.0574906	.0533322	.0105522	.0531235
.7	.1482769	.1371079	.1626568	.1680353	.1496750	.0555524	.1470288
.8	.3770204	.3407131	.4027063	.4225670	.3704681	.2643210	.3576139
.9	.7811348	.7321002	.7785587	.8112719	.7543486	.8201309	.7201511
.95	.9536628	.9335328	.9384753	.9563333	.9323257	.9729475	.9080047
.96	.9737323	.9606726	.9603024	.9734083	.9568249	.9848795	.9378690
.97	.9877573	.9807804	.9776152	.9860032	.9760643	.9926410	.9631327
.98	.9960033	.9933876	.9900911	.9942970	.9896622	.9971960	.9827207
.99	.9994514	.9990380	.997551	.9987262	.9975331	.9994048	.9954458

Table 4 Lower two-terminal bounds for the 1979 Arpanet, $s = 5$, $t = 56$

Thus α indicates the order of magnitude of the relative error between two adjacent bounds. This is reported as "=" when the bounds agree to 45 digit precision.

To permit other comparisons, we also present the numerical results themselves to six digit accuracy, although all were computed to 45 digits of accuracy. We suppress the results for $p = .1$ and $p = .2$ in the tables.

Computational results have indicated that the best direct two-terminal lower bound techniques are series-parallel (spl) [1] and edge-disjoint paths (edp) [9, 2]. Hence these are the two lower bounds we include in our tables. There are several competitive upper bound techniques which we have applied to the dual to obtain lower bounds indirectly. These include Shanthikumar (dual nc2) [14, 15], series-parallel upper (dual spu) [1], non-crossing k-cutsets (dual nck) [5], edge-disjoint cutsets (dual edc) [4, 10, 11], and *a priori* renormalization (dual ren) [7]. We include these results in our tables as well.

Acknowledgments

Thanks to Charles Colbourn for guidance in doing this research and writing this paper.

REFERENCES

[1] H.M. AboElFotoh, C.J. Colbourn, "Series-Parallel Bounds for the Two-Terminal Reliability Problem", *ORSA J. Comp* (1988), 209-222.

[2] T.B. Brecht, C.J. Colbourn, "Lower Bounds on Two-Terminal Network Reliability", *Discrete Appl. Math 21* (1988), 185-198.

[3] C.J. Colbourn, *The Combinatorics of Network Reliability*, Oxford University Press, Oxford, 1987.

[4] C.J. Colbourn, "Edge-Packings of Graphs and Network Reliability", *Discrete Math. 72* (1988), 49-61.

[5] C.J. Colbourn, L.D. Nel, T.B. Boffey, D.F. Yates, "Network Reliability and the Probabilistic Estimation of Damage from Fire Spread", *Annals Oper. Res.* 50 (1994) 173–185. itted for publication (1991).

[6] J.L. Gross, T.W. Tucker, *Topological Graph Theory*, John Wiley and Sons, New York, 1987.

[7] D.D. Harms, C.J. Colbourn, "Renormalization of Two-Terminal Reliability", *Networks* 23 (1993) 289–298.

[8] D.D. Harms, *A Symbolic Algebra Environment for Research in Network Reliability* Ph.D. Thesis, Simon Fraser University (1992).

[9] V.A. Kaustov, Y.I. Litvak, I.A. Ushakov, "The Computational Effectiveness of Reliability Estimates by the Method of Nonedge-Intersecting Chains and Cuts", *Soviet J. Comput. Systems Sci. 24* (1986), 70-73.

[10] L.D. Nel, C.J. Colbourn, "Locating a Broadcast Facility in an Unreliable Network", *INFOR 28* (1990), 363-379.

[11] L.D. Nel, H.J. Strayer, "Two-Terminal Reliability Bounds based on Edge Packing by Cutsets", *J. Comb. Math. Comb. Comp.* 13 (1993), 3–22.

[12] J.S. Provan, "The Complexity of Reliability Computations in Planar and Acyclic Graphs", *SIAM J. Computing 15* (1986), 694-702.

[13] B Richter, H. Shank, "The Cycle Space of an Embedded Graph", *J. Graph Theory 8* (1984), 365-369.

[14] J.G. Shanthikumar, "Reliability of Systems with Consecutive Minimal Cutsets", *IEEE Trans. on Reliab. 36* (1987), 546-550.

[15] J.G. Shanthikumar, "Bounding Network Reliability Using Consecutive Minimal Cutsets", *IEEE Trans. on Reliab. 37* (1988), 45-49.

[16] A.W. Shogan, "Sequential Bounding of the Reliability of a Stochastic Network", *Operations Research 34* (1976), 1027-1044.

[17] S. Stahl, "Generalized Embedding Schemes", *Journal of Graph Theory 2* (1978), 41-52.

[18] H. Strayer, C. Colbourn, "Bounding Network Reliability via Surface Duality", preprint.

DEFINING SETS FOR BLOCK DESIGNS: AN UPDATE

Anne Penfold Street

Centre for Combinatorics, Department of Mathematics
The University of Queensland, Brisbane 4072, AUSTRALIA

ABSTRACT

This paper deals with the following question: how many, and which, blocks of a design with given parameters must be known before the remaining blocks of the design are uniquely determined? We survey the theoretical background on such defining sets, some specific results for smallest and other minimal defining sets for small designs and the techniques used in finding them, the few known results on minimal defining sets for infinite classes of designs, and the conjectures on minimal defining sets for some classes of Hadamard designs.

1 INTRODUCTION

A *block design* $D = (V, B)$ is a set $V$of v elements (that is, a v–set) together with a set B of b subsets of V, called blocks, such that every block contains exactly k elements, and every element occurs in exactly r blocks. Here we are interested only in block designs which are *incomplete*, that is, having $k < v$. If every subset of t elements is contained in exactly λ_t blocks, the design is called a *t-design* and its parameters are indicated as t–(v, k, λ). An example is the 2–$(7, 3, 1)$ design F, with blocks

$$B_F = \{124, 235, 346, 457, 561, 672, 713\}.$$

If $t \geq 2$, the design is generally called a *balanced* incomplete block design; when $v = b$, it is said to be *symmetric*. The design F given above is a symmetric 2–$(7, 3, 1)$ design. In particular, it is an example of both a *Hadamard* design,

C. J. Colbourn and E. S. Mahmoodian (eds.), Combinatorics Advances, 307–320.

that is, a symmetric $2\text{–}(4h-1, 2h-1, h-1)$ design, and a *Steiner triple system*, that is, a $2\text{–}(v, 3, 1)$ design.

A set of blocks which is a subset of a unique $t\text{–}(v, k, \lambda)$ design is said to be a *defining set* of the design, and will be denoted by $d(t-(v, k, \lambda))$. A *minimal defining set*, denoted by $d_m(t-(v, k, \lambda))$, is a defining set, no proper subset of which is a defining set. A *smallest defining set*, denoted by $d_s(t-(v, k, \lambda))$, is a defining set such that no other defining set has smaller cardinality; the number of blocks it contains is denoted by s. Every t-design has itself as a defining set and hence must have a smallest defining set. A $d(t-(v, k, \lambda))$ defining set consisting of blocks of a particular $t\text{–}(v, k, \lambda)$ design D will often be abbreviated to dD, and a smallest defining set to d_sD. Definitions and theoretical results relating to defining sets in general were first given by Gray [7], [8], [9], although Curtis [3] had already shown that there is a set of eight blocks that defines the $5\text{–}(24, 8, 1)$ design.

The set of blocks $R = \{457, 713, 672\}$ is a subset of two 2-(7,3,1) designs. These designs are F and G where $B_F = R \cup T_1$, $B_G = R \cup T_2$, $T_1 = \{124, 235, 346, 156\}$ and $T_2 = \{125, 234, 356, 146\}$. Hence R is not a defining set of either design. But the set of blocks $S = \{124, 235, 346\}$ can be completed to a 2-(7,3,1) design only by adjoining the blocks $\{457, 561, 672, 713\}$. Hence S is a defining set of the design F.

The sets T_1 and T_2 are *mutually balanced*; that is, they contain the same t–sets ($t = 2$) the same number of times. Further, they are disjoint sets, forming a *trade*. We say that a trade is *minimal* if none of its proper subsets is a trade. Trades and defining sets are closely related, as shown by some of the results in the next section. Although strictly speaking, a trade consists of two collections of blocks, we frequently just refer to the collection contained in B to identify a trade.

We consider two interesting examples of designs containing repeated blocks. First, suppose that D, D_1 and D_2 are designs such that $B = B_1 \& B_2$, that is, the collection of blocks of both B_1 and B_2 with multiplicities counted. If S is a defining set of D, then S must contain defining sets of D_1 and D_2 respectively. But if S_1 and S_2 are defining sets of D_1 and D_2 respectively, then $S_1 \& S_2$ need not be a defining set of $D_1 \& D_2$. For example, if $D_1 = F = D_2$, we may choose $S_1 = \{124, 235, 346\}$ and $S_2 = \{457, 561, 672\}$. Then $S_1 \& S_2$ forces the last block of F, that is, 713. However it puts no restriction on the choice of the remaining seven blocks, other than that they must form a $2\text{–}(7, 3, 1)$ design, so there are 30 completions possible.

Next, consider the six 2–(7, 3, 1) designs with the following sets of blocks:

$$B_{11} = \{123, 145, 167, 246, 257, 347, 356\},$$

$$B_{12} = \{123, 146, 157, 247, 256, 345, 367\},$$

$$B_{13} = \{123, 147, 156, 245, 267, 346, 357\},$$

$$B_{21} = \{123, 145, 167, 247, 256, 346, 357\},$$

$$B_{22} = \{123, 146, 157, 245, 267, 347, 356\},$$

$$B_{23} = \{123, 147, 156, 246, 257, 345, 367\}.$$

The 2–(7, 3, 3) design E may be written as either

$$B_{11}\&B_{12}\&B_{13} \text{ or } B_{21}\&B_{22}\&B_{23}.$$

If we now choose defining sets S_{ij} for D_{ij}, where

$$S_{11} = \{123, 257, 347\}, S_{12} = \{123, 256, 345\},$$

$$S_{13} = \{123, 256, 346\}, S_{21} = \{123, 256, 346\},$$

$$S_{22} = \{123, 267, 347\}, S_{23} = \{123, 257, 345\},$$

then $S_{11}\&S_{12}\&S_{13} = S_{21}\&S_{22}\&S_{23}$ is a defining set for E.

From the small designs considered so far, it appears that the number of blocks in a smallest defining set of a 2–design is usually between a third and a half of the total number of blocks in the design, though these bounds are not exact. For instance, the non–cyclic 2–(13, 3, 1) design can be defined by only eight of its 26 blocks (Greenhill [12]), and the 2–(31, 15, 7) design developed from the quadratic residues by only 10 of its 31 blocks (Moran [18]), whereas the 2–(15, 7, 3) design associated with $PG(3, 2)$ needs 9 of 15 blocks to define it (Gray and Street [10], [11]).

Among non–isomorphic designs with the same parameters, it appeared at first that the number of blocks in the smallest defining sets was non–decreasing as the order of the automorphism group of the design increased. But among the 2–(9, 4, 3) designs there are several contradicting this idea [18].

2 SOME THEORETICAL RESULTS

Lemma 2.1 *[7] Let D be a t-(v, k, λ) design, with minimal trades*

$$T_1, T_2, \cdots, T_n \subseteq B$$

and let $S \subseteq B$. Then S is a defining set of the design D if and only if $S \cap T_i \neq \emptyset$, for each $i = 1, \cdots, n$.

Lemma 2.2 *If $T \subseteq B$ contains a block of every minimal defining set of D, then T contains a minimal trade.*

Proof: Since $B \setminus T$ contains no defining set of D, it can be completed in at least two ways to designs with the same parameters as D : first, to D itself by taking $(B \setminus T) \cup T$; secondly, to D' by taking a new set of blocks T', where $B' = (B \setminus T) \cup T'$. But now T and T' are mutually balanced and contain a minimal trade. □

If the t–(v, k, λ) design D is based on the set V, and $\rho \in Sym(V)$, then $\rho(D)$ is a design with the same parameters as D. For instance, if we choose $D = F$ and $\rho = (12)$, then $\rho(B_F) = \{124, 135, 346, 457, 562, 671, 723\}$. If $S \subseteq B$ and $S = dD$, then $\rho(S)$ is a defining set of $\rho(D)$ (Gray [9]). If $\rho(D) = D$, then ρ is an *automorphism* of D. For example, if $D = F$ as before, and $\rho = (24)(3675)$, then

$$\rho(B_F) = \{142, 463, 627, 235, 371, 754, 516\} = B_F.$$

The *automorphism group* of D, denoted by $Aut(D)$, is defined by

$$Aut(D) = \{\rho \in Sym(V) \mid \rho(D) = D\}.$$

Lemma 2.3 *[7] If S is a defining set of D, and $\rho \in Aut(D)$, then $\rho(S)$ is a defining set of $\rho(D) = D$.*

For instance, with $D = F$ as before, $S = \{124, 235, 346\}$ and $\rho = (24)(3675)$, then $\rho(S) = \{142, 463, 627\}$, again a defining set of F.

Lemma 2.4 *[7] If S is a defining set of the design D, then $Aut(S) \leq Aut(D)$.*

Lemma 2.5 *[7] $Aut(D) = \{\rho \mid \rho(S) \subseteq D\}$.*

We say that a group is *single-transposition-free* (STF) if none of its elements is a single transposition (ij). We say that a design is STF if its automorphism group is STF, and most further results concern such designs. This class of designs includes, among many others, all t–(v,k,λ) designs with $\lambda = 1$ and all symmetric t–(v,k,λ) designs with $t = 2$ and $k > 2$.

Lemma 2.6 *[9] Let S be a defining set of a STF t–(v,k,λ) design D. Then at least $v-1$ elements occur in the blocks of S and, if the elements i and j appear only once each in the blocks of S, they must appear in different blocks.*

These results enable us to find various lower bounds on the number of blocks in S, when D is an STF design [9]. For instance, if $S = dD$ consists of s blocks, then at most s elements occur once each in S, and at least $v-1-s$ elements occur at least twice each in S. Since only ks entries can occur in S, we have $s + 2(v-1-s) \le ks$, so that $2(v-1)/(k+1) \le s$. Other bounds may be found in a similar way.

A t–(v,k,λ) design is said to be *simple* if it contains no repeated blocks. For a simple STF design, the number of isomorphic copies of a defining set contained in the design is easily related to the sizes of the automorphism groups of the defining set and of the design.

Lemma 2.7 *[9] Suppose that the simple design D is also STF, and that S is a defining set of D. Then the number of isomorphic copies of S contained in D is $|Aut(D)|/|Aut(S)|$.*

There are also some results relating defining sets of a design to defining sets of its complementary design [7], and of its extensions (if any) [17], [19]. Suppose that the design D has blocks $B_1, \cdots, B_b$. Its *complementary design* $\tilde{D}$ has blocks $V \setminus B_1, \cdots, V \setminus B_b$, and if D is a t–design, so is $\tilde{D}$. If S is a defining set of D, then $\tilde{S}$ is a defining set of $\tilde{D}$. Also in forming extensions of D, we write $D(x)$ to mean the set of blocks $B_1 \cup \{x\}, \cdots, B_b \cup \{x\}$.

Lemma 2.8 *[7], [17] If a t–(v,k,λ) design D has exactly one extension to a $(t+1)$–$(v+1,k+1,\lambda_t)$ design, $E = D(x) \cup M$, and if S is a defining set of D, then $S(x)$ is a defining set of E. Hence $|d_s D| \ge |d_s E|$.*

Lemma 2.9 *[17] If D is a t-design with even t, and if $E = D(x) \cup \widetilde{D(x)}$, then* $|d_{\bullet}D| \leq |d_{\bullet}E|$.

Combining Lemma 2.8 and Lemma 2.9, we obtain a result which applies to all Hadamard designs.

Lemma 2.10 *[18] If D is a t-design with even t, and if $E = D(x) \cup \widetilde{D(x)}$ is its only extension, then $|d_{\bullet}D| = |d_{\bullet}E|$.*

There is also a particularly useful result for the case where the design E can be written as an extension by complementation in two ways.

Lemma 2.11 *[17] Suppose that D_1, D_2 are t-designs, that t is even, and that $E = D_1(x) \cup \widetilde{D_1(x)} = D_2(y) \cup \widetilde{D_2(y)}$. Then $|d_{\bullet}D_1| = |d_{\bullet}D_2|$, and D_1 and D_2 have the same number of smallest defining sets.*

3 FINDING SMALLEST DEFINING SETS IN SMALL DESIGNS

From the results in Section 2, Greenhill [12] developed the following implementable algorithm for finding smallest defining sets of simple STF t–(v, k, λ) designs.

1. Estimate a lower bound, l, on $|dD|$.
2. Generate all the l–subsets of blocks of D and sort them into isomorphism classes, say, $C_1, \cdots, C_m$. By Lemma 3, either every set in an isomorphism class is a defining set, or none of them is. Thus it is sufficient to consider a transversal of the classes, that is, a set $\{S_i \mid S_i \in C_i, i = 1, \cdots, m\}$, and to decide for each member of the transversal whether or not it is a defining set.
3. If S_i contains fewer than $v - 1$ distinct elements in its blocks, then it is not a defining set, by Lemma 6.
4. For each remaining class, determine both $|Aut(S_i)|$ and n_i, the number of sets in C_i.

5. Test $S_1, \cdots, S_m$ in turn and eliminate S_i if *either* $n_i \times |Aut(S_i)| \neq |Aut(D)|$ (contradicting Lemma 7) *or* if $Aut(S_i)$ is not a subgroup of $Aut(D)$ (contradicting Lemma 5). The remaining sets (if any) are said to be *feasible.*

6. If no feasible sets remain, increase l to $l+1$ and start again at Step 2.

7. If any feasible sets remain, list all known minimal trades in D, using $Aut(D)$ to generate more trades from known ones if possible.

8. Test the feasible sets and if $S_i \cap T = \emptyset$ for any S_i and minimal trade $T \subseteq B$, then by Lemma 1, S_i is not a defining set. If all S_i are eliminated, increase l to $l+1$ and start again at Step 2.

9. If any feasible sets remain, find all completions to t–(v, k, λ) designs of each one. If some S_i completes uniquely, then it is a defining set of D. If l was found from a lower bound (such as the one given in Section 2) then S_i is a smallest defining set of D, and the calculation is complete when all remaining sets of this cardinality have been dealt with. If l was estimated by some other means, decrease l to $l-1$ and start again at Step 2.

10. If *no* S_i completes uniquely, use the different completions to find more trades since the set of blocks $B \setminus S_i$ must contain a minimal trade. Then increase l to $l+1$ and start again at Step 2.

The lengthy computation of Step 2 can be carried out using nauty, a computer program developed by McKay for testing isomorphism [16]. Step 4, also lengthy, can be carried out using *CAYLEY*, developed by Cannon for algebraic computation [2]. Sometimes we may avoid the time–consuming computations of Step 9; see Gray and Street [11]. For if all designs with parameters t–(v, k, λ) are known, and if only one such design contains an l–subset isomorphic to S_i, then S_i is a defining set of that design. That is, in this case, a feasible set is a defining set [9]. When Step 9 is unavoidable, it can be carried out using either software of Sharry and Elmes [22] or a more recent version of Delaney [4].

Integer programming techniques have been applied by Khodkar [14] to find smallest defining sets of designs which are not necessarily simple and STF. As usual, let D be a t–(v, k, λ) design. For each $i \in I$, let

$$T_i = \{B_{i1}, B_{i2}, \cdots, B_{in_i}\} \subseteq D$$

be a trade. We have a family of trades $\{T_i \mid i \in I\}$, such that $\cup_{i \in I} T_i = \{B_j \mid j \in J\}$. The trade T_i corresponds to the inequality

$$x_{i1} + x_{12} + \cdots + x_{in_i} \geq 1. \qquad (i)$$

We minimize $\sum_{j\in J} x_j$ subject to the constraint that $x_j = 0$ or 1 for all $j \in J$ and the inequalities (i) for all $i \in I$.

If m is the optimal integer programming objective value for this system, then $|d_s D| \geq m$, and the solution exhibits a set of blocks S giving the value m. It remains to check that S is in fact a defining set, since in general we do not know that the family of trades that we have used includes all minimal trades. We can use a direct computer search to show that there is only one completion of the set of blocks that we have found, or in some cases we can argue from the properties of trades. The method was applied [14] to show that, of the 36 non-isomorphic 2-(9, 3, 2) designs, 10 have smallest defining sets of eight blocks each, 21 of nine blocks each, and 5 of ten blocks each.

v	b	r	k	λ	s	Reference
7	7	3	3	1	3	[7]
9	12	4	3	1	4	[8]
13	13	4	4	1	6	[13]
6	10	5	3	2	3	[8]
16	20	5	4	1	7	[8], [13]
21	21	5	5	1	8	[8], [13]
11	11	5	5	2	5	[8], [13]
13	26	6	3	1	9 (cyclic), 8 (non-cyclic)	[12]
7	14	6	3	2	6	[7]
10	15	6	4	2	8, 6, 5	[13]
16	16	6	6	2	9, 7, 7	[13]
15	35	7	3	1	≤ 16	[4], [19]
8	14	7	4	3	6	[7]
15	15	7	7	3	7 (three designs), 8 (one design)	
					9 ($PG(3,2)$)	[10], [11]
9	24	8	3	2	10 (five designs), 9 (21 designs)	
					8 (ten designs)	[14]
9	18	8	4	3	8 (nine designs), 6 (two designs)	[17]
7	21	9	3	3	9 (reducible), 7 (irreducible)	[7]
19	19	9	9	4	8	[18]
23	23	11	11	5	8 (quadratic residues)	[18]
27	27	13	13	6	≤ 11 (quadratic residues)	[18]
31	31	15	15	7	≤ 10 (quadratic residues)	[18]

Table 1 The sizes of smallest defining sets for some small 2–designs

Tables 1 and 2 show some of the results obtained for small designs.

t	v	b	r	k	λ	s	Reference
3	8	14	7	4	1	3	[7]
	12	22	11	6	2	5	[18]
	10	30	12	4	1	4	[13]
	8	28	14	4	2	6	[7]
	16	30	15	8	3	7 (three designs)	
						8 (one design)	
						9 ($PG(3,2)$)	[18]
	10	36	18	5	3	8 (four designs)	
						6 (three designs)	[17]
	20	38	19	10	4	8	[18]
	8	42	21	4	3	9 (from reducible)	
						7 (from irreducible)	[7]
	22	77	21	6	1	8	[19]
	24	46	23	12	5	8 (from quadratic residues)	[18]
	28	54	27	14	6	$\leq$ 11 (from quadratic residues)	[18]
	32	62	31	16	7	$\leq$ 10 (from quadratic residues)	[18]
4	11	66	30	5	1	5	[13]
	23	253	77	7	1	8	[19]
5	12	132	66	6	1	5	[13]
	24	759	253	8	1	8	[3], [19]

Table 2 The sizes of smallest defining sets for some small t–designs, $t \geq 3$

4 DEFINING SETS IN SOME INFINITE CLASSES OF DESIGNS

A different approach has been taken by Gower [5], [6], who has found minimal (but not smallest) defining sets for two infinite classes of Steiner triple systems, that is, for 2–$(v, 3, 1)$ designs. The particular classes are those which arise as point–line designs in either the projective geometry $PG(d, 2)$ or the affine geometry $AG(d, 3)$.

First consider a vector space V of dimension $d + 1$ over the field $GF[2]$. Its 1–dimensional subspaces may be considered as the points of $PG(d, 2)$, its 2–dimensional subspaces as the lines, its $r+1$–dimensional subspaces as r–dimensional projective objects and, finally, its d–dimensional subspaces as the hyperplanes. The point–line design in $PG(d,2)$ is a Steiner triple system on $2^{d+1}-1$ points, that is, a 2–$(2^{d+1}-1,3,1)$ design. In $PG(d,2)$, with $d \geq 2$, a set of $d+1$ hyperplanes such that no point of $PG(d,2)$ is incident with all of them is said to be of *type* $\mathcal{H}$. In such a set, there is precisely one point incident with none of them.

For example, we take the 2–$(15,7,3)$ design associated with a $PG(3,2)$ and developed from the difference set $\Delta_0 = \{0,1,2,4,5,8,10\}$. Since $d=3$ a set of type $\mathcal{H}$ must contain four blocks. We can choose Δ_0 itself, together with Δ_0+1, Δ_0+6 and Δ_0+7, which have no point common to all four of them, and all avoid the point 12.

The lines of $PG(3,2)$ are the intersections of its (hyper)planes; we now take all of the lines contained by these four blocks. In our particular example, these are

$$\{0,1,4\},\{0,2,8\},\{0,5,10\},\{0,6,13\},\{0,7,9\},\{1,2,5\},$$
$$\{1,3,9\},\{1,6,11\},\{1,7,14\},\{1,8,10\},\{2,3,6\},\{2,4,10\},$$
$$\{2,9,11\},\{3,5,11\},\{4,5,8\},\{5,6,9\},\{5,7,13\},\{6,7,10\},$$
$$\{6,8,14\},\{7,8,11\},\{9,10,13\},\{10,11,14\},$$

and these 22 lines form a minimal defining set for the 2–$(15,3,1)$ design associated with the $PG(3,2)$. Note however that this minimal set is not a smallest defining set, since this design has two non–isomorphic defining sets of 16 blocks each; see Delaney [4] and Moran [19].

In general, the hyperplanes of a set of type $\mathcal{H}$ cover a total of $(\frac{2}{3})4^d - (\frac{1}{2})3^d - 2^d + (\frac{5}{6})$ lines which form a minimal defining set.

Next, consider a vector space V of dimension d over the field $GF[3]$. Its vectors may be considered as the points of $AG(d,3)$, the cosets of its 1–dimensional subspaces as the lines of $AG(d,3)$, the cosets of its 2–dimensional subspaces as the planes, the cosets of its i–dimensional subspaces as the i–flats and, finally, the cosets of its $(d-1)$–dimensional subspaces as the hyperplanes. Cosets of the same i–dimensional subspace are parallel i–flats, forming a parallel class. In $AG(d,3)$ there are three hyperplanes in a parallel class. For example, if $d=2$ the four parallel classes of three lines each in $AG(2,3)$ are as follows.

$$123, 456, 789; 147, 258, 369; 159, 267, 348; 168, 249, 357.$$

The point–line design in the geometry $AG(d,3)$ is a Steiner triple system on 3^d points or 2–$(3^d,3,1)$ design.

We choose $2d$ hyperplanes of $AG(d,3)$ so that:

- they belong to d parallel classes with precisely two hyperplanes from each of these classes;
- no line of $AG(d,3)$ is incident with any d of them;
- there is a point of $AG(d,3)$ incident with none of them.

For example, in the copy of $AG(2,3)$ given above, the lines 123, 456, 147, 258, none of which is incident with the point 9, satisfy these requirements, and in general, working over $GF[3]$, we may take the set of lines

$$x_{d-1}=0, x_{d-1}=2, \cdots, x_1=0, x_1=2, x_0=0, x_0=2,$$

none of which is incident with the point $(1,1,\cdots,1)$. Then the set of all lines covered by these $2d$ hyperplanes forms a minimal defining set of the Steiner triple system associated with $AG(d,3)$. The set contains $[3^d(3^d-1)-7^d+1]/6$ lines. Thus for instance the Steiner triple system on 27 points associated with $AG(3,3)$ has 117 lines, 60 of which form a minimal defining set.

The other infinite class of designs whose defining sets have been studied are the Hadamard designs 2–$(4h-1, 2h-1, h-1)$, developed from the quadratic residues, where $v=4h-1$ is prime or prime–power. Seberry [21] has an interesting conjecture related to these designs, as follows.

Conjecture. *Let Q be the set of quadratic residues in the field $GF[v]$, say, $Q=\{q_1,\cdots,q_{2h-1}\}$. Then the set of blocks $\mathcal{Q}=\{Q+q_1,\cdots,Q+q_{2h-1}\}$ forms a defining set of the unique Hadamard design developed from the difference set Q.*

This conjecture has been verified for $v=4h-1=7,11,19,23,\cdots,67$, but has not been settled in the general case. Other work on this and related conjectures appears in Kunkle and Seberry [15] and Sarvate and Seberry [20]. However it has been shown [21] that $\mathcal{Q}$ does determine the residual 2–$(2h, 4h-2, 2h-1, h, h-1)$ of the Hadamard design developed from Q. As an example, we consider the 2–$(11,5,2)$ design, that is, the design with $h=3$. Here $Q=\{1,3,4,5,9\}$ and the blocks we consider are $Q+1$, $Q+3$, $Q+4$, $Q+5$, $Q+9$. Table 3 shows the incidence array of these five blocks, together with Q. Now reading the columns as blocks and the rows as elements 1 to 6, we find we have the 2–$(6,3,2)$ design with blocks

$$\{256, 135, 456, 126, 136, 124, 235, 234, 346, 145\},$$

that is, the residual of the 2–(11, 5, 2) design developed from Q. This is an example of a general result [21], but the question of whether $\mathcal{Q}$ is always a defining set of the Hadamard design is still open.

	1	2	3	4	5	6	7	8	9	10	
$Q+1$	0	1	0	1	1	1	0	0	0	1	1
$Q+3$	1	0	0	1	0	1	1	1	0	0	2
$Q+4$	0	1	0	0	1	0	1	1	1	0	3
$Q+5$	0	0	1	0	0	1	0	1	1	1	4
$Q+9$	1	1	1	0	0	0	1	0	0	1	5
Q	1	0	1	1	1	0	0	0	1	0	6

Table 3 Incidence array of $\mathcal{Q} \cup \{Q\}$ in the case $h = 3$

In conclusion, we note that defining sets in designs are in many ways analogous to critical sets in Latin squares [23] and are also related to unique completion sets in Latin squares [1].

Acknowledgments

I thank the Mathematics Department, Sharif University of Technology, Tehran, for their warm hospitality during and after the conference at which this paper was presented. This work was supported by ARC grant A49130102 and by an Australian Senior Research Fellowship.

REFERENCES

[1] Peter J Cameron, *How few entries determine any Latin square?* Bulletin of the Institute of Combinatorics and its Applications **10** (1994), 63–65.

[2] John Cannon, *A language for group theory,* Department of Pure Mathematics, University of Sydney, Sydney, Australia (1987).

[3] R T Curtis, *Eight octads suffice,* Journal of Combinatorial Theory, Series A **36** (1984), 116–123.

[4] Cathy Delaney, private communication, (1994).

[5] Rebecca A H Gower, *Minimal defining sets in a family of Steiner triple systems,* Australasian Journal of Combinatorics **8** (1993), 55–73.

[6] Rebecca A H Gower, *Minimal defining sets for another infinite family of Steiner triple systems,* submitted for publication.

[7] Ken Gray, *On the minimum number of blocks defining a design,* Bulletin of the Australian Mathematical Society **41** (1990), 97–112.

[8] Ken Gray, *Further results on smallest defining sets of well known designs,* Australasian Journal of Combinatorics **1** (1990), 91–100.

[9] Ken Gray, *Defining sets of single-transposition-free designs,* Utilitas Mathematica **38** (1990), 97–103.

[10] Ken Gray and Anne Penfold Street, *Smallest defining sets of the five non-isomorphic* 2–$(15,7,3)$ *designs,* Bulletin of the Institute of Combinatorics and its Applications **9** (1993), 96–102.

[11] Ken Gray and Anne Penfold Street, *The smallest defining set of the* 2–$(15,7,3)$ *designs associated with* $PG(3,2)$ *: a theoretical approach,* Bulletin of the Institute of Combinatorics and its Applications **11** (1994), 77–83.

[12] Catherine S Greenhill, *An algorithm for finding smallest defining sets of t–designs,* Journal of Combinatorial Mathematics and Combinatorial Computing **14** (1993), 39–60.

[13] Catherine S Greenhill and Anne Penfold Street, *Smallest defining sets of some small t–designs and relations to the Petersen graph,* Utilitas Mathematica, to appear.

[14] A Khodkar, *Smallest defining sets for the* 36 *non–isomorphic twofold triple systems of order nine,* Journal of Combinatorial Mathematics and Combinatorial Computing, to appear.

[15] Thomas Kunkle and Jennifer Seberry, *A few more small defining sets for* $SBIBD(4t-1, 2t-1, t-1)$, submitted for publication.

[16] Brendan D McKay, *nauty* User's Guide (Version 1.5), Australian National University Computer Science Technical Report TR–CS–90–02.

[17] Tony Moran, *Smallest defining sets of* 2–$(9,4,3)$ *and* 3–$(10,5,3)$ *designs,* Australasian Journal of Combinatorics **10** (1994), 265–288.

[18] Tony Moran, *Defining sets of the* 2–$(19,9,4)$ *designs and a class of Hadamard designs,* submitted for publication.

[19] Tony Moran, private communication, (1993).

[20] Dinesh Sarvate and Jennifer Seberry, *A note on small defining sets for some* $SBIBD(4t-1, 2t-1, t-1)$, Bulletin of the Institute of Combinatorics and its Applications **10** (1994), 26–32.

[21] Jennifer Seberry, *On small defining sets for some* $SBIBD(4t-1, 2t-1, t-1)$, Bulletin of the Institute of Combinatorics and its Applications **4** (1992), 58–62; *corrigendum,* **6** (1992), 62.

[22] Martin J Sharry and Sylvia H Elmes, *The program COMPLETE,* private communication, (1992).

[23] Anne Penfold Street, *Defining sets for t–designs and critical sets for Latin squares,* New Zealand Journal of Mathematics **21** (1992), 133–144.

Question: Determine (or get good bounds for)

$$\max_{L \in \mathcal{L}_n} \min_{S \in \mathcal{S}_{a,b}(L)} \sigma(S).$$

Problem Number 4. Proposed by *C. J. Colbourn*

Let $q = mt + 1$ be a prime power, and w be a primitive element of $GF(q)$. Let $\mathcal{S}_i = \{w^{am+i} \mid 0 \leq a < t\}$. A $V(m,t)$ vector is a vector $(a_1, \ldots, a_{m+1})$ with entries from $GF(q)$ so that for each d satisfying $1 \leq d \leq m+1$,

$$\{a_{d+i} - a_i : 1 \leq i \leq m+1, d \neq m+2-i\}$$

(computing subscripts modulo $m+2$) represent each of the sets $\mathcal{S}_0, \ldots, \mathcal{S}_{m-1}$.

Question: Find an infinite family of $V(m,t)$ vectors with $m \geq 4$. *Note:* The larger m is, the more useful $V(m,t)$ is.

Problem Number 5. Proposed by *M. R. Emamy K.*

Let C^n= The n-dimensional cube, E_n be the set of all edges of C^n, and B^n be $\{0,1\}^n$ = the set of all vertices of C^n.

For $x \in B^n$, denote by $\bar{x}$ the opposite vertex (antipode) of x (the i-th component of $\bar{x}$ is 0 if and only if i-th component of x is 1).

Let $x\bar{x}$ denote a diameter. Define H_x to be the hyperplane passing through 0, the center of C^n, and orthogonal to $x\bar{x}$.

Question. For n odd, what is the minimum number of H_x's, that can intersect all edges of E_n? When n is even, for any x consider $H_x(\epsilon)$ which is a small translation of H_x, and now answer the same question for $H_x(\epsilon)$'s.

Problem Number 6. Proposed by *R. K. Guy*

A 20-years-old problem of Allen Freedman is the following. Let n points inside the unit square be given where one of them is supposed to be the origin. Show that you can always cover at least half of the area of the square with n non-overlapping rectangles, each with sides parallel to the sides of the square, and each having one of the given points as its "southwest"-corner, or find a counter-example.

Example for the case $n = 4$:

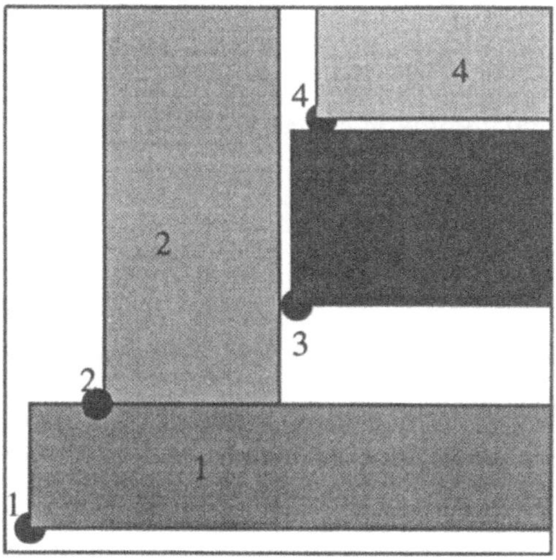

Figure 1

Problem Number 7. Proposed by *H. Kharaghani*

Show that all orthogonal designs $OD(2^{2^n-1}; a, a, \ldots, a)$, $a = 2^{2^n-n-2}$ in 2^{n+1} variables ($n > 2$), exist. In particular settle this conjecture for $n = 3$, i.e., show that there is an $OD(2^7; 8, 8, \ldots, 8)$ in 16 variables.

Problem Number 8. Proposed by *E. S. Mahmoodian*

For this problem there is set a prize of 100,000 Iranian Rials.

Let $r \leq s \leq t$ be three natural numbers. We conjecture that necessary and sufficient conditions to decompose the complete tripartite graph $K_{r,s,t}$ into 5-cycles are:

(i) r, s, and t all are even or all are odd;

(ii) $5 \mid rs + rt + st$;

(iii) $t \leq 4rs/(r + s)$.

Problem Number 9. Proposed by *E. S. Mahmoodian*

Prize: 10,000,000 Iranian Rials will be paid by M. Behzad to Iranian citizens.

Settle TCC; the total chromatic conjecture of M. Behzad. (For the definitions see his paper in this volume).

Problem Number 10. Proposed by *A. P. Street*

1. Explore defining sets for directed designs or group divisible designs.
2. Obtain bounds or other theoretical results for designs whose automorphism group contains a single transposition.

Problem Number 11. Proposed by *A. P. Street*

A $K_4 - e$-design of order n is an edge-disjoint decomposition of the complete graph K_n into copies of $K_4 - e$. They exist for all $n \equiv 0$ or $1 \pmod 5$, $n \geq 6$.

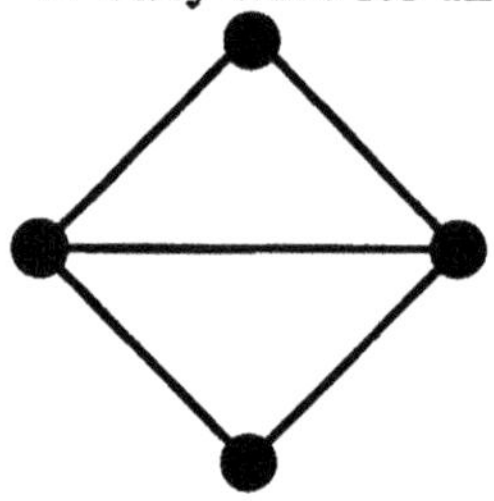

Figure 2 $K_4 - e$

1. Do resolvable $K_4 - e$-designs exist? (Need $n \equiv 16 \pmod{20}$) [*Editors' Note: C.J. Colbourn, D.R. Stinson and L. Zhu, "More frames with block size four", have obtained examples where* $n = 60t + 16$ *for all* $t \geq 1$ *except for* $t = 8, 12$.]
2. Do large sets of $K_4 - e$-designs exist?
3. Do overlarge sets of $K_4 - e$-designs exist?

Other *Mathematics and Its Applications* titles of interest:

I.H. Dimovski: *Convolutional Calculus*. 1990, 208 pp. ISBN 0-7923-0623-6

Y.M. Svirezhev and V.P. Pasekov: *Fundamentals of Mathematical Evolutionary Genetics*. 1990, 384 pp. ISBN 90-277-2772-4

S. Levendorskii: *Asymptotic Distribution of Eigenvalues of Differential Operators*. 1991, 297 pp. ISBN 0-7923-0539-6

V.G. Makhankov: *Soliton Phenomenology*. 1990, 461 pp. ISBN 90-277-2830-5

I. Cioranescu: *Geometry of Banach Spaces, Duality Mappings and Nonlinear Problems*. 1990, 274 pp. ISBN 0-7923-0910-3

B.I. Sendov: *Hausdorff Approximation*. 1990, 384 pp. ISBN 0-7923-0901-4

A.B. Venkov: *Spectral Theory of Automorphic Functions and Its Applications*. 1991, 280 pp. ISBN 0-7923-0487-X

V.I. Arnold: *Singularities of Caustics and Wave Fronts*. 1990, 274 pp. ISBN 0-7923-1038-1

A.A. Pankov: *Bounded and Almost Periodic Solutions of Nonlinear Operator Differential Equations*. 1990, 232 pp. ISBN 0-7923-0585-X

A.S. Davydov: *Solitons in Molecular Systems. Second Edition*. 1991, 428 pp. ISBN 0-7923-1029-2

B.M. Levitan and I.S. Sargsjan: *Sturm-Liouville and Dirac Operators*. 1991, 362 pp. ISBN 0-7923-0992-8

V.I. Gorbachuk and M.L. Gorbachuk: *Boundary Value Problems for Operator Differential Equations*. 1991, 376 pp. ISBN 0-7923-0381-4

Y.S. Samoilenko: *Spectral Theory of Families of Self-Adjoint Operators*. 1991, 309 pp. ISBN 0-7923-0703-8

B.I. Golubov A.V. Efimov and V.A. Scvortsov: *Walsh Series and Transforms*. 1991, 382 pp. ISBN 0-7923-1100-0

V. Laksmikantham, V.M. Matrosov and S. Sivasundaram: *Vector Lyapunov Functions and Stability Analysis of Nonlinear Systems*. 1991, 250 pp. ISBN 0-7923-1152-3

F.A. Berezin and M.A. Shubin: *The Schrödinger Equation*. 1991, 556 pp. ISBN 0-7923-1218-X

D.S. Mitrinovic, J.E. Pecaric and A.M. Fink: *Inequalities Involving Functions and their Integrals and Derivatives*. 1991, 588 pp. ISBN 0-7923-1330-5

Julii A. Dubinskii: *Analytic Pseudo-Differential Operators and their Applications*. 1991, 252 pp. ISBN 0-7923-1296-1

V.I. Fabrikant: *Mixed Boundary Value Problems in Potential Theory and their Applications*. 1991, 452 pp. ISBN 0-7923-1157-4

Other *Mathematics and Its Applications* titles of interest:

A.M. Samoilenko: *Elements of the Mathematical Theory of Multi-Frequency Oscillations.* 1991, 314 pp. ISBN 0-7923-1438-7

Yu.L. Dalecky and S.V. Fomin: *Measures and Differential Equations in Infinite-Dimensional Space.* 1991, 338 pp. ISBN 0-7923-1517-0

W. Mlak: *Hilbert Space and Operator Theory.* 1991, 296 pp. ISBN 0-7923-1042-X

N.Ja. Vilenkin and A.U. Klimyk: *Representation of Lie Groups and Special Functions. Volume 1: Simplest Lie Groups, Special Functions, and Integral Transforms.* 1991, 608 pp. ISBN 0-7923-1466-2

N.Ja. Vilenkin and A.U. Klimyk: *Representation of Lie Groups and Special Functions. Volume 2: Class I Representations, Special Functions, and Integral Transforms.* 1992, 630 pp. ISBN 0-7923-1492-1

N.Ja. Vilenkin and A.U. Klimyk: *Representation of Lie Groups and Special Functions. Volume 3: Classical and Quantum Groups and Special Functions.* 1992, 650 pp. ISBN 0-7923-1493-X
(Set ISBN for Vols. 1, 2 and 3: 0-7923-1494-8)

K. Gopalsamy: *Stability and Oscillations in Delay Differential Equations of Population Dynamics.* 1992, 502 pp. ISBN 0-7923-1594-4

N.M. Korobov: *Exponential Sums and their Applications.* 1992, 210 pp. ISBN 0-7923-1647-9

Chuang-Gan Hu and Chung-Chun Yang: *Vector-Valued Functions and their Applications.* 1991, 172 pp. ISBN 0-7923-1605-3

Z. Szmydt and B. Ziemian: *The Mellin Transformation and Fuchsian Type Partial Differential Equations.* 1992, 224 pp. ISBN 0-7923-1683-5

L.I. Ronkin: *Functions of Completely Regular Growth.* 1992, 394 pp. ISBN 0-7923-1677-0

R. Delanghe, F. Sommen and V. Soucek: *Clifford Algebra and Spinor-valued Functions. A Function Theory of the Dirac Operator.* 1992, 486 pp. ISBN 0-7923-0229-X

A. Tempelman: *Ergodic Theorems for Group Actions.* 1992, 400 pp. ISBN 0-7923-1717-3

D. Bainov and P. Simenov: *Integral Inequalities and Applications.* 1992, 426 pp. ISBN 0-7923-1714-9

I. Imai: *Applied Hyperfunction Theory.* 1992, 460 pp. ISBN 0-7923-1507-3

Yu.I. Neimark and P.S. Landa: *Stochastic and Chaotic Oscillations.* 1992, 502 pp. ISBN 0-7923-1530-8

H.M. Srivastava and R.G. Buschman: *Theory and Applications of Convolution Integral Equations.* 1992, 240 pp. ISBN 0-7923-1891-9

Other *Mathematics and Its Applications* titles of interest:

A. van der Burgh and J. Simonis (eds.): *Topics in Engineering Mathematics*. 1992, 266 pp. ISBN 0-7923-2005-3

F. Neuman: *Global Properties of Linear Ordinary Differential Equations*. 1992, 320 pp. ISBN 0-7923-1269-4

A. Dvurecenskij: *Gleason's Theorem and its Applications*. 1992, 334 pp. ISBN 0-7923-1990-7

D.S. Mitrinovic, J.E. Pecaric and A.M. Fink: *Classical and New Inequalities in Analysis*. 1992, 740 pp. ISBN 0-7923-2064-6

H.M. Hapaev: *Averaging in Stability Theory*. 1992, 280 pp. ISBN 0-7923-1581-2

S. Gindinkin and L.R. Volevich: *The Method of Newton's Polyhedron in the Theory of PDE's*. 1992, 276 pp. ISBN 0-7923-2037-9

Yu.A. Mitropolsky, A.M. Samoilenko and D.I. Martinyuk: *Systems of Evolution Equations with Periodic and Quasiperiodic Coefficients*. 1992, 280 pp. ISBN 0-7923-2054-9

I.T. Kiguradze and T.A. Chanturia: *Asymptotic Properties of Solutions of Non-autonomous Ordinary Differential Equations*. 1992, 332 pp. ISBN 0-7923-2059-X

V.L. Kocic and G. Ladas: *Global Behavior of Nonlinear Difference Equations of Higher Order with Applications*. 1993, 228 pp. ISBN 0-7923-2286-X

S. Levendorskii: *Degenerate Elliptic Equations*. 1993, 445 pp. ISBN 0-7923-2305-X

D. Mitrinovic and J.D. Kečkić: *The Cauchy Method of Residues, Volume 2*. Theory and Applications. 1993, 202 pp. ISBN 0-7923-2311-8

R.P. Agarwal and P.J.Y Wong: *Error Inequalities in Polynomial Interpolation and Their Applications*. 1993, 376 pp. ISBN 0-7923-2337-8

A.G. Butkovskiy and L.M. Pustyl'nikov (eds.): *Characteristics of Distributed-Parameter Systems*. 1993, 386 pp. ISBN 0-7923-2499-4

B. Sternin and V. Shatalov: *Differential Equations on Complex Manifolds*. 1994, 504 pp. ISBN 0-7923-2710-1

S.B. Yakubovich and Y.F. Luchko: *The Hypergeometric Approach to Integral Transforms and Convolutions*. 1994, 324 pp. ISBN 0-7923-2856-6

C. Gu, X. Ding and C.-C. Yang: *Partial Differential Equations in China*. 1994, 181 pp. ISBN 0-7923-2857-4

V.G. Kravchenko and G.S. Litvinchuk: *Introduction to the Theory of Singular Integral Operators with Shift*. 1994, 288 pp. ISBN 0-7923-2864-7

A. Cuyt (ed.): *Nonlinear Numerical Methods and Rational Approximation II*. 1994, 446 pp. ISBN 0-7923-2967-8

Other *Mathematics and Its Applications* titles of interest:

G. Gaeta: *Nonlinear Symmetries and Nonlinear Equations.* 1994, 258 pp.
ISBN 0-7923-3048-X

V.A. Vassiliev: *Ramified Integrals, Singularities and Lacunas.* 1995, 289 pp.
ISBN 0-7923-3193-1

N.Ja. Vilenkin and A.U. Klimyk: *Representation of Lie Groups and Special Functions.* Recent Advances. 1995, 497 pp. ISBN 0-7923-3210-5

Yu. A. Mitropolsky and A.K. Lopatin: *Nonlinear Mechanics, Groups and Symmetry.* 1995, 388 pp. ISBN 0-7923-3339-X

R.P. Agarwal and P.Y.H. Pang: *Opial Inequalities with Applications in Differential and Difference Equations.* 1995, 393 pp. ISBN 0-7923-3365-9

A.G. Kusraev and S.S. Kutateladze: *Subdifferentials: Theory and Applications.* 1995, 408 pp. ISBN 0-7923-3389-6

M. Cheng, D.-G. Deng, S. Gong and C.-C. Yang (eds.): *Harmonic Analysis in China.* 1995, 318 pp. ISBN 0-7923-3566-X

M.S. Livšic, N. Kravitsky, A.S. Markus and V. Vinnikov: *Theory of Commuting Nonselfadjoint Operators.* 1995, 314 pp. ISBN 0-7923-3588-0

A.I. Stepanets: *Classification and Approximation of Periodic Functions.* 1995, 360 pp. ISBN 0-7923-3603-8

C.-G. Ambrozie and F.-H. Vasilescu: *Banach Space Complexes.* 1995, 205 pp.
ISBN 0-7923-3630-5

E. Pap: *Null-Additive Set Functions.* 1995, 312 pp. ISBN 0-7923-3658-5

C.J. Colbourn and E.S. Mahmoodian (eds.): *Combinatorics Advances.* 1995, 338 pp.
ISBN 0-7923-3574-0

GPSR Compliance

The European Union's (EU) General Product Safety Regulation (GPSR) is a set of rules that requires consumer products to be safe and our obligations to ensure this.

If you have any concerns about our products, you can contact us on ProductSafety@springernature.com

In case Publisher is established outside the EU, the EU authorized representative is:

Springer Nature Customer Service Center GmbH
Europaplatz 3
69115 Heidelberg, Germany

Batch number: 10682323

Printed by Printforce, the Netherlands